Biological
Environmental
Science

Biological Environmental Science

William V. Dashek
Retired from Adult Degree Program
Mary Baldwin College
Richmond/Staunton, Virginia, USA

Editor
David E. McMillin
Mathematics and Natural Science Division
Brewton Parker College
Mount Vernon, Georgia, USA

Science Publishers

Enfield (NH)   Jersey   Plymouth

Science Publishers   *www.scipub.net*

Post Office Box 699
Enfield, New Hampshire 03748
United States of America

General enquiries : *info@scipub.net*
Editorial enquiries : *editor@scipub.net*
Sales enquiries   : *sales@scipub.net*

Published by Science Publishers, Enfield, NH, USA
An imprint of Edenbridge Ltd., British Channel Islands
Printed in India

ISBN 978-1-57808-536-1

Library of Congress Cataloging-in-Publication Data

Biological environmental science/William V. Dashek, editor. David McMillin.
   p. cm.
 Includes bibliographical references and index.
 ISBN 978-1-57808-536-1 (pbk.)
 1. Pollution--Environmental aspects. 2. Environmental sciences. I. Dashek, William V. II. McMillin,
David (David E.).
 QH545.A1B563 2008
 363.7--dc22

                                                                        2008039696

Preface

Bioligical Environmental Science is textbook for 'upper' undergraduates who desire a *biological* and, to a lesser extent, *chemical* introduction to environmental science. Thus, the book is not suited to environmental science courses whose thrusts are earth science, physical science or chemistry. The book assumes that the student has taken one semester of biology and chemistry.

The book utilizes the original research literature and possesses numerous references. The environmental research literature is vast and, therefore, the references are copious but not comprehensive. In addition to the references, the book cites pertinent websites.

A unique feature of this book is the inclusion of appendices centering about the scientific method, the metric system, statistical tests, spreadsheets, ecological and environmental journals and exam questions. One of the appendices lists laboratory manuals.

Following introductory chapters describing the scopes of ecology and environmental science, the book concentrates on global warming, energy, pollution, population growth and speciation, and concludes with a chapter regarding environmental law, management and organizations.

Finally, the book does not offer opinion or social policy. The student can analyze the provided data and read the cited research literature and, subseqently, draw conclusions regarding the significance of the topics to the environmental quality of Planet Earth.

Contents

Acknowledgements

I am grateful to the late Drs. J.E. Varner, W.G. Rosen and W.F. Millington as well as Dr. D.T.A. Lamport for my scientific traning. I am especially grateful to Dr. Rosen who tought me the value of preserving a life-sustaining environment for Planet Earth. I thank Professor Cathy McPherson for the opportunity to teach environmental science at Mary Baldwin College's Adult Degree Program, Richmond Regional Center. I am indebted to Ms. Nina Smith for her clerical assistance and to Mr. Terry Barham of Staples Copy Center, Richmond, Va for media preparation.

I acknowledge the financial support of The National Institutes of Health. The National Science Foundation, The United States Department of Energy and the United States Department of Agriculture-Forest Service for enabling a lengthy career of academic research.

I dedicate this volume to Kristin Dashek Peyton and Karin Dashek Bryant who granted me the gift of time. Finally, I appreciate my parents' efforts for providing the opportunity to pursue higher education.

CHAPTER 1

What is Ecology?

Ecology is the science of the relationships of organisms to the surrounding world. This discipline is concerned with the interaction of organisms and their environment (Scott, 1996). Ecology is often divided into physiological ecology, population biology, community ecology and ecosystem ecology. Organisms are dependent upon their ·environments, e.g. they must convert a source of energy, ingest water and minerals, dispose of wastes and regulate temperature. Table 1.1 presents significant ecological and environmental terms used throughout this text.

What are Biogeochemical Cycles and What is their Significance?

For nutrient cycles, consumption must balance production, e.g., nutrients can be recycled from dead biomass through inorganic forms and back into living organisms. Of the 80 plus naturally occurring elements approximately 40 are used by living systems to sustain life. Many of the 40 are required in trace amounts (see chapter on Metal Pollution for discussion of major and minor nutrients). The organic world is composed of Carbon (C), Oxygen (O_2), and Hydrogen (H_2) to a large extent. Nitrogen (N_2) and Sulfur (S) are also important elements for living systems.

Oxygen Cycle (Fig. 1.1)

Much of the Earth's Oxygen is in the form of molecular oxygen. This element is a component of CO_2, H_2O, ions (e.g., NO_3 or CO_3) and many organic compounds. Ozone (O_3) is a derivative of O_2 and forms a protective layer in the atmosphere shielding Earth's organisms from harmful ultraviolet (UV) radiation (see chapter on global warming).

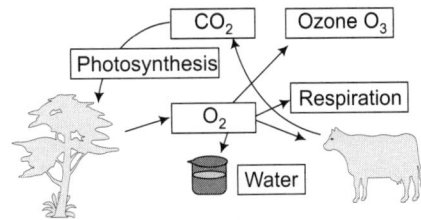

Fig. 1.1 Oxygen cycle
Adapted from: *http://geog.byu.edu/oghart/214pres/geochem/sed010.htm*

Oxygen is utilized in respiration, which consists of glycolysis and the Kreb's cycle and some ancillary processes (Bowlby, 2006). Glycolysis is the conversion of starch or glucose to pyruvic acid. This compound is transformed to acetyl CO-A which enters the Kreb's cycle. Electrons and hydrogens derived from respiration are utilized in

TABLE 1.1 Significant Ecological and Environmental Terms

Ecological Term	Definition
Ecosystem	An association of plants and animals and the physical factors of their environment (Scott, 1996). Inhabitants of ecosystems are producers, consumers and decomposers (Table 1.4).
Community	Spatial subsystems within ecosystems.
Biosphere	Totality of living organisms on the Earth (Scott, 1996).
Biomes	Large stable, terrestrial ecosystems (Turk and Turk, 1997).
Biomass	Total mass of organic matter present at one time in an ecosystem (Turk and Turk, 1997).
Energy	Capacity to do work and transfer heat when a body is forced to move.

Types:

Potential	Stored energy, i.e., energy available to do work.
Kinetic	Energy used to perform work.

Kinetic energy and potential energy are freely interconvertible. A discussion of these forms of energy and the Laws of Thermodynamics is presented in Hokikian (2002) and Jorgensen and Svirezkev (2004).

Importance of energy to organisms: Many biochemical reactions which occur in organisms require energy. The flow of energy in ecosystems occurs through food webs (interconnected feeding relationships in a food chain occurring in a particular place and time – www.arcytech.org/java/population/foods-foodchain.html).

Organisms in a food web (Neutel et al., 2002) and types of nutrition.

Omnivores	Organisms which consume both plants and animals
Herbivores	A plant-eating animal
Carnivores	Flesh-feeding animals, as well as plants that supplement their diets by capturing animals (Scott, 1996)
Saprophytes	Organisms which 'live-off' of decaying organic material
Parasites	Organisms which live at the expense of another organism
Ectoparasites	Associated with the exterior of the host.
Endoparasites	Live within their hosts
Detritus Feeders	Organisms which utilize non-living organic matter

Nutrition

Autotrophs vs. Heterotrophs

Autotrophs	Organisms which synthesize organic compounds from simple inorganic compounds
Heterotrophs	Organisms which cannot perform photosynthesis or chemosynthesis;they have a nutritional dependency upon organic compounds.

Trophic Levels (Tscharntke and Hawkins, 2002)

First Level	Autotroph
Second Level	Heterotrophs which are primary consumers – obtain their energy directly from Autotrophs.
Third Level	Heterotrophs which are secondary to consumers – obtain their energy indirectly from Autotrophs.
	These levels form an ecological food pyramid with the producers at the base and primary, secondary and tertiary consumers at successive less abundant levels. There are many diagrams of ecological pyramids of biomass in textbooks and on the Internet.

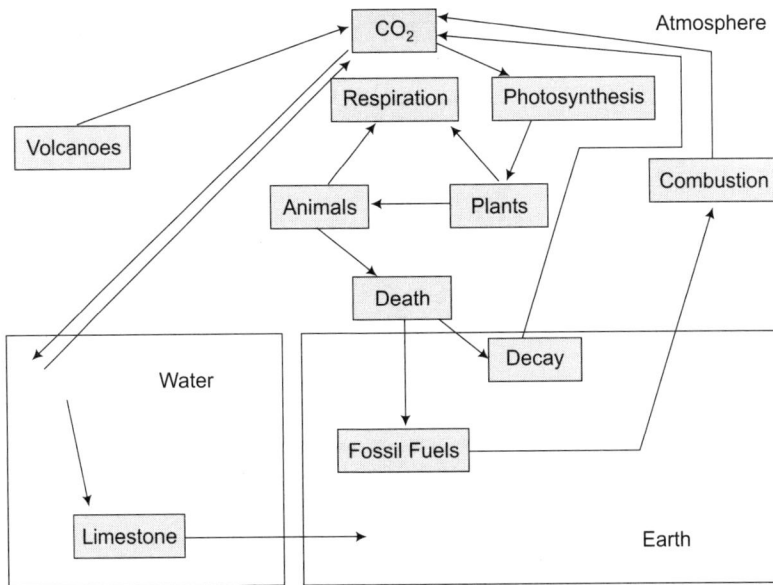

Fig. 1.2 Carbon cycle
From: *http://www.marietta.edu/~biol/102/ecosystem.html*

electron transport to yield ATP (adenosine triphosphate). In contrast to respiration, photosynthesis evolves O_2.

There are three types of photosyntheses: *Calvin cycle* which occurs in C4 plants, *Hatch and Slack cycle* which is a characteristic of C3 plants and *Crassulacean metabolism* which is prevelant in succulents (Trabalha and Reichle (1986). Less than one half of the total C cycling occurs through biological or related organic pathways. Atmospheric CO_2 (0.03%) dissolves in water and some dissolved CO_2 escapes from the sea to the air (volcanic eruption) while other CO_2 molecules react to yield calcium carbonates in the sediment. The carbon cycle is very complex in that $CaCO_3$ occurs in the oceans as well as terrestrial and aquatic environments. A comparison of Figs 1.1 and 1.2 reveals that the carbon and oxygen cycles are closely related.

Nitrogen Cycle (Fig. 1.3)

Nitrogen, a constituent of proteins, comprises the amino group (NH_2) of amino acids. Amino acids are linked to each other via peptide bonds yielding peptides. Proteins are polypeptides consisting of many amino acids in peptide linkage. There are both structural and enzymatic (organic catalysts) proteins. Nitrogen is also a constituent of other biologically important compounds (Table 1.2).

Lightning and certain photochemical reactions can transform molecular nitrogen to derivatives which are usable by organisms. Once fixed, N_2 is assimilated by plants which can be consumed by animals and enter the heterotrophic food web. A food web is an interrelationship of organisms in an ecological community in which energy is transferred (Baron, 2003). Food webs per-

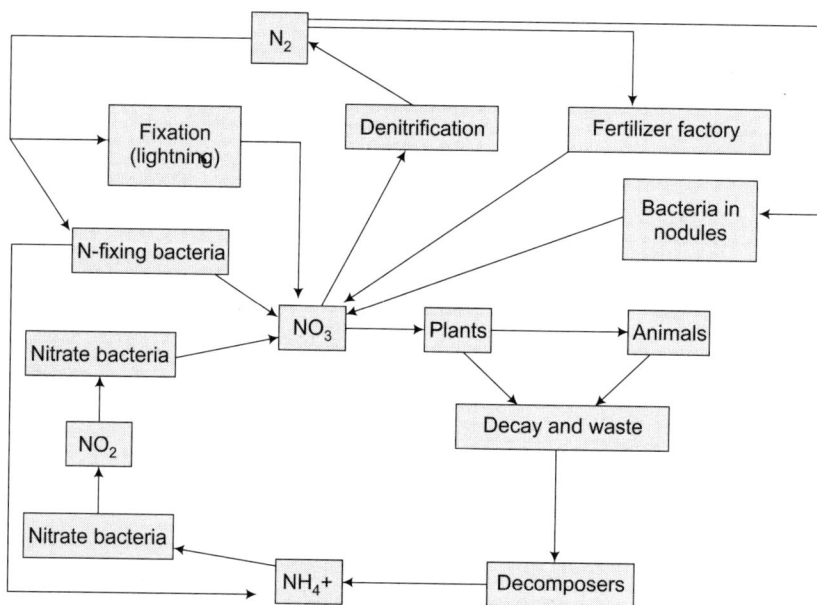

Fig. 1.3 Nitrogen cycle
From: *http://www.marietta.edu/~biol/102/ecosystem.html*

TABLE 1.2	Biologically Important Compounds Which Contain Nitrogen
Class of Compounds	*Generalized Structure*
Amides	$\begin{matrix} O \\ R-C-NH_2 \end{matrix}$
Amines	$\begin{matrix} H \\ N \quad R \\ R_1 \end{matrix}$
Betaines	$\begin{matrix} CH_3 \\ H_3C\ N^+ \quad CH_2COO \\ CH_3 \end{matrix}$
Alkaloids	Dashek (1997)[a]
Porphyrins	Dashek (1997)
Vitamins	Dashek (1997)

[a]There are diverse classes of alkaloids, porphyrins and vitamins. Thus, the student is referred to Dashek, W.V. (1997). *Methods in Plant Biochemistry and Molecular Biology.* CRC Press, Boca Raton, FL. Nucleic acids, DNA and RNA, contain nitrogen in purines and pyrimidines (Fig. 1.4). Nitorgen is 4X as abundant in the atmosphere as O_2.

petuate the flow of energy and the cycle of nutrients resulting in ecosystem homeostasis (balance of nature). Nitrogen in organic compounds (Table 1.2) can be returned to the soil through decomposition or as animal feces and urine.

Sulfur Cycle (Fig. 1.5)

Plants can take up sulfur from the soil and water as the sulfate (SO_4^{2-}) anion. Sulfur in plants can be incorporated into animals via ingestion. Sulfur occurs in various compounds in both plants and animals. Through decomposition and weathering the sulfur in plants and animals can be returned to the soil. Sulfur present in fossil fuels can be released during burning of the fuel thereby releasing SO_4^{2-} into the atmosphere. Volcanic activity releases SO_2 which can be

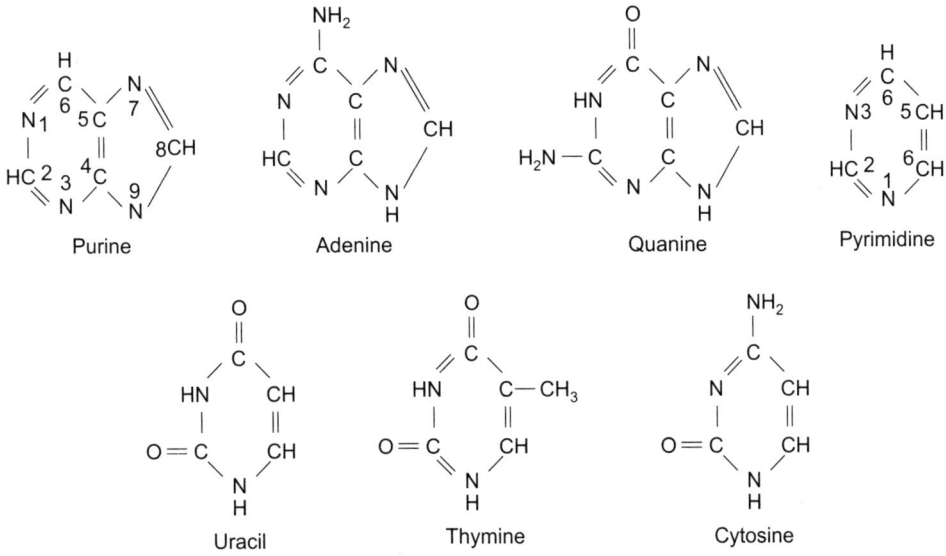

Fig. 1.4 Structure of purines and pyrimidines

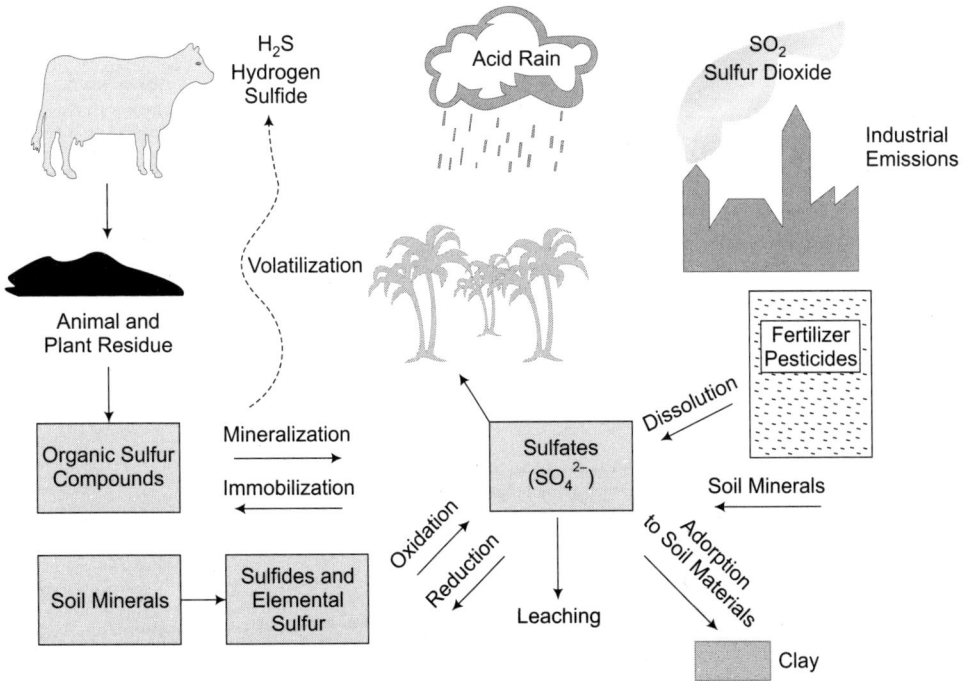

Fig. 1.5 Sulfur cycle
Adapted From: *http://www.agnr.umd.edu/users/agron/nutrient/factshee/sulfur/Sulfur.html*

transformed into SO_4^{2-}. Atmospheric SO_4^{2-} can be transported and through chemical reactions be converted to sulfuric acid, a destructive component of acid rain (see chapter 5).

In addition to O_2, C, N_2 and S other elements, e.g., phosphorus (Fig. 1.6), calcium (Ca^{++}, sodium (Na^{++}), magnesium (Mg^{++}) or iron (Fe^{++} or Fe^{++}) can cycle. However, the available pools of these elements are not as abundant as those for O_2, C, N_2 and S (Salomons et al., 1995). Phosphorus is significant because it is a constituent of fertilizers. which can run-off into waterways (see chapter 11).

What are the Major Ecosystems of the Biosphere? These are shown in Table 1.4.

What are the Ecological Sampling Methods Beneficial to Environmental Science?

There are many relevant ecological methods. Some ecological methods other than the Quadrat Procedure are presented in Table 1.5.

Table 1.6 presents kinds of plants and animals in a particular habitat and how many there are of each species.

The conclusion to the next chapter presents the student with references to both Ecology and Environmental Science books. These can be explored to various depths depending upon the student's interest and/or the class' orientation. However, if the student is an Environmental Science major,

---

TABLE 1.4 Ecosystems of the Biosphere

Systems	Characteristics
Aquatic	
Oceans (Scott, 1996)	*Euphotic zone* illuminated zone to ~ 200 meters, varies considerably in turbidity; notable organisms are primary autotrophs such as phytoplankton; also present are zooplankton (primary consumers) and omnivores and predators; bottom sediment contains benthic organisms.
Freshwater (Carpenter et al, 1992)	Involves ecological relationships similar to those in the oceans.
Terrestrial (Walker and Steffen, 1997; Barbour et al, 1980; Agren and Bosatta, 1996; Gooran and Bosatta,1996)	Biomes are large, stable terrestrial ecosystems resulting from the interactions of many environmental, biological and evolutionary factors.

TABLE 1.5 Some Ecological Methods Other than the Quadrat Procedure[a]

Estimating Abundance

Mark-Recapture Techniques
Removal Methods and Resight Methods
Line Transects and Distance Methods

Estimating Community Patterns

Similarity Coefficients and Cluster Analysis
Species Diversity Measures
Niche Measures

[a]These methods are discussed in Krebs (1999) and Southwood and Henderson (2006).

TABLE 1.6	Kinds of Plants and Animals in a Particular Habitat and How Many there are of Each Species?
Difficulty	How can one count each and every species?
Solution	Collect a number of species from randomly-selected regions of the habitat
Assumption	Samples are representative of the habitat
Sampling Method	QUADRAT
	Def. A square frame
	Purpose: A square frame enables
	comparable samples to be obtained from areas of consistent size and shape
	Type of Quadrat
	Standard Sampling Unit
	Traditional square frame although a circular frame is possible; square frame allows for comparison with existing data
	Quadrat size can vary
	Small quadrats: provide a quick survey but are less reliable (see *http://www.offwell.free-online. co.uk/howto.htm* for a discussion of quadrat size.

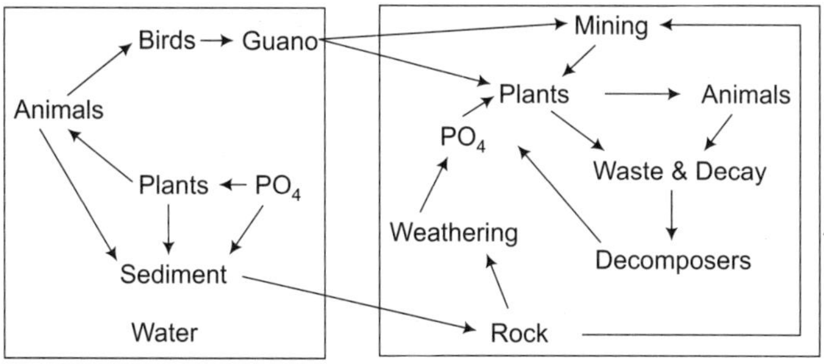

Fig. 1.6 Phosphorus Cycle
From: *http://.www.marietta.edu/~bioil/102/ecosystem.html*

one should read one or more books from both the Ecology and Environmental Science list.

References

Agren, G.I. and Bosatta, E. 1996. *Theoretical Ecosystem Ecology: Understanding Element Cycles*. Cambridge Univ. Press, New York, NY, USA.

Barbour, M.B., Burk, J.H. and Pitts, W.D. 1980. *Terrestrial Plant Ecology*. The Benjamin Cummings Co., Menlo Park, CA, USA.

Baron, J.S. 2003. *Rocky Mountain Features: An Ecological Perspective*. Island Press, Covelo, CA, USA.

Bowlby, N. 2006. Plant metabolism-respiration In: *Plant Cell Biology* (Dashek, W.V. and Harrison, M., eds.). Science Publishers, Enfield, NH, USA, pp. 359-398.

Carpenter, S.R. , Fisher, S.G., Grimm, N.B. and Kitchell, J.F. 1992. Global change and freshwater ecosystems. *Ann. Rev. Ecology and Systematics* 23: 119-139.

Gooran, A and Bosatta, E. 1996. *Theoretical Ecosystems Ecology: Understanding Element*

Cycles. Cambridge Univ. Press, Cambridge, UK.

Hokikian, J. 2002. *The Science of Disorder: Understanding the Complexity and Pollution in Our World.* Los Feliz Pub., Los Angeles, CA, USA.

Jorgensen, S.E. and Svirezhev, Y.M. 2004. *Towards a Thermodynamic Theory for Ecological Systems.* Pergamon (Elsevier), New York, NY, USA.

Krebs, C.J. 1999. *Ecological Methods.* Addison Wesley and Benjamin Cummings, Menlo Park, CA, USA.

Neutel, Anje-Margriet, Heesterbeek, J.A.P. and de Ruiter, P.C. 2002. Stability in real food webs: weak links in long loops. Science 296: 1120-1123.

Osborne, P.L. 2000. *Tropical Ecosystems and Ecological Concepts.* Cambridge Univ. Press, Cambridge, UK.

Salomons, W., Forstner, U. and Maden, P. 1995. *Heavy Metals.* Springer-Verlag, New York, NY, USA.

Scott, T.A. 1996. *Concise Encyclopedia of Biology.* Walter de Gruyter, Berlin, Germany.

Southwood, R. and Henderson, P.A. 2006. 'Ecological Methods'. *Blackwell Science,* Boston, MA, USA.

Tscharntke, T. and Hawkins, B.A. 2002. *Multitrophic Level Interactions.* Cambridge Univ. Press, Cambridge, UK.

Trabalha, J.R. and Reichle, D.E. 1986. *The Changing Carbon Cycle: A Global Analysis.* Springer Verlag, New York, NY, USA.

Turk, J. and Turk, A. 1997. *Environmental Science.* Harcourt-Brace College Publishers, Orlando, FL, USA.

Walker, B. and Steffen, N. 1997. Management ecosystem. An overview of the implications of global change for natural and conservation. Ecology [online] 1(2):2.

CHAPTER 2

Environmental Science

INTRODUCTION TO ENVIRONMENTAL SCIENCE AND ITS RELATIONSHIP TO ECOLOGY

Environmental science, which is closely related to Ecology, is concerned with global environmental change which can occur in both aquatic and terrestrial ecosystems (Chapter 1, Table 1.2). This change can occur via both natural and anthropogenic (man-made) means.

Many of these changes are from non-point and point pollution. Two major concerns of contemporary environmentalists are global climate change and pollution. Whereas the former centers around global warming, the latter involves detecting, analyzing and reducing air, ground, noise, thermal and water pollutions. Other concerns are population growth and species extinction. The scope of environmental science also includes law and management. To this end, environmentalists often use ecological methods (Table 2.1). Environmental Science can be approached from a variety of viewpoints, e.g., atmospheric sciences, biology, chemistry, economics, engineering, geology, management, mathematics, physics and policy decisions. Thus, environmental science is often more encompassing than ecology, which in its pure form is a branch of Biology. In popular usage, ecology has been mistakenly equated with environmental activism.

A specific Environmental Sampling Method Monitoring Carbon Dioxide (A Greenhouse Gas)

There is a variety of methods for analyzing carbon dioxide (Csuros, 1997). These include: solid electrolyte CO_2 gas sensors, a non-dispersive infrared gas analyzer, and diffusion CO_2 monitor Hoskin's diagnostic kits. In addition, a simple bioassay employing leaf discs from sun and shade plants (SAPS Student Sheet 3, 1990) is available (*http://www.sapsplantsci.com.oc.uk/worksheet/disc.htm*).

Gas-liquid chromatography (Barnola et al., 1983, 1987; Gravestein et al., 1987.) injects a vaporized sample into the head of a column which consists of a liquid stationary phase absorbent onto the surface of an inert solid (Fig. 2.2). An inert, mobile gaseous phase transports the sample through the column. Table 2.2 summarizes the components of a gas chromatography system.

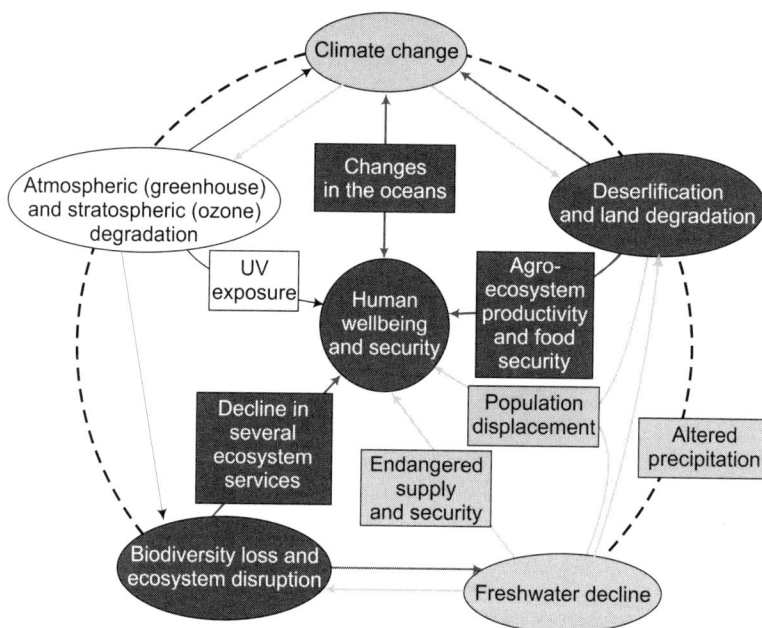

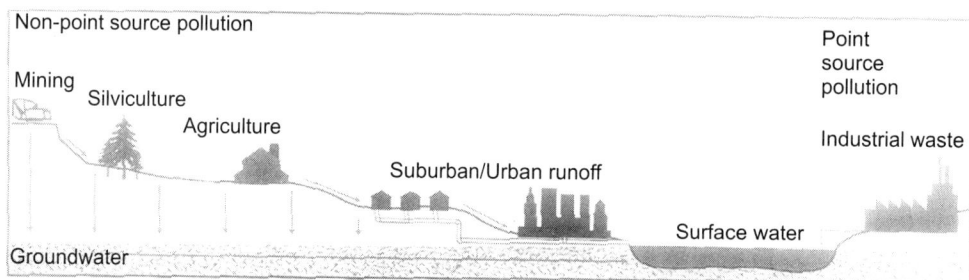

Point

Stationary sources that can be identified individually by name and location, such as electric power plants. A point source may be designated a major source if its emissions exceed a level set by Federal regulations.

Non-point

Small stationary sources that are not identified individually, such as neighborhood dry cleaners; and diffuse stationary sources, such as wildfires and agricultural tilling. (This category was previously called area sources.) Although each non-point source's emissions may be small, their aggregate emissions can be significant.

Fig. 2.1 Global Environmental Change and Comparison of Non-point and Point Pollution. From: *http://www.uvic.ca/clear/*. Adapted from: *http://www.who.int/global change*. The WHO diagram is less detailed than that in Uvic.ca/cler/scopl oragion. Comparison of point and non-point pollutions. From: *http://www.hsrc.ssw.org/update25.pdf*.

TABLE 2.1   Comparison of Ecological and Environmental Sampling Methods[f]

Ecological Methods[a]	Application	Environmental Methods[b]	Application
Mark-Recapture Technique	Estimating abundance	Microscopic techniques	Analytical methods (See Tables 2.4 and 2.5)
Removal and Resight Methods		Immunoassays	
Quadrat Counts[d]		Spectroscopic Methods Atomic Absorption Inductively Coupled	
		Plasma (ICP) Infrared Ion Selective Electrodes Mass-Spectrometry Raman Spectroscopy	
Sampling Coefficients and Cluster Analysis[e]	Estimating community patterns	Well installation[c] Soil cores	Field methods
Species Diversity		Dredges Kemmerers	
Niche Measures		ISCO Autosamplers Secchi Discs	

[a]Discussed in Krebs (1999) and Southwood and Henderson (2000).

[b]Described in Green and Harrison (1979), Keith (1996) and Gallagher, doug@vt.edu.

[c]US Water News Online – Environmental Sampling and Analysis (2002). *http://www.uswaternews.com.books/bksbycategory/ 4dENVChesamp/4dENVChe Envsamp.*

[d]Reviewed in Sorrells and Glenn (1991). Review of sampling techniques used in studies of grassland plant communities. Proceedings of the Oklahoma Academy of Science, Volume 71. (see literature cited).

[e]*http://www.fiu.edu/~ecosyst/lab%20manual/lab%20info/lab2.htm.* See Data Analysis Technique.

[f]See appended references

Gallagher, D., doug@vt.edu EMBL Data Bases and Sample Archives.

Green, R. and Harrison, R. 1979. *The Sampling Design and Statistical Method for Environmental Biologists.* Wiley, New York, NY, USA.

Keith, L.H. 1996. *Principles of Environmental Sampling.* 2nd Ed., Oxford University Press, Oxford, England.

Krebs, C.J. 1999. *Ecological Methods.* Addison Wesley and Benjamin Cummings, Menlo Park, CA, USA.

Southwood, T. and Henderson, P.A. 2000. *Ecological Methods.* 3rd Ed., Blackwell Science, Oxford, England.

TABLE 2.2   Gas Chromatography System[a]

Component	Characteristics
Carrier gas	Chemically inert; such as nitrogen, helium, argon and carbon dioxide; gas is chosen on the basis of detector used.
Sample injection port	Uses small sample (1-20 µl) introduced onto the packed column as a 'plug' of vapor via a microsyringe; sample's temperature is maintained a t~50°C > the boiling point of the least volatile component of the sample.
Columns	There are two types, packed or capillary columns (open tubular); capillary columns are wall-coated open tubular or support-coated open–tubular (see *http:// www.shu.oc.uk/schools/sci/chem./Tutorials/chrom/gaschem.htm for details*).
Column Temperature	Optimum column temperature is dependent upon the boiling point of the sample
Detectors	Non-selective detector responds to all compounds except the carrier gas; Selective detector responds to a range of compounds with a common physical or chemical property; Specific detector responds to a single chemical compound. Detectors can be grouped into concentration dependent factors and mass flow dependent detectors; the former does not destroy the sample and is related to the concentration of

(Contd.)

(Contd.)

soluble molecules in the detector; the latter destroys the sample and the signal is related to the rate at which soluble molecules enter the detector. The available detectors are flame ionization, thermal conductivity, electron capture, nitrogen-phosphorus, flame photometric, photoionization and Hall electrolytic conductivity.

[a]Adapted from Sheffield Hallom University School of Science and Mathematics, *http://www.sju.oc.uk/schools/sci/chem./tutorials/chrom/gaschem.htm.*

Barnola, J.M., Raynaud, D., Neftel, A. and Oeschager, H. 1983. Comparison of CO_2 measurements by two laboratories on air from bubbles in polar ice. Nature 303: 410-413.

Barnola, J.M., Raynaud, D., Korotkevich, Y.S. and Lorius, C. 1987. Vostok ice core provides 160,000 year record of atmospheric CO_2. Nature 329: 408-414.

Csuros, M. 1997. *Environmental Sampling and Analysis Laboratory Manual.* CRC Press, Boca Raton, FL, USA.

Gravenstein, J.S., Paulus, A. and Hayes, T.J. 1995. Gas Monitoring in Clinical Practice. Butterworth-Heineman, Boston, MA, USA.

TABLE 2.3 Summary of an HPLC Method[a]

Process

- Injection of a solution of compounds into a loop calibrated to contain a specific volume;
- Transfer from the loop onto a column packed with an appropriate stationary phase;
- Eluent is delivered from a pump at a constant rate and pressure adequate to overcome the column's backpressure; High pressures are necessary to force a liquid through a column tightly packed with small particles.
- Differential migration occurs because of differences in partitioning of compounds between the mobile phase (eluent) and stationary phase (column packing).
- Retention time is the time for a compound to pass through the column and is an indicator of a compound's identity; Quantitative analysis is derived from the area or height of a peak generated by a detector; Calibration curves can be constructed using a series of standards.

[a]Adapted from: *http://pages.towson.edu/larkin/210DOCS/exp8.pdf.* The student is encouraged to read this source and that of Neue (1997). 'HPLC Coulmns Theory, Technology and Practice.' Wiley-VCH, New York, NY, USA.

In addition to gas chromatography, high performance liquid chromatography (HPLC) is frequently employed in environmental analyses. This method (Fig. 2.3) is utilized when a sample to be analyzed cannot easily be converted to a gas phase for GC. Table 2.3 presents a summary of an HPLC method.

Another chromatographic technique useful to environmental scientists is ion chromatography (Fig. 2.4). This methodology (Weiss and Weiss, 2005) is valuable for determining inorganic and organic ions (Shaw and Haddad, 2004) as well as sulfur dioxide (ISO TC 146/SCI, 2007). The method is dependent upon ion exchange on stationary phases containing functional groups.

In contrast to chromatographies, electrophoresis involves the differential migration of charged molecules in an electric field. For capillary electrophoresis (Fig. 2.5.) , the following steps are employed: 1. Electrophoresis occurs in buffer-filled capillary tubes of 25-100 um internal diameter., 2. Anode and cathode electrodes are positioned in a solution containing ions., 3. High voltage is applied across the electrodes. and 4. Anions and cations migrate through the solution towards electrodes of

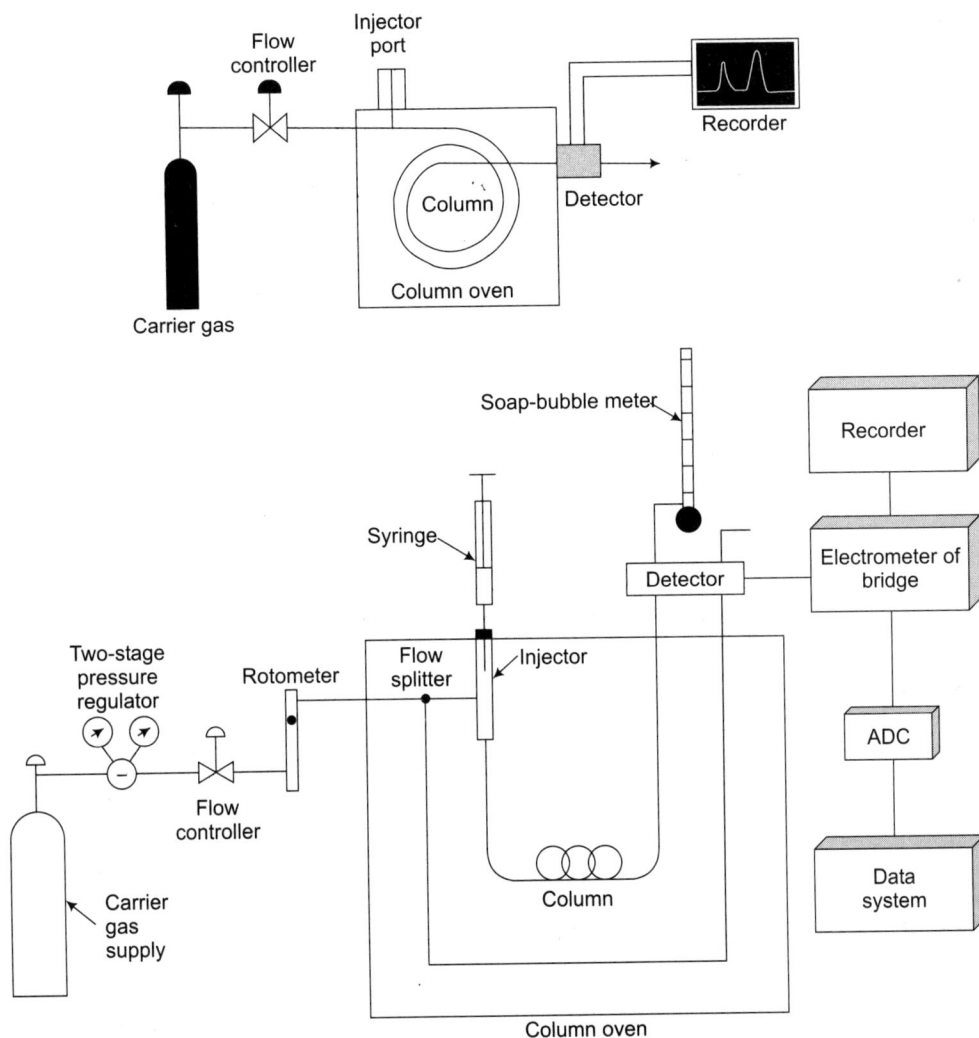

Fig. 2.2   Schematic of a gas chromatograph. From: *http://www.shu.oc.uk/schools/sci/chem./tutorials/chrom/ gaschem.htm*

opposite charge. Capillary electrophoresis is useful for the speciation of arsenic, chromium and cobalt in soil solutions (Shintani and Polonski, 1996).

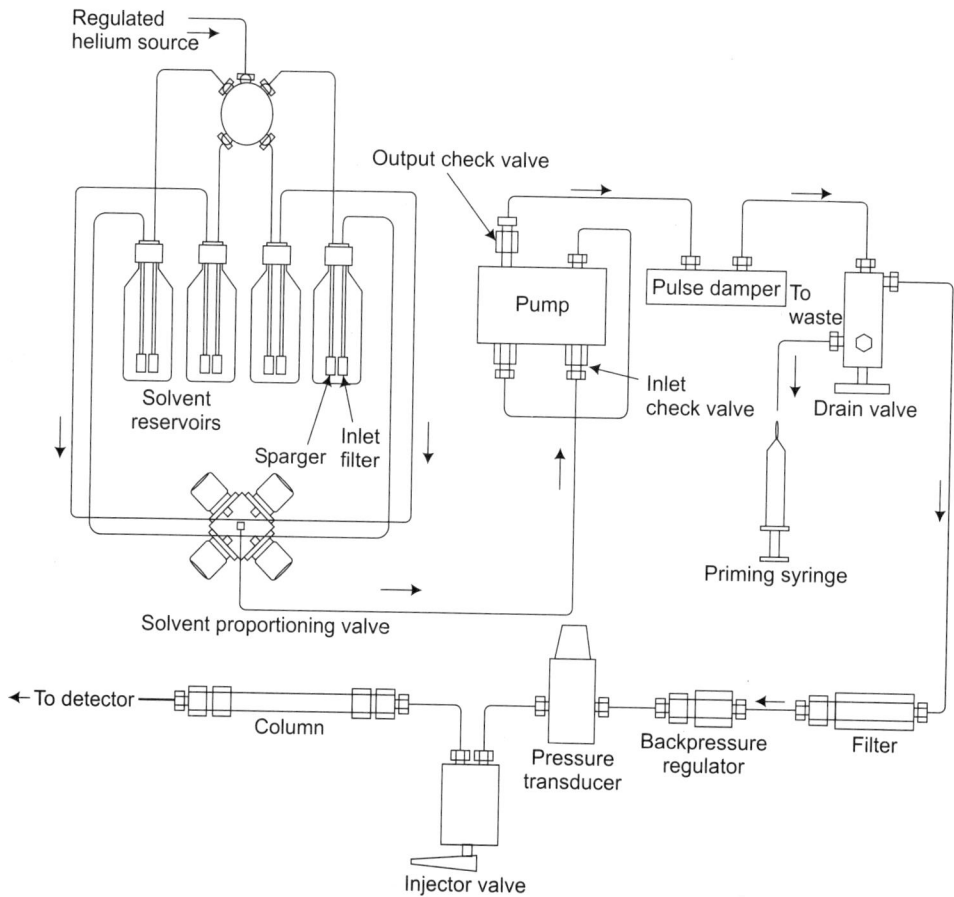

Fig. 2.3 HPLC System. From: pages *http://towson.edu/larkin/210DOCS/exp.8.pdf*

TABLE 2.4 Summary of Light and Electron Microscopies Useful to Environmental Scientists

Light Microscopy		*Electron Microscopy and Ancillary Methods*	
Technique	Application	Technique	Application
Bright field	Conventional microscopy	Atomic force	Mapping of surfaces to an atomic scale
Confocal scanning optical	Examination of cells in live tissue in bulk samples	Cryoelectron	Imaging of biological macromolecules in the absence of specimen dehydration and staining
Confocal fluorescence	DNA labeled with more than one fluorescent tag		
Dark field	Visualization technique for ashes produced by micro-incineration and	Electron systems imaging EM shadowing	Detection, localization and quantitation of light elements. Structural

(Contd.)

(Contd.)

	fluorescence microscopy; useful for low-contrast subjects		information from ordered arrays of macromolecules
Reflection contrast	Quantification in gap between light and EM microscopies	Immunoelectron	Localization of cellular antigens
Reflection-imaging	Useful for imaging highly reflective particles such as silver grains in autoradiographs		
Field ion	Atomic structure of crystals		
Nearfield scanning optical	Determination of single molecules on surface	Negative staining	Useful for detergent-extracted cytoskeletons, membrane fractions, isolated organelles
Nuclear magnetic resonance	High-resolution 3-D imaging of living plants; forms images if H_2O in the body; water distribution and binding in transpiring plants and H_2O transport in plants with light-stressed foliage	Scanning electron	Surface topography
Nomarski differential interference contrast	Reveals edges in biological structures, e.g., organelle and nuclear boundaries, cell walls; also images fibrous subcellular components, e.g., microtubules	Scanning tunneling	Surface topography, image internal structure of macromolecules such as proteins, liquid crystals, and DNA
Phase contrast	Produces visible differences in retardation of light waves, useful for biological material which possesses limited inherent direct contrast	Transmission electron	Subcellular morphology
Polarization	Most useful for highly birefringent objects, e.g., cellulosic microfibrils in cell walls and distinguishing crystalline and non-crystalline inclusions	X-ray microanalysis	Detection, localization and quantitation of elements
Raman	Analysis of bioaccumulations in plant vacuoles		
Tomographic phase	Creates 3-D images of organelles, cells and small organisms		
Scanning confocal near-infrared	Images thin sections of thick tissue specimens non-destructively		

(Contd.)

(Contd.)

Atom probe field ion	Images solid surfaces with atomic resolution
Scanning force	Visualizes DNA and protein complex
Fluorescent *in situ* hybridization	Localizes presence or absences of specific DNA
X-ray microscopy	Characterizes radioactive fuel particles from reactors
Multiphoton fluorescence	Generates confocal-like images

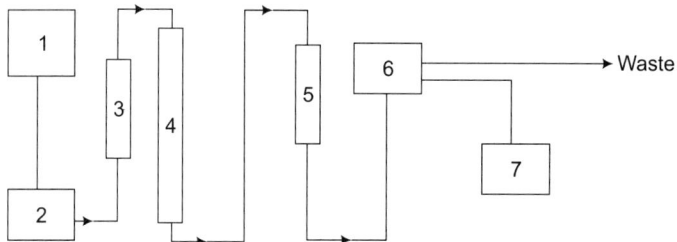

(1) Eluent reservoir
(2) Pump
(3) Precolumn
(4) Separator column
(5) Suppressor column
(6) Detector
(7) Recorder or integrator, or both

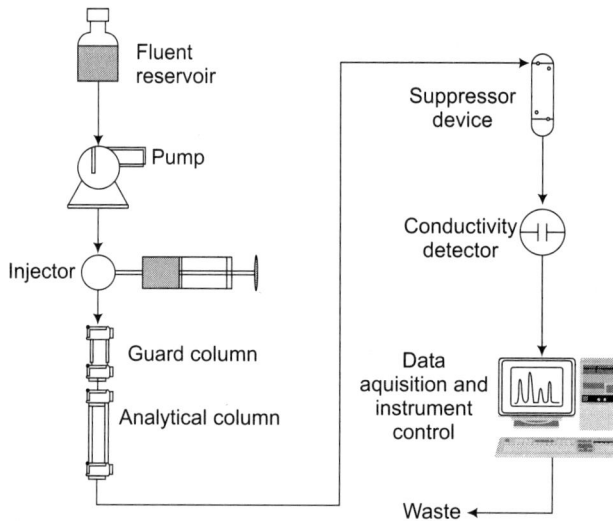

Fig. 2.4 Depiction of an ion chromatograph. From: *http://www.epa.gov/SW-846/pdfs/9056.pdf.*

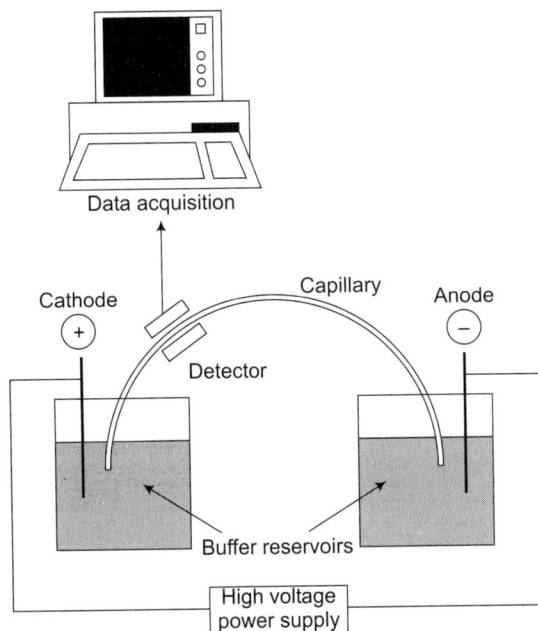

Fig. 2.5   Illustration of a capillary electrophoresis 'set-up.' From: *http:www.rsc.org/pdf/books/caplelectrosc.pdf.* Other useful diagrams occur at: *http://www.shsu.edu/~chemtgc/sounds/flashfiles/CE.swf and http:// www.beckman.com/literature/BioResearch/360643-CEPrimer1/pdf.*

Microscopy References

Bronnikov, A.V. 2002. Theory of quantitative phase contrast computed tomography. *J. Opt. Soc. Am.* 19: 472-480.

Brown, L.M. 1997. A synchrotron in a microscope. In: *Electron Microscopy and Analysis.* IOP Publishing, Bristol, UK pp. 17-28.

Burrells, W. 1977. *Microscope Technique. A Comprehensive Handbook for General and Applied Microscopy.* Halsted Press, Wiley, New York, NY, USA.

Callaghan, P. 1994. *Principles of Nuclear Magnetic Resonance Microscopy.* Oxford Univ. Press, Oxford, UK.

Chen, Y., Shah, N., Huggins, F.E. and Huffman. 2006. Microanalysis of ambient particles from Lexington, KY by electron microscopy. *Atoms. Environ.* 40: 651-663.

Cherry, R.J. 1991. *New Techniques of Optical Microscopy and Microspectroscopy.* CRC Press, Boca Raton, FL, USA.

Choma, M.A., Ellerbee, A.K., Yang, C. and Izatt, J.A. 2005. Spectral-domain phase microscopy. *Opt. Lett.* 30:1162-1163.

Croft, W.J. 2006. 'Under the Microscope: A Brief History of Microscopy.' *World Scientific,* Singapore, Malaysia.

Diver, S. 2004. [Compostteas] Fluorescence microscopy for visualization of soil microorganisms – a review. *Biol. Fert. Soils* 39: 301-311.

Egerton, R.F. 2008. *'Physical Principles of Electron Microscopy. An Intoduction to TEM, SEM and AEM.'* Springer, New York, NY, USA.

Egerton, R.F. and Malac, M. 2005. EELS in the TEM. *J. Electron Spectrosc.* 143:45-52.

Eggleston, C.M., Higgins, S.R. and Maurice, P.A. 1998. Scanning probe microscopy of environmental interfaces. *Environ. Sci. Technol.* 32: 456A-459A.

Filler, T.J. and Peuker, E.T. 2000. Reflection contrast microscopy (RCM) a forgotten technique. *J. Pathol.* 190:635-638.

TABLE 2.5 Summary of an Immunoassay and Spectroscopic Methods Useful to Environmental Scientists[a,b]

Technique	Application
Spectroscopy	
Atomic Absorption (AA)	Detection of alkali and alkaline earth metals
Differential Optical Absorption (DOAS)	Measures amounts of atmospheric trace gases
Fourier Transform Infrared Red (FTIR)	Air contaminant analyses
Gas Chromatography/Mass Spectrometry and Gas Chromatography	Detection of semivolatiles insoluble in a water matrix but soluble in a non-polar organic solvent
Inductively Coupled Argon Plasma Emission (ICP or ICAP)	Examination of elements; low levels of heavy metals can be determined by ICP/MS
Infrared (IR)	Analyses of petroleum and air contaminants products, hydrocarbons and volatile organic compounds
Mass Spectrometry (MS)	Detection of ppbv concentrations of volatile organics
Ion Trap Mass Spectrometer	
Tandem Source Glow Discharge Quadrupole Mass Spectrometer	
Nuclear Magnetic Resonance (NMR)	Characterization of nutrients and natural organic matter
Raman	Certain pollutants in aqueous systems
Spectroscopies based on flame diagnostics, e.g., atomic emission (AES)	Establish amounts of certain metals
X-ray fluorescence	Detection of major and trace elements is solids
Immunoassay	
Enzyme-linked Immunoabsorbent (ELISA)	Detection of Pesticides

[a]Thorough accounts of environmental analytical methods can be found in Kebbekus and Mitra (1997) and Smith (1997). Drum et al. (2001) review field analysis methods for water, solids, gels, oils , non-aqueous substances, air and gas pollutants, as well as residential noise and thermal pollutions. Other useful tests regarding environmental analysis are in Patnark (1997), Telliard (1999), Reeve (2002), Dean (2003) and Tanner (2005).

[b]Photothermal, Mössbauer and infrared vibration spectroscopies are also useful for environmental analyses. Dean, J.R. 2003. *Methods for Environmental Trace Analysis.*

Finzi, L. and Dunlap, D.D. 2001. *'Polarized Light Microscopy. Encyclopedia of Life Sciences.'* Wiley, New York, NY, USA.

Gersh, I. 1973. *Submicroscopic Cytochemistry.* Academic Press, New York, NY, USA.

Goldstein, J.L. 1981. *Scanning Electron Microscopy and X-ray Microanalysis: A Text for Biologists. Material Scientists and Geologists,* Plenum Press, New York, NY, USA.

Hall, J.L. 1978. *Electron Microscopy and Cytochemistry of Plant Cells.* Elsevier North Holland Biomed. Press, Amsterdam.

Harder, D.P. 1992. *Image Analysis in Biology.* CRC Press, Boca Raton, FL, USA.

Harris, N. and Oparka, K.J. (eds.). 1994. *Plant Cell Biology.* IRL Press, Oxford, UK, pp. 156-157.

Hawes, C. 1994. Electron microscopy. In: *Plant Cell Biology.* Harris, N. and Oparka, K.J. (eds.). IRL Press, Oxford, UK, pp. 52-68.

Hayat, M.A. 1980. *X-ray Microanalysis in Biology.* Univ. Park Press, Baltimore, MD, USA.

Herman, B. and LeMasters, J.J. 1993. *Optical Microscopy.* Academic Press, San Diego, CA, USA.

Inoue, S and Spring, R. 1997. *Video Microscopy. The Fundamentals.* Springer, Berlin, Germany.

Inoue, S. 2006. Foundations of confocal scanned imaging in light microscopy. In: *Handbook of Biological Confocal Microscopy.* (Pawley, J.B., ed.). Springer, New York, NY, USA, pp. 1-19.

Jahne, B. 1997. *Practical Handbook on Image Processing for Scientific Applications.* CRC Press, Boca Raton, FL, USA.

Jones, C., Mulloy, B. and Thomas, H. 1994. *Optical Spectroscopy and Macroscopic Techniques.* Humana Press, Totowa, NJ, USA.

Juniper, B.C., Cox, C.C., Gilchrist, A.J. and Williams, P.R. 1970. *Techniques for Plant Electron Microscopy.* Blackwell Sci. Publ. Oxford, UK.

Kawata, S. 2001. *Near-Field Optics and Surface Plasmon Polaritons.* Springer, Berlin, Germany.

Kaupp, G. 2006. *Atomic Force Microscopy, Scanning Nearfield Optical Microscopy and Nanoscratching Application to Rough and Natural Surfaces.* Springer-Verlag, Berlin, Germany.

Koenraad, P.M. and Kemerink, M. 2003. Scanning tunneling microscopy/spectroscopy and related techniques. *AIP Conference Proceedings* 696: 179p, Eindhoven, The Netherlands.

Marmasse, C. 1980. *Microscopes and Their Uses.* Gordon and Breach, New York, NY, USA.

MedGadget. 2007. Tomographic phase microscopy: A new imaging modality for cells. *http://medgadget.com/archives/2007/08/tomographic_phase_microscopy-a-new-imaging.*

Meyer-llse,W., Warwick, T. and Attwood, D. 2000. 'X-ray Microscopy.' American Insitute of Physics, Melville, NY, USA.

Mohanty, S.B. 1982. *Electron Microscopy for Biologists.* Charles Thomas, Springfield, IL, USA.

Morita, S. 2007. 'Roadmap of Scanning Probe Microscopy.' Springer Verlag, Berlin, Germany.

Morrison, G. and Buckley, C.H. 1992. *X-ray Microscopy.* Springer Verlag, Berlin, Germany.

Muller, M. 2005. *Introduction to Confocal Fluorescence Microscopy.* SPIE Publications, Bellingham, WA, USA.

Murthy, M.S.N., Jones, M.G., Kulka, J., Davies, J.D., Halliwell, M., Jackson, P.C. and Bull, D.R. 1995. Scanning confocal near-infa-red microscopy. A new microscopy technique for three dimensional histopathology. *Eng. Sci. Educ. J.* 4: 223-230.

Othmar, M. and Amrein, M. 1993. *STM and SRM in Biology.* Academic Press, San Diego, CA, USA.

Panitz, J.A. 1982. Field ion microscopy-a review of basic principles and selected applications. *J. Phys. E. Sci. Instrum.* 15:1281-1294.

Posteck, M.A., Howard, K.S., Johnson, A.H. and mcMichael, K.L. 1980. *Scanning Electron Microscopy: A Student's Handbook.* Ladd Res. Indus. Inc., Burlington, VT, USA.

Prins, F.A. 2006. Reflection contrast microscopy: the bridge between light and electron microscopy. In: *Cell Imaging Techniques Methods and Protocols.* (Taatjes, D.J. and Mossman, B.T., eds.). Humana Press, Totowa, NJ, USA.

Reed, S.J.B. 1993. *Electron Microprobe Analysis.* Cambridge Univ. Press, Cambridge, MA, USA.

Rost, F.W.D. 1992. *Fluorescence Microscopy.* Cambridge Univ. Press, Cambridge, UK.

Russ, J.C. 1995. *The Image Processing Handbook.* CRC Press, Boca Raton, FL, USA.

Sako, Y., Sekihato, A., Yanagisaiwa, Y., Yamamoto, M., Shimata, Y, Opaki, K. and Kusimi, A. 1995. Comparison of two-photon excitation laser scanning microscopy with uv-confocal laser scanning microscopy in three-dimensional calcium imaging using the fluorescence indicator Indo-1. *J. Micro.* 185: 9-20.

Samson, E.C. and Blanca, C.M. 2007. Dynamic contrast enhancement in widefield microscopy using projector-generated illumination patterns. *New J. Phys.* 9: paper 363. pp. 1-14.

Satoshi, U. and Ewing, R.C. 2003. Application of high-angle annular dark field transmission electron microscopy, scanning transmission electron microscopy-energy dispersive x-ray spectrometry and energy-filtered transmission electron microscopy to the characterization of nonparticles in the environment. *Environ. Sci. Technol.* 37: 786-791.

Shaw, P.J. and Rawlins, D.J. 1994. An introduction to optical microscopy for plant cell biology. In: *Plant Cell Biology, A practical Approach.* (Harris, N. and Oparka, K.J., eds.). Oxford Univ. Press, Oxford, UK. pp. 1-26.

Shotton, D. (ed.). 1993. *Electronic Light Microscopy. Modern Biomedical Microscopy.* Wiley-Liss, New York, NY, USA.

Sigee, D.C. 1993. *X-ray Microanalysis in Biology: Experimental Techniques and Applications.* Cambridge Univ. Press, Cambridge, UK.

Slayter, E.M. 1993. *Light and Electron Microscopy.* Cambridge Univ. Press, Cambridge, UK.

Smith, R.F. 1994. *Microscopy and Photomicrography.* CRC Press, Boca Raton, FL, USA.

Snigireva, I. and Snigirev, A. 2006. X-ray microanalytical techniques based on synchrotron radiation. *J. Environ. Monit.* 8: 33-42.

Stephens, D.J. and Allan, V.J. 2003. Light microscopy techniques for live cell imaging. *Science* 300: 82-86.

Swatland, H.J. 1997. *Computer Operations for Microscope Photometry.* CRC Press, Boca Raton, FL, USA.

Tribe, M.A., Evant, R.M. and Snook, R.K. 1975. *Electron Microscopy and Cell Structure.* Cambridge Univ. Press, Cambridge, UK.

Tsong, T.T. 2005. *Atom-Probe Field Ion Microscopy Field Ion Emission, Surfaces and Interfaces at Atomic Resolution.* Cambridge Univ. Press, Cambridge, UK.

Tsong, T.T. 2005. Some aspects of filed emission and field ion microscopy. *J. Electron Micr. Tech.* 2:229-245.

Turrell, G. and Corset, J. 1996. *Raman Microscopy Developments and Applications.* Academic Press, New York, NY, USA.

Webb, R.H. 1999. Theoretical basis of confocal microscopy. Methods in Enzymology. 370: 3-20.

Weis, D.G., Maile, W., Wick, R.A. and Steffen, W. 1999. Video microscopy. In: *Light Microscopy in Biology. A Practical Approach.* (Lacey, A.J., ed.). Oxford Univ Press, Oxford, England. pp. 73-149.

Williams, P.M., Cheema, M.S., Davies, M.C., Jackson, D.E. and Tedler, S.J.B. 1994. Methods in Molecular Biology, vol. 22, *Microscopy, Optical Spectroscopy and Microscopic Techniques.* (Jones, C., Mulloy, B. and Thomas, A.H., eds.). Humana Press, Totowa, NJ, USA.

Dean, J.R. 2003. *Methods for Environmental Trace Analysis.* Wiley, New York, NY, USA.

Drum, D.A., Bouman, S.L. and Shugar, G.J. 2001. *Environmental Field Testing and Analysis. Ready Reference Handbook.* McGraw Hill, New York, NY, USA.

Kebbekus, B.B. and Mitra, S. 1997. *Environmental Chemical Analysis.* CRC Press, Boca Raton, FL, USA.

Patnark, P. 1997. *Handbook of Environmental Analysis. Chemical Pollutants in Air, Water, Soil and Solid Wastes.* Lewis Publishers, Boca Raton, FL, USA.

Reeve, R.N. 2002. *Introduction to Environmental Analysis.* Wiley, New York, NY, USA.

Smith, Roy-Keith. 1997. *Guide to Environmental Analytical Methods.* Genium Publishing Corporation, Amsterdam, NY, USA.

Tanner, P.A. 2005. Application of spectroscopic methods to environmental problems. *Spectrosc. Lett.* 38: 211-212.

Telliard, W.A. 1999. EPA Analytical methods for the determination of pollutants in the environment. *Crit. Rev. Anal. Chem.* 29: 249-257.

Spectrometry References

Almond, D. and Patel, P. 1996. *Photothermal Science and Techniques.* Chapman and Hall, London, England.

Beckhoff, B. 2006. *Handbook of Practical X-ray Fluorescence Analysis.* Springer, Berlin, Germany.

Bialkowski, S.E. 1996. *Photothermal Spectroscopy Methods for Chemical Analysis.* Chemical Analysis: A Series of Monographs on Analytical Chemistry and Its Applications.' Voulme 134. Wiley and Sons, New York, NY, USA.

Bloemen, H., J. Th. and Burn, J. 1993. *Chemistry and Analysis of Volatile Organic Compounds in the Environment.* Blackie Academic, London, England.

Bonnell, D. 2000. *Scanning Probe Microscopy and Spectroscopy. Theory, Technique and Applications.* Wiley-VCH, New York, NY, USA.

Buhrke, V.E., Jenkins, R. and Smith, D.K. 1997. *A Practical Guide for the Preparation of Specimens for X-Ray Fluorescence and X-Ray Diffraction Analysis*. Wiley, New York, NY, USA.

Burlingame, A.L. 2005. *Biological Mass Spectrometry*. Elsevier, Amsterdam, The Netherlands.

Clark, R.J.H. and Hester, R.E. 1995. *Spectroscopy in Environmental Science*. Wiley, New York, NY, USA.

Cresser, M. 1994. *Flame Spectrometry in Enviromental Chemical Analysis: A Practical Guide*. Royal Society of Chemistry, London, England.

Dean, J.R. 2007. *Bioavailability, Bioaccessibilty and Mobility of Environmental Contaminants*. Wiley New York, NY, USA.

Downard, K. 2004. *Mass Spectrometry: A Foundation Course*. Royal Society of Chemistry, London, England.

Giannakopulos, A., Thomas, B., Colburn, A.W., Reynolds, D.J., Raptakis, E.N., Markarov, A. and Derrick, P.J. 2002. Tandem time-of-flight mass spectrometer (TOF-TOF) with a gquadratic-filed-ion mirror. Review of Scientific Instruments. 73:115-2123.

Grieken, R.E. and Markowicz, A.A. 2002. *Handbook of X-ray Spectrometry*. Marcel Dekker, New York, NY, USA.

Griffith, P. 2007. *Fourier Transform Spectrometry*. Wiley, Hoboken, NJ, USA. Hammes, G.C. *Spectroscopy for the Biological Sciences.'* Wiley, Hoboken, NJ,USA.

Hill, S.J. 2007. *'Inductively Coupled Plasma Spectrometry and Its Applications*. Blackwell, Oxford, England.

Hirokawa, K., Yasuda, K. and Takada, K. 1992. *Graphite Furnace Atomic Absorption Spectrometry and Phase Diagrams of Alloy*. Anal. Sci. 8: 411-417.

Hites, R. 2007. *Elements of Environmental Chemistry*. Wiley, New York, NY, USA.

Hufner, S. 2007. *'Very High Resolution Photoelectron Spectroscopy.'* Springer, Berlin, Germany.

Jacobsen, N.E. 2007. *'NMR Spectroscopy Explained. Simplified Theory, Applications and Examples For Organic Chemistry and Structural Biology.'* Wiley Interscience, Hoboken, NJ, USA.

Karasasek, F. and Clement, R.E. 1991. *Basic Gas Chromatography-Mass Spectrometry*. Elsevier, Burlington, MA, USA.

Lajunen, L.H.J. and Peramaki, P. 2004. *Spectrochemical Analysis by Atomic Absorption and Emission*. Royal Society of Chemistry, Cambridge, England.

Lakowicz, J.R. 2006. *Principles of Fluorescence Spectroscopy*. Springer, New York, NY, USA.

Leclercq, P.A. and Cramers, C.A. 1998. High speed GC-MS. Mass Spectrometry Reviews. 17:37-49.

Lungbland, R. 2007. *Biochemistry and Molecular Biology Compendium*. CRC Press, Boca Raton, FL, USA.

March, R.E. and Todd, J.F.J. 2005. *Quadrupole Ion Trap Mass Spectrometry*. Wiley, Hoboken, NJ, USA.

Murad, E. and Cashion, J. 2003. *Mossbauer Spectroscopy of Environmental Materials and Their Industrial Utilization*. Kluwer, Norwell, Mass., USA.

Nanny, M.A., Minear, R.A. and Leenheer, J.A. 1997. *'Nuclear Magnetic Resonance Spectroscopy in Environmental Chemistry.'* Oxford Univ. Press, Oxford, England.

Patnaik, P. 1997. *A Handbook of Environmental Analysis. Chemical Pollutants in Air, Soil and Solid Wastes*. Lewis Publishers, Boca Raton, FL, USA.

Pellizzari, E.D. and Bursey, J.T. 1985. Gas Chromatography/mass spectrometry in water pollution studies. In: *Mass Spectrometry in Environmental Science*. (Karasek, F.W., Hutzinger, O. and Safe, S., eds.). Plenum Press, New York, NY, USA, pp. 139-158.

Platt, U. and Stutz, J. 2007. *Differential Optical Absorption Spectroscopy: Principles and Applications*. Springer, Berlin, Germany.

Ramamoorthy, A. 2006. *NMR Spectroscopy of Biological Solids*. Taylor and Francis, Boca Raton, FL, USA .

Reeve, R.N. 2002. *Introduction to Environmental Analysis*. Wiley, New York, NY, USA.

Schlag, E.W. 1994. *Time-of-Flight Spectrometry and its Applications*. Elsevier, Amsterdam, The Netherlands.

Smith, E. and Dent, G. 2005. *Modern Raman Spectroscopy: A Practical Approach.* Wiley, New York, NY, USA.

Smith, Roy-Keith. 1997. *Guide to Environmental Analytical Methods.* 5th Ed. Genium Publishing Corporation, Amsterdam, New York, USA.

Speight, J.G. 2005. *Environmental Analysis and Technology for the Refining Industry.* Wiley-Interscience, Hoboken, NJ, USA.

Stuart, B.H. and Ando, D.J. 1997. *Biological Applications of Infrared Spectroscopy.* Wiley, New York, NY, USA.

Tanner, P.A. 2005. *Application of Spectroscopic Methods to Environmental Problems.* Spectrosc. Lett. 38: 211-212.

Telliard, W.A. 1999. *EPA Analytical Methods for the Determination of Pollutants in the Environment.* Crit. Rev. Anal. Chem. 29: 249-257.

Tranter, G.E., Holmes, J.L. and Lindon, J.C. 2000. *Encyclopedia of Spectroscopy and Spectrometry.* Academic Press, New York, NY, USA.

Van Emon, J.M. 2006. *Immunoassays and Other Bioanalytical Techniques.* CRC Press, Boca Raton, FL, USA.

Van Hare, D.R., Carreira, L.A. and Rogers, L.B. 1986. Quantitative measurements of individual polycylic aromatic hydrocarbons in a mixture by coherent anti-stokes Raman Spectroscopy. In: *Air and Water Analysis: New Techniques and Data.* (Frei, R.W. and Albaiges, eds.). *Current Topics in Environment and Toxicological Chemistry.* Gordon and Breach, New York, NY, USA. pp. 59-71.

Venables, J.A. and Liu, J. 2005. High spatial resolution studies of surfaces and small particles using electron beam techniques. J. Electron. Spectrosc. 143: 207-220.

Verma, H.R. 2007. '*Atomic and Nuclear Analytical Methods: XRF, Mossbauer, XPS, NAA and Ion Beam Spectroscopic Techniques.* Springer, Berlin, Germany.

Westwood, S.A. 1993. *Supercritical Fluid Extraction and Its Use in Chromatographic Sample Preparations.* Chapman and Hall, New York, NY, USA.

Wise, M.B., Hurst, G.B., Thompson, C.V., Buchannan, M.V. and Guerin, M.R. 1991. Screening volatile organics by direct ion trap and glow discharge mass spectrometry. In: *Proceedings of the Second International Symposium on Field Screening Methods for Hazardous Wastes and Toxics.* Las Vegas, NV, USA. pp. 273-288.

General References for Chapter 1 and 2

Akcakapr, H.R., Burgman, M.A. and Ginzburg, L.R. 2002. *Applied Population Ecology.* Romas Ecolab., Setauket, NY, USA.

Ablers, P.H., Heinz, G.H. and Ohlendorf, H.M. 1998. *Environmental Contaminants and Terrestrial Vertebrates: Effects on Populations, Communities and Ecosystems.* SETAC, Pensacola, FL, USA.

Allaby, M. 1996. *Basics of Environmental Science.* Routledge, New York, NY, USA.

Andersa, J.M. 1981. *Ecology For Environmental Sciences: Biosphere, Ecosystem and Man.* Wiley, New York, NY, USA.

Arms, K. 1996. *Environmental Science.* H. Holt and Co., New York, NY, USA.

Barbour, M.B., Burk, J.H. and Pitts, W.D. 1980. *Terrestrial Plant Ecology.* The Benjamin Cummings Co., Menlo Park, CA, USA.

Bazzaz, F.A. 1996. *Plants in Changing Environments: Linking Physiological, Population and Community Ecology.* Cambridge Univ. Press, Cambridge, England.

Beacham, W. 1993. *Beacham's Guide to Environmental Issues and Sources.* Beacham Pub., Washington, DC, USA.

Begon, M., Harper, J.L. and Towsend, C.R. 1986. *Ecology: Individuals, Populations and Communities.* Blackwell Scientific Publications, Oxford, England.

Berg, P. and Raven, L. 2000. *Environment.* Wiley, New York, NY, USA.

Bernards, N. 1991. *The Environmental Crisis.* Greenhaven Press, San Diego, CA, USA.

Botkin, D.B. and Keller, E. 2007. *Environmental Science: Earth as a Living Planet*. Wiley, New York, NY, USA.

Bramwell, A. 1989. *Ecology in the 20th Century: A History*. Yale Univ. Press, New Haven, CT, USA.

Bush, M. 2003. *Ecology of a Changing Planet*. 3rd ed., Prentice Hall, Upper Saddle, NJ, USA.

Camp, W.G. and Donahue, R.L. 1994. *Environmental Science for Agriculture and Life Sciences*. Delmar, Albany, NY, USA.

Chapman, J.L. and Reiss, M.J. 1998. *Ecology Principles and Applications*. Cambridge Univ. Press, Cambridge, UK.

Chiras, D.D. 2006. *Environmental Science*. Jones and Bartlett, Sudbury, MA, USA.

Cotgreve, P. and Forseth, I. 2002. *Introductory Ecology*. Blackwell Science Publications, Oxford, England, UK.

Crawley, M.J. and Crawley, M. 1997. *Plant Ecology*. Blackwell Scientific Publications, Oxford, England, UK.

Cunningham, W.P. and Saigo, B. 1996. *Environmental Science: A Global Concern*. Wm. C. Brown Publishers, Dubuque, Iowa, USA.

Cunningham, W.P. and Cunningham, M.A. 2006. *Principles of Environmental Science: Inquiry and Application*. McGraw Hill, New York, NY, USA.

Dale, M.R.T. 1999. *Spatial Patterns in Plant Ecology*. Cambridge Univ. Press, Cambridge, England.

Deington, J.M. and Edington, M.A. 1977. *Ecology and Environmental Planning*. Chapman and Hall, London, England, UK.

Delmer, E. 1999. *Discoveries of the Environment*. Blackwell, Oxford. Malden, MA, USA.

Dubey, S.K. 2006. *A Textbook of Ecology*. Dominant Publishers, New Delhi, India.

Enger, E. and Smith, B.F. 2006. *Environmental Science – A Study of Interrelationships*, 5th Ed. McGraw Hill, Dubuque Iowa, USA.

Ennos, A.R. and Bailey, S.E.R. 1995. *Problem Solving in Environmental Biology*. Longman Scientific and Technical, Wiley and Sons, New York, NY, USA.

Etherington, J.R. 1982. *Environment and Plant Ecology*. Wiley and Sons, Chichester, England.

Fowden, L., Mansfield, T. and Stoddart, J. 1993. *Plant Adaptation to Environmental Stress*. Chapman and Hall, London, England.

Franck, I. and Brownstone, D. 1992. *The Green Encyclopedia*. The Prentice Hall General Reference. New York, NY, USA.

Frankhan, R. and Kingsoeven, J. 2004. Responses to Environmental Change Adaptation or Extinction In: *Evolutionary Conservation Biology* (Ferrier, R., Diechmann, U. and Convet, D. (eds.). Cambridge Univ. Press, Cambridge, England, pp. 85-100.

Gloss, G., Downes, B. and Boulton, A. 2003. *Freshwater Ecology*. Blackwell Science Publications, Oxford, England.

Goldfrank, W.L., Goodman, D. and Szasz, A. 1999. *Ecology and the World-System*. Greenwood Press, Westport, CT, USA.

Gooran, A. and Bosatta, E. 1996. *Theoretical Ecosystems Ecology: Understanding Element Cycles*. Cambridge Univ. Press, Cambridge, UK.

Greeshoff, R.M. 1993. *Plant Responses to the Environment*. CRC Press, Boca Raton, FL, USA.

Greig-Smith, P. 1983. *Quantitative Plant Ecology*. University of California Press, Berkeley, CA, USA.

Gurevitch, J., Scheiner, S.M. and Fox, G.A. 2002. *The Ecology of Plants*. Sinauer, Sunderland, MA, USA.

Hale, M.G. and Orcutt, D.M. 1987. *The Physiology of Plants Under Stress*. Wiley, New York, NY, USA.

Hall, J.A. and Morrison, J.W. 1978. *Environmental Studies. A Field and Laboratory Approach*. Acro Publishing Co., New York, NY, USA.

Hairston, N.G. 1989. *Ecological Experiments. Purpose, Design and Execution*. Cambridge Univ. Press, Cambridge, UK.

Henry, J.G and Heinke, G. W. 1996. *Environmental Science and Engineering*. Prentice Hall, Upper Saddle, NJ, USA.

Hokikian, J. 2002. *The Science of Disorder: Understanding the Complexity, Understanding and*

Pollution in Our World. Los Feliz Pub., Los Angeles, CA, USA.

Honari, M. and Boleyn, T. 1999. *Health Ecology: Health, Culture and Human Environment.* Routledge, London, England.

Howell, D.J. 1994. *Ecology For Environmental Professionals.* Quorum Books, Westport, CT, USA.

Hughes, J. 2000. *Ecology and Historical Materialism Physiological Plant Ecology.* Cambridge Univ. Press, Cambridge, UK.

Indergit, K.M., Dakshini, M. and Foy, C.L. 1999. *Principles and Practices in Plant Ecology: Allelochemical Interactions.* CRC Press, Boca Raton, FL, USA.

ISO TC 146/SCI. 2007. ISO 11632:1998. *Stationary Source Emissions-Determination of Mass Concentrations of Sulfur Dioxide-Ion Chromatography Method.* American National Standards Institute. Washington, DC, USA.

Jackson, L.E. 1997. *Ecology in Agriculture.* Academic Press, San Diego, CA, USA.

Jjemba, P.K. 2004. *Environmental Microbiology Principles and Applications.* Science Publishers, Enfield, NH, USA.

Jorgensen, S.E., Hallig-Sorenson, B. and Nielsen, S.N. 1996. *Handbook of Environmental and Ecological Modeling.* Lewis Publishers, Boca Raton, FL, USA.

Kemp, D. 1994. *Global Environmental Issues.* Taylor and Francis, London, England.

Kennedy, D. 2006. Science Magazine's State of the Planet. Island Press, Washington, DC, USA.

Kershaw, K.A. 1985. *Quantitative and Dynamic Plant Ecology.* Edward Arnold, London, England.

Krebs, J.R., Davies, N.B. and Parrr, J. 1993. *An Introduction to Behavioral Ecology.* Blackwell Science Publications, Oxford, England.

Kumar, A. 2004. *A Textbook of Environmental Science.* APH Publishing Corporation, New Delhi, India.

Lambers, H., Chapin, F.S., III and Pons, T.L. 1998. *Plant Physiological Ecology.* Springer, New York, NY, USA.

Lean, G. 1994. *Atlas of the Environment.* ABC-CLIO. Santa Barbara, CA, USA.

Lerner, H.R. 1999. *Plant Responses to the Environmental Stress.* Marcel Dekker, New York, NY, USA.

Leveque, C. 2003. *Ecology From Ecosystem to Biosphere.* Science Publishers, Enfield, NH, USA.

Levin, M.A. and Israeli, E. 1996. *Engineering Organisms in Environmental Settings.* CRC Press, Boca Raton, FL, USA.

Mackenzie, F.T. 1998. *Our Changing Planet: An Introduction to Earth System Science and Environmental Change.* Prentice Hall, Upper Saddle River, NJ, USA.

Marina, M.L., Rios, A. and Valcarcel, M. 2005. *Analysis and Detection by Capillary Electrophoresis. Volume XLV (Comprehensive Analytical Chemistry).* Elsevier, New York, NY, USA.

McConnell, R.L. and Abel, D.C. 2000. *Environmental Issues: An Introduction to Sustainability.* 3E, Prentice Hall, Upper Saddle River, NY, USA.

McIntosh, R.P. 1985. *The Background of Ecology: Concept and Theory.* Cambridge Univ. Press, New York, NY, USA.

McKinney, M.L. *et al.* 2003. *Environmental Science: Systems and Solutions.* Jones and Bartlett. Sudbury, MA, USA.

Meyers, J.C. 1997. *The Geostatistical Error Management Quantifying Uncertainty for Environmental Sampling and Mapping.* Van Nostrand, Reinhold, New York, NY, USA.

Middleton, G.V. and Wilcock, P.R. 1994. *Mechanics in the Earth and Environmental Sciences.* Cambridge Univ. Press, Cambridge, England.

Miller, G.T. 2005. *Environmental Science: Working With the Earth.* Brooks Cole, Belmont, CA, USA.

Molles, M.C. 2006. *Ecology.* McGraw Hill, New York, NY, USA.

Mongkolsuik, S., Lovett, P.S. and Trempy, J.E. 1993. *Biotechnology and Environmental Science. Molecular Approaches.* Plenum Press, New York, NY, USA.

Morgan, S. 1995. *Ecology and Environment. The Cycles of Life.* Oxford University Press, New York, NY, USA.

National Research Council. 1999. *Global Environmental Change: Research Pathways for the Next Decade.* National Academy Press, Washington, DC, USA.

Nazaroff, W.W. and Alvarez-Cohen, L. 2001. *Environmental Engineering Science.* Wiley, New York, NY, USA.

Nobel, P.S. 1983. *Biophysical Plant Physiology and Ecology.* W.H. Freeman, San Francisco, CA, USA.

Odum, E. and Barrett, G. W. 2004. *Fundamentals of Ecology.* Brooks Cole, Belmont, CA, USA.

Oldfield, F. 2005. *Environmental Change. Key Issues and Alternative Approaches.* Cambridge Univ. Press, Cambridge, England, USA.

O'Niell, J. 1993. *Ecology, Policy and Politics: Human Well-Being and the Natural World.* Routledge, London, UK.

Ogren, G.I. and Bosatta, E. 1996. *Theoretical Ecosystem Ecology Understanding Element Cycles.* Cambridge Univ. Press, New York, NY, USA.

Osborne, P.L. 2000. *Tropical Ecosystems and Ecological Concepts.* Cambridge Univ. Press, Cambridge, England.

Pearcy, R.W. 1989. *Plant Physiological Ecology.* Chapman and Hall, London, England.

Percival, R.V., Schroeder, C.H., Miller, A.S. and Leapli, J.P. 2003. *Environmental Regulation: Law, Science and Policy.* Aspen Publishers, New York, NY, USA.

Pollock, S. 1993. *Ecology.* Dorling Kindersley, London, England.

Press, M.C., Scholes, J.D. and Baker, M.G. 1999. *Physiological Plant Ecology.* Blackwell Science, Oxford, England.

Ranta, E. Kaitala, V. and Lundberg, P. 2005. *Ecology of Populations.* Cambridge Univ. Press, Cambridge, England.

Remment, H. 1980. *Ecology, A Textbook.* Springer-Verlag, Berlin, Germany.

Rosenthal, D.B. 1994. *Environmental Science Activities.* Wiley and Sons, New York, NY, USA.

SAPS Student Sheet 3 1990. Investigating the behavior of leaf discs *http://www.saps.plantsci.com.oc.uk/worksheets/disc.htm.*

Schneider, D.C. 1994. *Quantitative Ecology: Spatial and Temporal Scaling.* Academic Press, New York, NY, USA.

Schmidt, G. 2001. *The Essential Guide to Environmental Chemistry.* Wiley, Chichester, West Sussex, UK.

Schoch, R.M., McKinney, M.L. and Yonavjak, L. 2007. *Environmental Science: Systems and Solutions.* Jones and Bartlett, Sudbury, MA, USA.

Shaw, M.J. and Haddad, P.R. 2004. The determination of trace metal pollutants in environmental matrices using ion chromatography. Environ. Int. 30:403-431.

Shintani, H. and Polonski, J. 1996. *Handbook of Capillary Electrophoresis Applications.* Springer, New York, NY, USA

Silvertown, J.W. 1993. *Introduction to Plant Population Biology.* Blackwell Scientific Publications, Oxford, England.

Sinclair, A.R.E., Fryfell, J.N. and Caughley, G. 2005. *Wildlife Ecology Conservation and Management.* Wiley, New York, NY, USA.

Sinclair, T.R. and Gardner, F.P. 1998. *Principles of Ecology in Plant Production.* CAB International, Oxon, New York, NY, USA.

Smith, T.M. and Smith, L.L. 2002. *Elements of Ecology Update.* Addison Wesley-Longman. Upper Saddle River, NJ,USA.

Spellerberg, I.F. 2005. *Monitoring Ecological Change.* Cambridge Univ. Press, Cambridge, UK.

Spiro, T.G. and Stigliani, W.M. 2003. *Chemistry of The Environment.* Prentice Hall, Upper Saddle River, NJ, USA.

Subramanian, V. 2002. *A Textbook in Environmental Science.* CRC Press, Boca Raton, FL, USA.

Trivedi, R. 2002. *A Textbook of Environmental Science.* Anmol Publication Pvt, New Delhi, India.

Uirzo, R. and Sarukhan, J. 1984. *Perspectives on Plant Population Ecology.* Sinauer Associates, Sunderalnd, MA, USA.

Underwood., A.J. 1997. *Experiments in Ecology. Their Logical Design and Interpretation Using Analysis of Variance.* Cambridge Univ. Press, Cambridge, UK.

Vesley, D. 1999. *Human Health and the Environment: A Turn of the Century Perspective.* Kluwer Academic, Boston, MA, USA.

Vander Moarel, E. 2004. *Vegetation Ecology.* Blackwell Science Publications, Oxford, England, UK.

Walker, B. and Steffen, W. 1996. *Global Change and Terrestrial Ecosystems.* Biodiversity Documentation Centre, Bangalore, India.

Watts, S. and Halliwell, L. 1996. *Essential Environmental Science Methods and Techniques.* Taylor and Francis, London, England.

Weiss, J. and Weiss,T. 2005. *Handbook of Ion Chromatography.* Wiley, New York, NY, USA.

Wilkinson, D. 2006. *Fundamental Processes in Ecology: An Earth Systems Approach.* Oxford University Press, Oxford, England.

Williams, G. 1987. *Techniques and Fieldwork in Ecology.* Bell and Hyman, London, England.

Willson, M.F. 1983. *Plant Reproductive Ecology.* Wiley, New York, NY, USA.

Withgoth, J.H. and Brennan, S.R. 2007. *Essential Environment: The Science Behind the Stories,* 2/5. Pearson Prentice Hall, Upper Saddle River, NJ, USA.

Wright, R.T. and Nevel, B.J. 2008. *Environmental Science: Toward A Sustainable Future.* Prentice Hall, Upper Saddle River, NJ, USA.

Zweers, W. 1994. *Ecology, Technology and Culture.* White House Press, Paul and Co., London, UK.

CHAPTER 3

Global Warming

Is Global Warming Occurring and What are the Causes?

This chapter is concerned with whether global climate change (Dow and Downing, 2006; Burrough, 2007) is occurring and, if so, what are the root causes (Crowley, 2000; Crowley and Bener, 2001; Maslin, 2007) and consequences (Linden, 2007).

Global climate is the spatial and temporal distribution patterns of temperature world-wide. To obtain additional information regarding alleged global climate change, the student can consult one or more of the following references: (Singer, 1999; Peters and Lovejoy, 1992; Newton, 1993; Paterson, 1996; Houghton et al., 2001; Balinus and Soon, 2001; Roleff et al., 1997; Tretenberg, 1997; Drake, 2000; Hansen et al., 2000; Stat et al., 2001; Berger and Lautre, 2002; Linden, 2007). Of special interest is the report of Ravindranath and Sathage (2002) who discuss climate change in developing countries. Ferrey and Cabraal (2005) consider renewable power in relation to global warming for Thailand, Indonesia, India, Sri Lanka and Vietnam.

What is the Evidence For and Against Global Climate Change? The Earth's Past and Current Temperatures.

Figure 3.1 presents variations in the Earth's surface temperature for the past 140 years (respectively). According to the USA's National Academy of Sciences, the Earth's surface temperature has risen 1° C in the past century. Figure 3.1 indicates that accelerated warming has occurred during the last two decades. Many scientists have expressed concern about the recent increase and there has been intense debate whether this rise is due to anthropogenic activities or, alternatively, to natural variations in climate. One of the reasons that a controversy exists involves a few inconsistencies in surface versus satellite temperature data (Fig. 3.2). Hansen et al. (1998) have attempted to reconcile the data. In their review, the discrepancy may be explainable by an oversight in the satellite data. The oversight appears to involve failure to account for decay of satellite data altitude caused by satellite drag. When this is compensated for, the satellite and surface data are in closer accord. In addition, a 2000 National Academies report suggested means to reconcile surface and atmospheric temperature measurements. The National Research Council of the National Academies asserts that better monitoring systems are required 'to ensure continuity

Variations of the Earth's surface temperature for

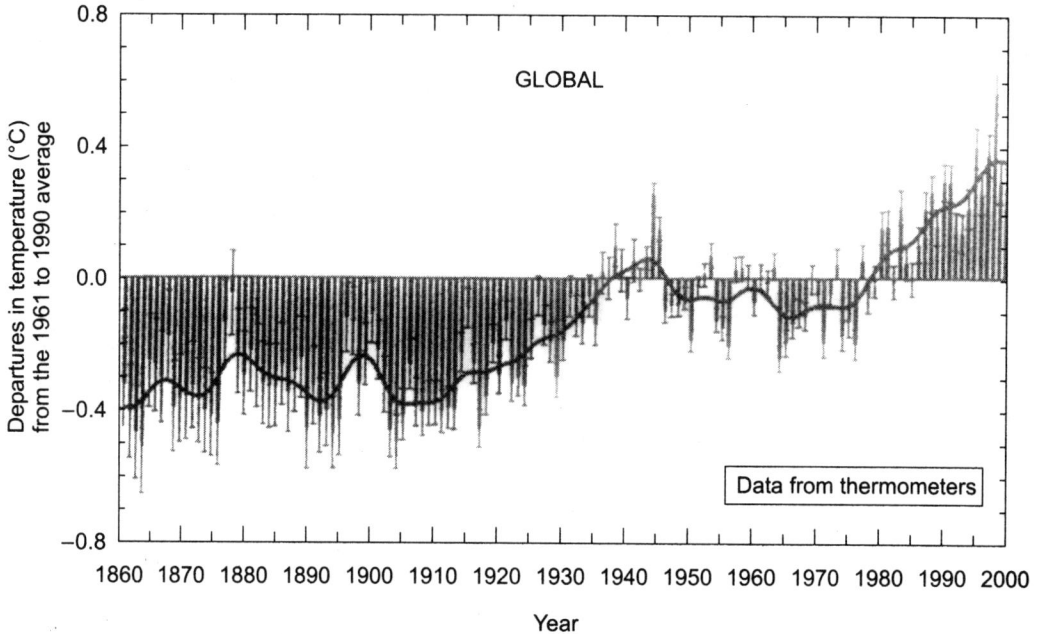

The past 140 years

Fig. 3.1 Variations of the Earth's temperature for the past 140 years. Earth's temperature is shown on a year by year basis and approximately decade by decade (a filtered annual curve suppressing fluctuations below near decadal time scales). Over the last 140 years..., the best estimate is that the global average temperature has increased by 0.6–0.2°C. From: A Report of Working Group 1 of the Intergovernmental Panel on Climate Change. Summary for policymakers. *http://www.ids.oc.uk/eldris/wgi.htm.*

and quality in data collection'. Table 3.1 presents systems other than temperature for monitoring global temperature change.

What are the Possible Causes of Global Warming?

Since the beginning of the industrial revolution atmospheric carbon dioxide, methane and nitrous oxide levels have been rising (Fig. 3.3). These and other gases (Table 3.2) have been termed greenhouse gases. Why? Energy from the sun influences the Earth's weather, climate and heats the Earth's surface. Some of this energy is radiated back into space (Fig. 3.4). Atmospheric greenhouse gases trap some of the radiated energy causing the Earth to retain heat.

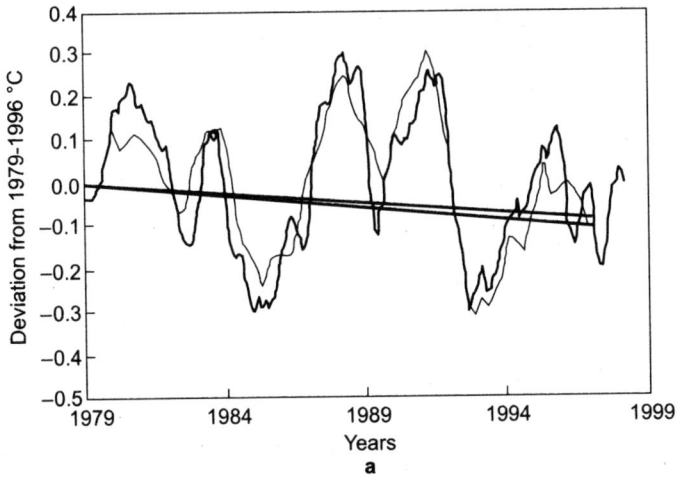

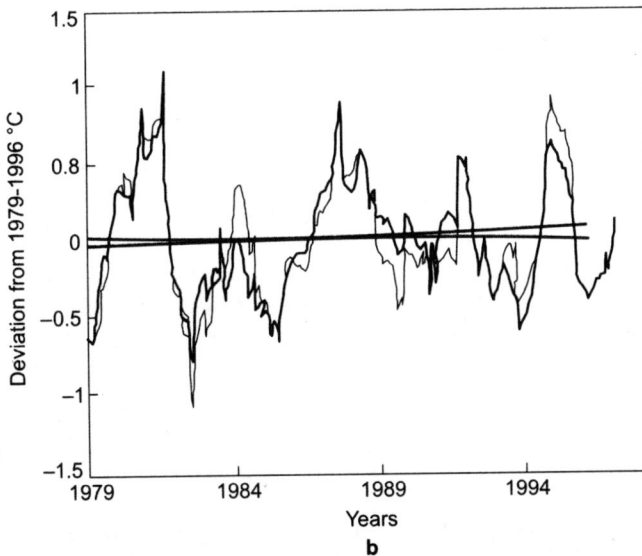

Fig. 3.2a,b Comparison of balloon temperature (light line) and global satellite temperature (dark line). (a) From: *www.oism.org/pproject/review.pdf.* Other diagrams comparing surface and satellite temperature can be found at *http:www.answers.com/main/ntqueyjessionid=/vsoyoy333tmjj* method = 4 dsname = wik and *http://www.asu.edu/lib/noble/earth/* warming.htm. (b) comparison of troposphere measurements by satellite from North America between 30°-70° N and 75°-125° W with the surface record. Source as in oism.org.

The sources of these gases are presented in Table 3.2. Of major concern is the increase in atmospheric carbon dioxide which appears to result from the burning of fossil fuels (see literature in Graves and Reavy, 1996), deforestation (Bonnie et al., 2000) and land use changes (Dale, 1994). Thus, the possibility of anthropogenic-driven global warming has been proposed (Karl and Trenberth, 1999). Some support for this proposal has been derived from superimposing time-dependent Earth's

TABLE 3.1 Monitors Other Than Temperature for Assessing Global Temperature Change

Parameter	Measurement
Ice Cores	Sediment Layer Thickness *http://www. St-edmunds.com.oc.uk/us/houghton/ lecture2.html* and Petit et al. (1999).
Tree Rings	Width of ring reflects precipitation levels
Coral growth	Reflects ocean temperatures as coral growth occurs at different rates

temperature changes and atmospheric CO_2 levels (Fig. 3.5) as well as comparing simulated and observed global mean surface temperature changes relative to natural and anthropogenic forcings (Fig. 3.6). Natural factors (Sailor, 1989) which could enhance the Earth's temperature are: solar luminosity, decreases in stratospheric aerosol loading, urban heat island effects, alterations in anthropogenic sulfur and carbon aerosols (*http://users.erols.com/dhotyl/amer110.htm*).

Therefore, skeptics (Moore, 1998; Singer, 1999; Allen et al., 2001) have questioned the view (IPCC, 2001) that current increases in global warming (Johansen, 2006) are, in the main, man-made (Table 3.3). Roleff et al. (1997) review opposing views in the global warming debate. Hansen et al. (2000) have concerned themselves with pointing out the key differences between global warming proponents and skeptics. These differences are discussed on the Internet *www.giss.nasa.gov/edu/gov/debate*. Allen et al. (2001) reviewed uncertainties in IPCC's third assessment report. Oreskes (2004) has reviewed the

TABLE 3.2 Sources of Greenhouse Gases[a]

Gas	Concentration	Source(s)	% Increase Since 1750
Carbon dioxide (CO_2)	350 ppm	Burning of fossil fuels	31
		Deforestation (Bonnie et al., 2000)	
		Land use changes	
Chloroflurocarbons[b] ($CFCl_3$ and CF_2Cl_2)		Aerosols	Increasing
Methane (CH_4)	1,750 ppb	Agricultural development (Keppler et al., 2006) Anaerobic respiration Oil/natural gas extraction	151
Nitrous oxide (N_2O)	280 ppb	~ 1/3 are anthropogenic emissions (agricultural and chemical industries)	17
Ozone (O_3)	4 – 65 ppb	Anthropogenic emissions of several O_3-forming gases	36
Sulphur dioxide (SO_2)		Coal fire	
		Plants	
Water vapor	0 – 4%, variable		
Sulfur hexafluoride (SF_6) or alternative SF_5CF_3 (Sturges *et al.* (2000)		Emitted during the generation, transmission and distribution of energy	

[a]Adapted from a report of working group 1 of the intergovernmental panel on climate change, Houghton, J.T., et al. (eds.) Climate Control 2001: The Scientific Basis. Cambridge Univ. Press, Cambridge, England.
[b]Substitutes for chlorofluorocarbon compounds are: CHF_2Cl, CF_3CH_2F, perfluorocarbons and sulphur hexafluoride.

Indicators of the human infuence on the atmosphere during the industrial era

Global atmospheric concentrations of three well mixed greenhouse gases

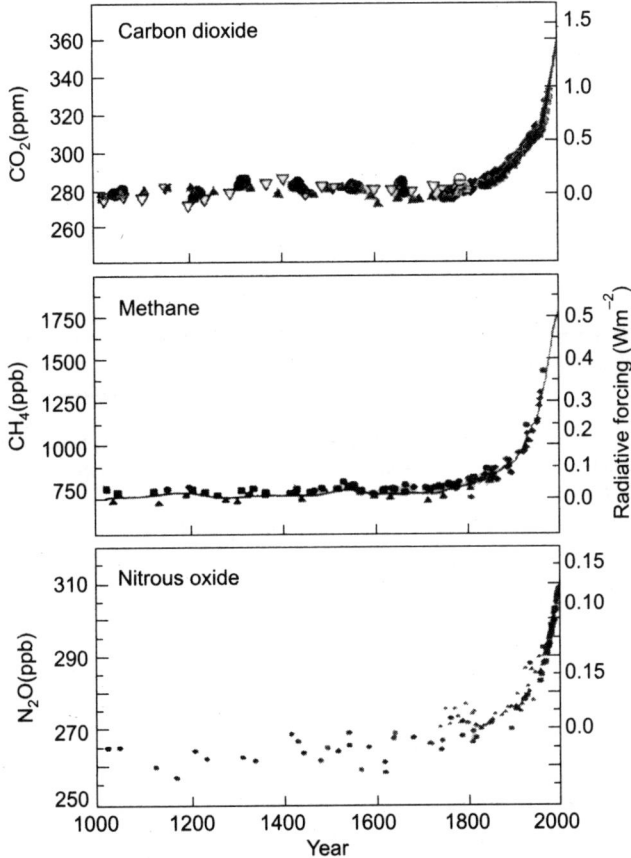

Fig. 3.3 Global atmosphere concentrations of three well mixed greenhouse gases. Atmospheric concentrations of carbon dioxide (CO_2), methane (CH_4) and nitrous oxide (N_2O) over the past 1000 years. The ice core data from several sites in Antarctica and Greenland are supplemented with data from direct atmospheric samples over the past few decades (shown by the line for CO_2 and incorporated in the curve representing the global average of CH_4). The estimated positive radiative forcing of the climate system from these gases is indicated on the right-hand scale. Since these gases have atmospheric lifetime of a decade or more, they are well mixed, and their concentrations reflect emissions from sources throughout the globe. All three records show effects of the large and increasing growth in anthropogenic emissions during the Industrial Era. *From*: A Report of Working Group 1 of the Intergovernmental Panel on Climate Change. Cicerone and Overland (1988) is the landmark paper concerning biogeochemical aspects of atmospheric methane. Recent analysis has revealed that global methane levels have been constant from 1999 to 2002 (*http://www.usgcrp.gov/usgcrp/library/vcp2004-5/ocp2004-5-hi-atmos.htm*)

existing literature concerning global climate change and concludes that the scientific community is in overwhelming agreement for evidence that the "Earth's climate is be-ing affected by human activities". Figure 3.10 provides IPCC's assessment on the level of scientific understanding of factors which can influence the Earth's temperature.

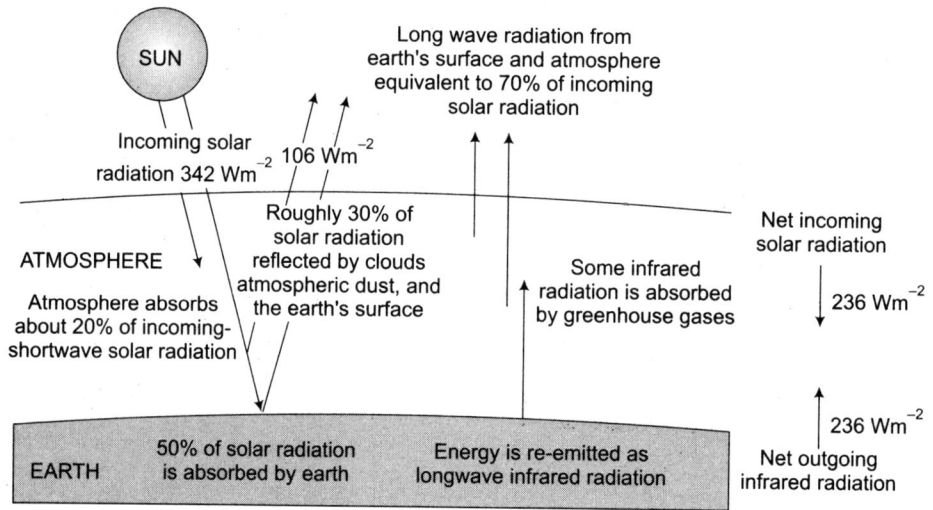

Fig. 3.4 Diagram of the greenhouse effect: From: *www.umich.edu/ngs265/society/greenhouse.htm*. Other diagrams occur at *http://yosemite.epa.gov/oar/globalwarmingnsf/content/climate.html* and *http://www.epa.gov/oar/agtrmd99/html/global.html*

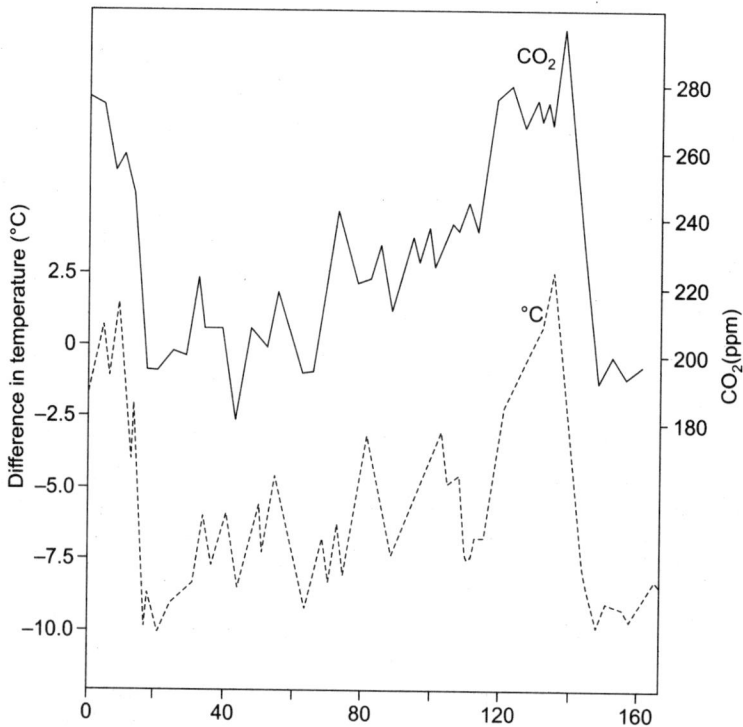

Fig. 3.5 Alteration in CO_2 concentration and temperature over the past 160,000 years. From: Barnola et al. (1987).

Simulated Annual Global Mean Surface Temperatures

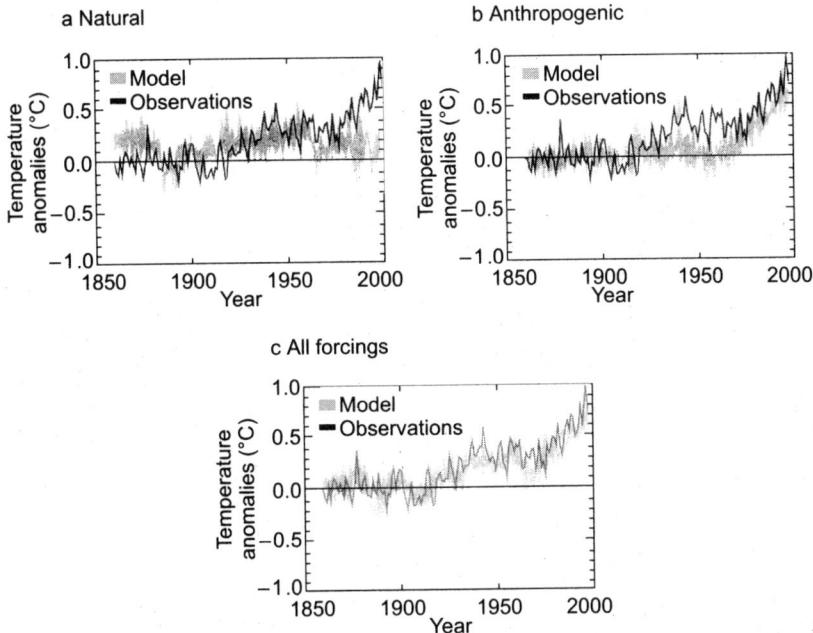

a Natural

b Anthropogenic

c All forcings

Fig. 3.6 Simulating the Earth's Temperature variations.
A climate model can be used to simulate the temperature changes that occur both from natural and anthropogenic causes. The simulations represented by the band in (a) were done with only natural forcings: Solar variations and volcanic activity. Those encompassed by the band in (b) were done with anthropogenic forcings included. From (b), it can be seen that inclusions of anthropogenic forcings provide a plausible explanation for a substantial part of the observed temperature changes over the past century, but the best match with observations is obtained in (c) when both natural and anthropogenic factors are included. These results show that the forcings included are sufficient to explain the observed changes, but do not exclude the possibility that other forcings may also have contributed. The bands of model results presented here are four runs from the same model. Similar results to those in (b) are obtained with other models with anthropogenic forcing.
From: A Report of Working Group 1 of the Intergovernmental Panel on Climate Change: Summary for Policymaker. *http://www.ids.uc.uk/eldris.wgi.htm* (with permission).

TABLE 3.3 Sampling of Skeptics' Views About Global Warming. These views can be found in summaries on global warming websites[a].

1. Earth will not become warmer uniformly, i.e., some areas will get warmer and some will become cooler. These alterations could result from localized evaporation as well as enhanced cloud cover. (Fig. 3.7)

2. Chlorofluorocarbons migrate to the atmosphere (Fig. 3.8) where they react with ozone (O_3). However, the concentration of chlorofluorocarbons is low enough to be a negligible contributor to O_3 depletion.

3. Global warming appears to be a natural geological process which could start to reverse itself in 10-20 years.

4. Global warming seems to be caused by changes in solar energy rather than anthropogenic activities (Fig. 3.9) and is dependent upon the Earth's orbital alteration (Benestad, 2002).

(Contd.)

(Contd.)

The Earth's axis is tilted at a 23.5 degree angle and the Earth revolves (obliquity) around the sun in an ecliptic (Fig. 3.9 a,b) of very low eccentricity. The result is that different regions of the Earth receive varying amounts of sunlight during the year yielding the seasons. The length of the seasons can alter the time between the vernal and autumnal equinoxes thus affecting solar energy reaching the Earth.

Solar radiation affects the Earth's temperature (Benestad, 2002). The Milankovic Cycles (d) produce an irregular but predictable level of incoming solar radiation. Sunspot number (c) is the number of individual sunspots and ten times the number of sunspot groups. Sunspots (dark magnetic blemishes on the face of the sun) can affect global warming if the average number increases. The height of the sunspot cycle is the solar max when the suns's magnetic field creates energy in the magnetic field. Then, solar flares, x-rays and a cloud of plasma are emitted (Suplee, 2004)[b]

5. Investigations which deposit an increase in global temperatures appear to be biased and unrepresentative resulting from measurements at urban weather stations which are often warmer.

6. Global temperature measurements remote from human habitation and activity do not reveal any evidence of warming during the last century. These sites include proxy measurements such as tree rings, marine sediments, ice cores, weather balloons, satellite measurements and many surface sites.

7. Oceans may play a role in CO_2 increase (Monnin et al., 2001)

[a]Hansen et al. (2000) contend that CO_2 buildup is of less importance to increases in global temperature than other greenhouse gases. Lindzen (1997) is a critic of the role of rising CO_2 in climate change as are Robinson et al. (1996).
[b]Suplee, C. 2004. A Stormy Star. National Geographic July pp. 2-33.
Lomborg, B. 2007. Cool It: The Skeptical Environmentalist's Guide To Global Warming. Alfred A. Knopf, New York, NY, USA.

Global dimming (Stanhill and Cohen, 2001) describes the gradual reduction in the level of total solar radiance at the Earth's surface from 1950's onward. This dimming causes a cooling effect which may have prompted scientists to underestimate the role of greenhouse gases in global warming (http://en.wikipedia.org/wiki/global-dimming).

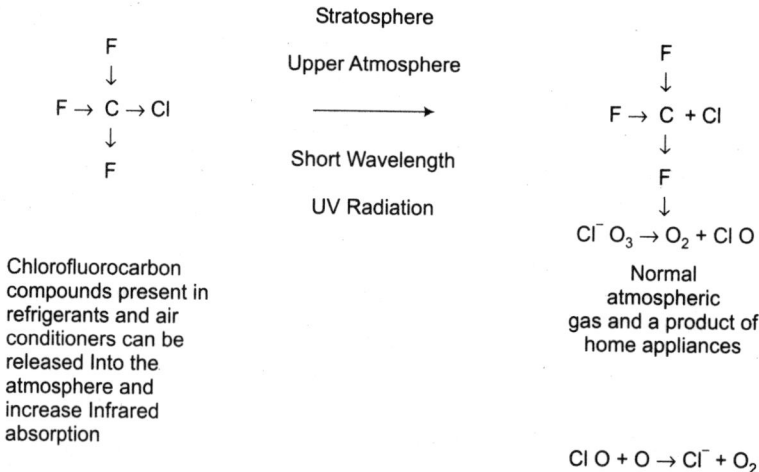

	Stratosphere	
F	Upper Atmosphere	F
↓		↓
F → C → Cl	⟶	F → C + Cl
↓		↓
F	Short Wavelength	F
	UV Radiation	↓
		$Cl^- O_3 → O_2 + Cl\,O$

Chlorofluorocarbon compounds present in refrigerants and air conditioners can be released Into the atmosphere and increase Infrared absorption

Normal atmospheric gas and a product of home appliances

$Cl\,O + O → Cl^- + O_2$

O_3 = Ozone; One Cl atom can destroy > 100,000 O_3 molecules. Decrease in atmospheric level in O_3 ("Ozone Hole") Permits More UV to Reach Earth's Atmospheric Results Increase Risk of Skin Cancer and Cataracts. Some examples of CFCS are CFC -11 ($CFCl_3$) and CFC-12 (CF_2Cl_2). CFCs warming can dissolve in seawater affecting marine ecosystems.

a

Physical Climate System

Atmospheric Physics/Dynamics

Climate Change

External forcing

Sun

Volcanoes

Stratospheric Chemistry/Dynamics

Ocean Dynamics

Terrestrial Energy/Moisture

Global Moisture

Soil

CO_2

Human Activities

Marine Biogeochemistry

Terrestrial Ecosystems

Land Use

Tropospheric Chemistry

CO_2

Biogeochemical Cycles

Pollutants

b

Changes in Solar inputs

Changes in the Atmosphere: Composition, Circulation

Changes in the Hydrological Cycle

Atmosphere

Clouds

N_2, O_2, Ar, H_2O, CO_2, CH_4, N_2O, O_3, O_3, etc. Aerosols

Volcanic Activity

Atmosphere Biosphere Interaction

Atmosphere-Ice Interaction

Precipitation Evaporation

Terrestrial Radiation

Glacier

Ice Sheet

Heat Exchange

Wind Stress

Human Influences

Biosphere

Land Atmosphere Interaction

Sea Ice

Hydrosphere: Ocean

Soil-Biosphere Interaction

Land Surface

Ice-Ocean Coupling

Hydrosphere: Rivers & Lakes

Cryosphere: Sea Ice, Ice Sheets, Glaciers

Changes in the Ocean: Circulation, Sea Level, Biogeochemistry

Changes in/on the Land Surface: Orography, Land Use, Vegetation, Ecosystems

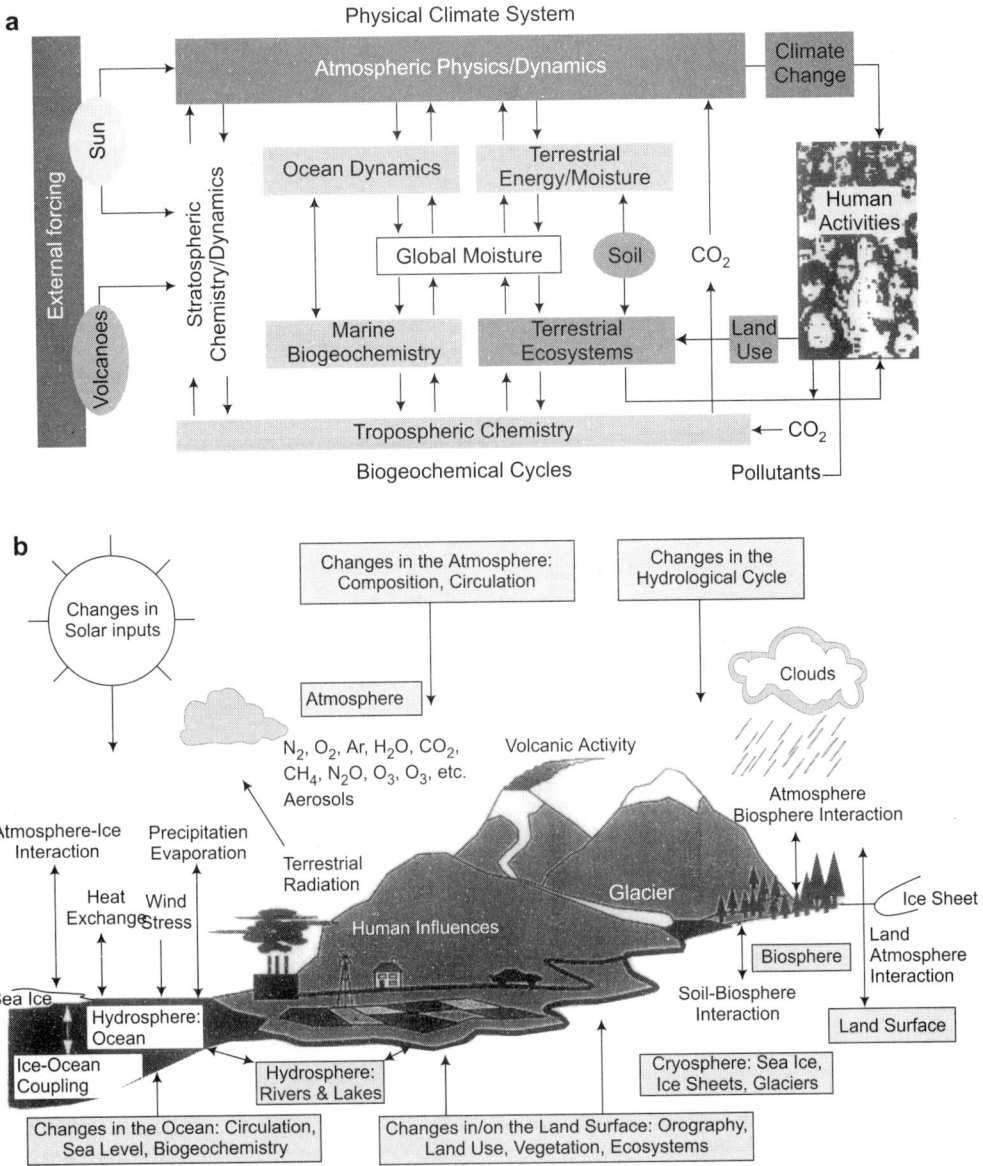

Fig. 3.7 Components of the Climate System. a. From: *http://gsfc.nasa.gov/sect16/sect16_4.html*. and b. Grace (1983) discusses plant-atmosphere relationships.

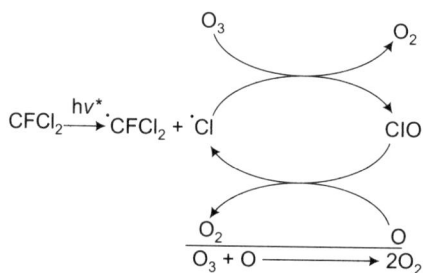

Ozone Destruction Caused by Manmade Compounds (e.g., CFCs)

Ultraviolet radiation from the sun strikes the CFC molecule and causes a chlorine atom to break away.

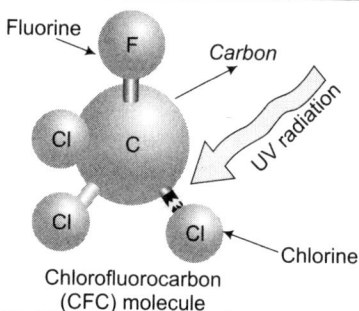

The chlorine atom reacts with an ozone molecule to form chlorine monoxide and diatomic oxygen.

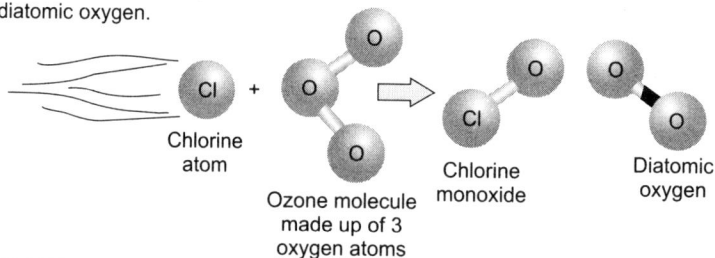

When a free atom of oxygen reacts with chlorine monoxide molecule, diatomic oxygen is formed and the chlorine atom is released to destroy more ozone.

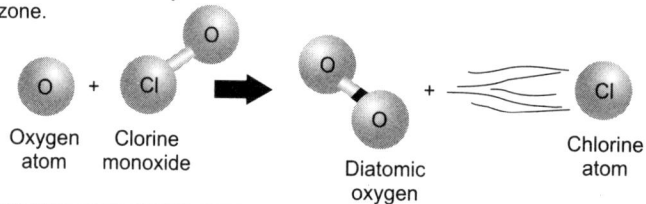

Fig. 3.8 Interaction of atmospheric chlorofluorocarbon compounds with ozone (O_3) From: *http://www.epa.gov/ord/webpubs/stratoz.pdf.*

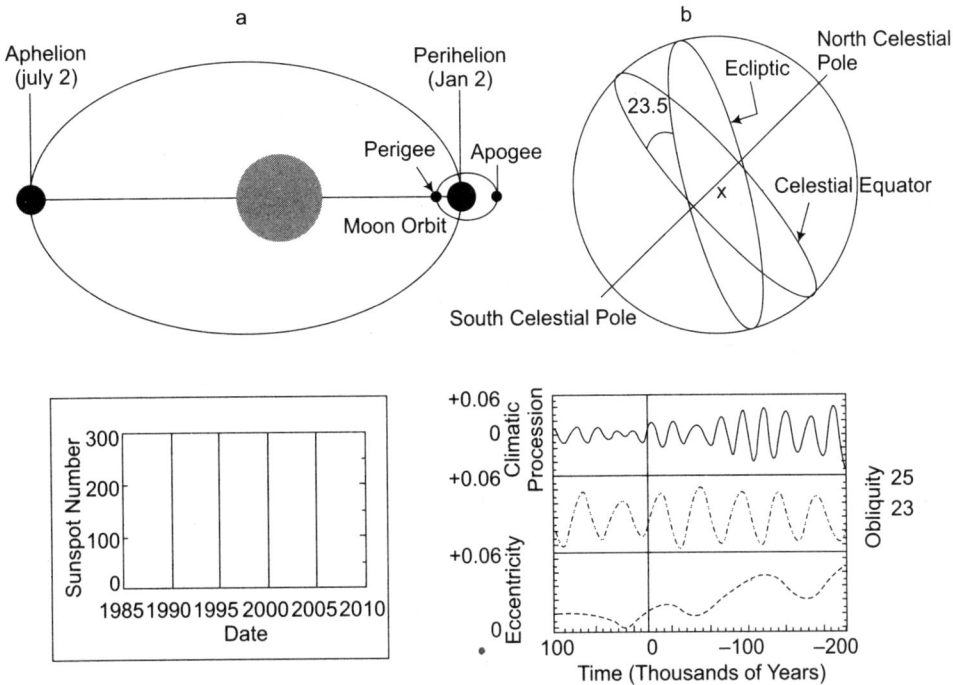

Fig. 3.9 Earth's orbit around the sun is elliptical (a, b) Perigree – Moon is closest to the Earth; Apogree – Moon is farthest from the Earth; Perihelion – Earth is closest to the Sun, Aphelion – Earth is farthest from the Sun.

EFFECTS OF CO_2 AND GLOBAL CLIMATE CHANGE ON BIOLOGICAL ORGANISMS (GATES, 1993)

Whole Plants

While Table 3.4 displays the general effects of elevated CO_2 on plants, Table 3.5 summarizes the CO_2 levels and increased temperature on individual plant species. Thorough reviews of this topic have been published by Badiani et al., 1999; Heath and Mansfield, 1999; Idsu and Idsu, 1994; Lemon, 1983; Lucht et al., 2002). Morrison and Morecroft (2006) discuss the effects of climate change on Plant growth.

Agriculture

Crop yields may be stunted by higher tem-

TABLE 3.4	Effects of Elevated CO_2 on Plants.
Plant Responses	*Effect*
Plants in General	Increases plant water acquisition via stimulating root growth; reduces water loss by constricting stomatal apertures; enhances plant photosynthetic rates and biomass production (Idsu and Idsu,1994 and journal refs in *http://www.CO₂salice.org/subject/g/summaries/growthwaterog.htm*).

Reviews regarding the responses of plants to CO_2 are: Amthor (1995), Bazzaz et al. (1996), Vivin et al. (1996), Linacre (1998) and Morrison and Morecroft (2006) Ehleringer et al. (2005).

peratures and damaged by enhanced difficulty in water retention. This could lead to regional famine. Because heat accelerates the growth cycle, crops may be less nutritious from failure to accumulate carbohydrates.

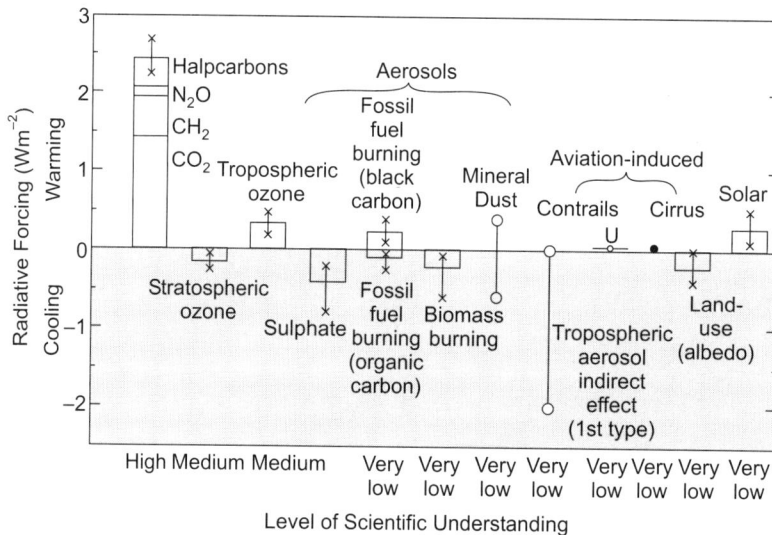

Fig. 3.10 The global mean radiative forcings of the climate system for the year 2000, relative to 1750. These radiative forcings originate from changes in atmospheric composition, alteration of surface reflectance by land use, and variation in the output of the sun. Except for solar variation, some form of human activity is linked to each. The rectangular bars represent estimates of the contributions of these forcings – some of which yield warming, and some cooling. Forcing due to episodic volcanic events, which lead to a negative forcing lasting only for a few years, is not shown. The indirect effect of aerosols shown is their effect on the size and number of cloud droplets. A second indirect effect of aerosols on clouds, namely their effect on cloud lifetime, which would also lead to negative forcing, is not shown. Effects of aviation on greenhouse gases are included in the individual bars. The vertical line about the rectangular bars indicates a range of estimates, guided by the spread in the published values of the forcings and physical understanding. Some of the forcings possess a much greater degree of certainty than others. A vertical line without a rectangular bar denotes a forcing for which no best estimate can be given owing to large uncertainties. The overall level of scientific understanding for each forcing varies considerably, as noted. Some of the radiative forcing agents are well mixed over the globe, such as CO_2, thereby perturbing the global heat balance. Others represent perturbations with stronger regional signatures because of their spatial distribution, such as aerosols. For this and others reasons, a simple sum of the positive and negative bars cannot be expected to yield the net effects on the climate system. The simulations of this assessment report indicate that the estimated net effect of these perturbations is to have warmed the global climate since 1750. Wigley and Raper (2001) conclude that there is 90% probability that warming for 1990 to 2100 will be 1.7° to 4.9°C. From: A report of Working group 1 of the Intergovernmental Panel on Climate Change. Summary for Policymakers. *http://www.ids.uc.uk/eldris.wgi.htm*
From: *http://en.wikipedia.org/wiki/image:recent_sea_level_rise.prg.*

The following can be consulted for in depth coverage of the effects of climate change and/or CO_2 on agriculture (Cure and Acock, 1986; Rosenzeweig and Parey, 1994; Solomon and Shugart, 1993; Decker et al., 1986; Fisvold and Kuhn, 1999; Kimball, 1982; Parry 1990; Roberston et al. 2000; Long et al. (2006) discuss lower than expected crop yield stimulation with rising CO_2 correlations.

Animals

A summary of effects of CO_2 on some animals is shown in Table 3.6. Some reviews pertinent to animals and elevated CO_2 levels and/or temperature are: Schaefer (1982), Weike and Mertens (1991) and World Health Organization (1990).

TABLE 3.5 Some Examined Plant Responses for CO_2-enriched and/or Temperature Increases

Environmental Parameter	Examined Plant Response	Reference (s)
Elevated CO_2	Photosynthesis and root function	Luo et al. (1994)
	Chilling; induced water stress and photosynthetic reduction	Wolfe-Bellin et al. (2006) Boese et al. (1997)
	Plant root responses	Rogers et al. (1992)
	Carbon flow in photosynthetic cells	Stitt (1991)
	Root growth and functioning	Stulen and Den Hertog (1983)
	Gas exchange in *Liquidanbar styracifluor* and *Pinus taeda* seedlings	Tolley and Strain (1985)
	Stomatal numbers	Woodward (1987)
	CO_2 exchange, carbon allocation and osmoregulation in *Quercus robur* seedlings	Vivin et al. (1996)
	Photosynthetic yield	Nijs et al. (1988)
	Mitochondrial respiration	Drake et al. (1991)
	Photosynthetic capacity of *Scirpus olneyi*	Arp and Drake (1991)
	Phosphorus requirements in pine species	Conroy et al. (1990)
	Stomatal numbers	Woodward (1987)
	Growth and partitioning of dry matter and nitrogen in wheat and maize	Hocking and Meyer (1991)
	Downward regulation of photosynthesis and growth	Idso and Kimball (1991)
	Quantum yield of photosynthesis in *Scirpus olneyi*	Long and Drake (1991)
	Growth and yield of CO_2-enriched wheat	Gifford (1979)
	Environmental productivity indices of *Opuntian ficus-indica*	Nobel (1991)
	Rice acclimation	Rowland-Bamford et al. (1991)
	Water economy	Heath and Mansfield (1999),] Law et al., (2001)
Elevated CO_2 and temperature	Biomass accumulation in white clover	Ryle and Powell (1992)
	Photosynthetic productivity	Long (1991), Long and Drake
	Evapotranspiration, Shifts in spring phenology	Allen (1991), Leon (2002) Wolfe et al. (2004)

Reviews regarding the responses of terrestrial plants are: Amthor (1995), Bazzaz et al. (1996) and Linacre (1998).

Populations and Communities

Table 3.7 offers some effects of elevated CO_2 levels on plant populations and communities. Pertinent review references regarding CO_2 and terrestrial ecosystems include: Amthor (1995), Bruce et al. (1998), Cav and Woodward (1998 a, b), Ehleringer et al. (2005), Koch and Mooney (1996), Schimel (1995), Walker and Steffen (1996) and Xiao et al. (1998).

Finally, Arnell (1996), Bousquet et al. (2002), Carpenter et al. (1992), Firth and Fisher (1992), Sweeney et al. (1992), Vorosmarty et al. (2000) and Wright et al. (1986) have discussed the effects of elevated CO_2 on global climate change on aquatic ecosystems.

TABLE 3.6 Summary of Effects of CO_2 on Animals[a]

Effect	Reference(s)
Hypercapnia increased ventilation constriction of pulmonary arteries reduced contraction of heart muscles fall in blood pressure	Brackett et al. (1969) VanYpersele de Strihou (1974) Turino et al. (1974)
% increase in ventilation 0-10% volume rose as concentration of CO_2 increase. The responses were human > hamster > manatee >, seal > echidna > gopher	Darden (1972) Luft et al. (1974)
Shifts in distribution of European cork borer	Porter et al. (1991)
Soil invertebrates exhibit several adaptations to high CO_2 e.g., low metabolic rate, better buffering capacity of the haemolymph and regulation of the balance between acids and bases	(see Graves and Reavey, 1996)
CO_2 is important for the activities such as host seeking, mating and even ventilation	Nicolas and Sillans (1989)
CO_2 can be an indicator of the presence of a possible food source	
CO_2 could also be a cue for females searching for a site for oviposition	Kobayashi et al. (1984)
Soil micro-arthropods increase as a consequence of CO_2-enhanced soil fungal systems	see literature in: http://www.co2science.org/subject/a/summaries/animals.htm
Northward expansion of certain butterflies and bug species	http://www.co2science.org/subject/a/summaries/animals.htm
Certain herbivorous insects may be inhibited	http://www.co2science.org/subject/a/summaries/animals.htm
Freshwater and marine fish thermal tolerance	Wood and McDonald (1997)

[a]See Schaefer (1982), Leaf (1989) and Weike and Mertens (1991), Wyman (1991) for potential health effects of global climate change.

TABLE 3.7 Effects of Elevated CO_2 Levels on Populations and Communities[a]

Plant Responses	Reference (s)
C_4 species 'lose out' to C_3 species as CO_2 concentrations rise when grown in competition. However, the impact of CO_2 on different plant communities is difficult to predict.	Arp et al. (1993), Koerner and Bazzaz (1996)
CO_2 concentrations can affect flowering timing	Bazzaz (1990)
Respiration of pine ecosystem	Law et al. (1999)
High CO_2 concentrations assist species to retain water during the onset of drought	
Plant interactions between C_3 and C_4 saltmarsh plant species	Kruper and Groth (2000)
Competition between C_3 plants	Arp et al. (1993)
High latitude vegetation greening	Lucht et al. (2002)
Plant-insect herbivore interactions	Lincoln et al. (1993)
Disease risks for terrestrial and marine biota	Harvell et al. (2002)

| Animal Responses | Schmidtmann and Miller (1989) |
| Horn fly and honeybee populations regulation of atmospheric CO_2 | Seely (1974) |

[a]Reviews pertinent to CO_2 and ecosystems are: Baldocchi et al. (1996), Bazzaz and Carlson (1984), Bruce et al. (1998), Cav and Woodward (1998 a,b), Koch and Mooney (1996), Norby et al. (2001), Schimel (1995), Walker and Steffen (1996) and Bousquet et al., (2002).

Physical Consequences of Global Warming

Sea Level Rises

Titus and Richman (2004) have published a U.S. State by State table as well as Sea Level Rise Maps depicting the areas of land on both the Atlantic and Gulf Coasts that are at or near sea level. A scientific consensus exists that there is a significant risk that sea level rise will increase during the 21st century (Fig. 3.11; Table 3.8). A one-meter rise in sea level is likely to take place over a two hundred year period but may occur as early as 2100 (see literature in Titus and Richman, 2004). The causes and consequences of sea level change are presented in Table 3.9.

In addition to sea level rise (McCarthy and McKenna, 2000; Titus and Richman, 2004; Chanton, 2002; Vinnikov, 1999, Carton et al., 2005, Hansen et al., 2006), other possible consequences of global warming

include: Permafrost degradation (Lawrence et al.) drought (Seager et al., 2007) and wildfire, famine, more intense rainstorms, deadly heat waves, the spread of disease (Cook, 1992; Harvell et al., 2002; IPCC, 2007), warming water and ecosystem disruption. These consequences are discussed in detail at *http://www.nrdc.org/global warming/fcors.asp.naturalresources defensecouncil.* The EPA expects "changing regional climate could alter forests, crop yields and water supplies. Deserts may expand into existing rangelands" *(http://www.yosemite. epa.gov/oar/globalwarming.nsf/content/ impacts.html).*

Projected Climate Change

The U.S. EPA stated *(http://yosemite/epa.gov/ oar/globalwarming.nsf/content/climate uncertainties.html)* that "scientists are more confident about their projections for large scale areas (e.g., global temperature and

TABLE 3.8	Estimated Sea Level Rise, 2000 – 2100, by Scenario (in cm, with inches in parentheses)				
Year	Conservative	Mid-Range Scenarios Moderate[a]	High	High Scenario	Historical Extrapolation
2000	4.8 (1.9)	8.8 (3.5)	13.2 (5.2)	17.1 (6.7)	2-3 (0.8-1.2)
2025	13.0 (5.1)	26.2 (10.3)	39.3 (15.5)	54.9 (21.6)	4.5-8.25 (1.8-3.2)
2050	23.8 (9.4)	52.3 (20.6)	78.6 (30.9)	116.7 (45.9)	7-12 (2.8-4.7)
2075	38.0 (15.0)	91.2 (35.9)	136.8 (53.9)	212.7 (83.7)	9.5-15.5 (3.7-6.1)
2100	56.2 (22.1)	144.4 (56.9)	216.6 (85.3)	345.0 (135.8)	12-18 (4.7-7.1)

From J. Hoffman, D. Keyes and J. Titus, 1983, Projecting Future Sea Level Rise: Methodology, Estimates to the Year 2100, and Research Needs, 2nd rev. ed., U.S. GPO No. 055-000-00236-3, Washington, DC: Government Printing Office.

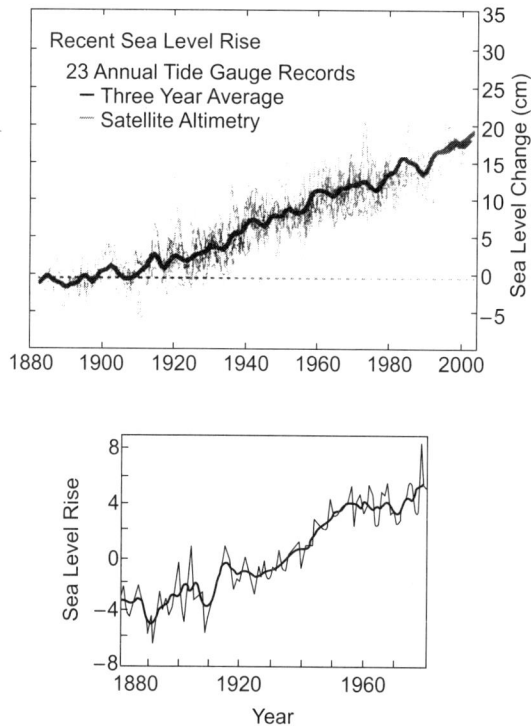

Fig. 3.11 Sea Level Rise. Hansen, J. Goddard Institute for Space Studies, NASA Space. Flight Center, USA noted that the rising trend of sea level closely parallels the alterations in average global temperatures. The estimated sea level rise for 2000 – 2100 is shown in Table 3.7.
Cazenave and Nerem (2004) consider current sea level changes.

TABLE 3.9 Causes and Consequences of Sea Level Change

Oceanic Alterations	Terrestrial Alterations
Possible Causes	Possible Causes
Surface and deep ocean circulation changes	Ground water extraction
(O'Brien et al., 1992)	Freshwater mass from continents
Storm surges (Trenberth, 2005)	Melting of mountain glaciers and arctic and Antartic
Exacerbate ocean desets (Hophis, 2008)	polar icecaps
Increased heat content causes water expansion	Land movements
(Carton et al., 2005; Cazenave and Nerem, 2004)	Tectonic displacements
	Runoff
	Water storage and land reservoirs (IPCC, 2001)

Consequences
(*http://www.cap.noaa.gov/stressors/climatechange/sea_level_rise.html.*) Accelerated Costal Retreat and Erosion Damange to Costal Infrastructure (O'Brien et al., 1992).
Expanded flooding during severe storms and high tides

Broad impacts in the coastal economy

Higher and more frequent flooding of wetlands and adjacent shores (McGranahan et al., 2007)

Increased wave energy in the near-shore area

Intrusion into coastal freshwater acquifiers

Satellite measurements during the past decade have revealed that both "Antarctica and Greenland are each less mass overall" (Serreze et al., 2007; Shepherd and Wingham, 2007).

precipitation change, average sea level rise) but less confident about the ones for small-scale areas (e.g., local temperature and precipitation changes, altered weather patterns, soil moisture changes." The EPA believes this to be due to the inability of computer models (Gates *et al*, 1992) to forecast climate-indirect alterations at smaller scales.

If the concentrations of greenhouse gases continue to increase, many scientists anticipate that the average global surface temperature may rise 0.6 – 2.5°C in the next 50 years and 2.2 - 10°C in the next century.In addition to the greenhouse gases, other previously mentioned factors can influence global temperature. Thus, these factors also need to be taken into account. Two of the factors which require further investigation are natural variations in the luminosity of the sun and volcanic activity.

It is very difficult to predict the future although certain scientific groups have attempted to do so. Generally, these predictions are based upon extrapolation of current trends and belief that the status quo will not change. The myriad of possible scenarios regarding future global temperatures and resultant consequences staggers the mind (Table 3.10). One can propose a multitude of questions concerning the future actions of mankind. For example, what will happen to the world's climate if the burning of fossil fuels continue? The costs of tackling climate change have been examined by the House of Lords Select Committee (2006).

What if many countries adhere to the Kyoto Protocol (see below) but the major polluting countries do not? What if new treaties are proposed to limit greenhouse gases? Will the World's political leaders head the advice of scientists? Will scientists develop new technologies to replace fossil fuel utilization? The list of possible questions is endless whether one is positive or negative about the future depends on one's view of human nature (King, 2004).

Greaves and Reavy (1996) have suggested the following effects of an increase of CO_2 from 350 to 700 ppm. These include: 1. no significant effects on humans, soil insects and bees, 2. no effect on organisms which acclimate to new CO_2 levels and 3. considerable impact on animals which require plants.

Attempts to Ameliorate Global Warming

There are many efforts going on to control global warming (Dolsak and Dunn, 2006) including existing mechanisms to reduce excess atmospheric CO_2 levels (Anderson and Newell, 2004), research programs (US Climate Action Report 2002) and treaties (Sandalow and Bowles, 2001). The mechanisms involve using soils (Wofsy, 2001) and vegetation as sinks (Pacala et al., 2001; Scholes and Noble, 2001), disposal of excess CO_2 in oceans (Tamburri et al., 2000) and geological sinks (Fig. 3.12) (Lombardi et al., 2006). The former has been discussed at length in the scientific literature while the latter has received less attention (Science

TABLE 3.10 Recent Trends, Assessment of Human Influences on the Trend, and Projections for Extreme Weather Events for Which There is an Observed Late 20th Century Trend (Tables 3.7, 3.8, 9.4, Sections 3.8, 5.5, 9.7, 11.2 – 11.9) a.)

Phenomenon[a] and direction of the trend	Likelihood that trend occurred in late 20th century (typically post 1960)	Likelihood of a human contribution to observed trend [b]	Likelihood of future trends based on projections for 21st century using SRES scenarios
Warmer and fewer cold days and nights over most land areas	Very likely [c]	Likely [e]	Virtually certain [e]
Warmer and more frequent hot days and nights over most land areas	Very likely [d]	Likely (nights) [e]	Virtually certain [e]
Warm spells/heat waves. Frequency increases over most land areas	Likely	More likely than not [f]	Very likely
Heavy precipitation events. Frequency (or proportion of total rainfall from heavy falls) increases over most areas	Likely	More likely than not [f]	Very likely
Area affected by droughts increases	Likely in many regions since 1970	More likely than not	Likely
Intense tropical cyclone activity increases	Likely in some regions since 1970	More likely than not	Likely
Enhanced incidence of extreme high sea level (excludes tsunamis) [g]	Likely	More likely than not [f, h]	Likely [i]

IPCC Working Group I Contribution to the Fourth Assesment Report. Climate Change 2007: The Physical Science Basis.

http://www.ipcc.ch/spm2feb07.pdf.

Notes:

[a]See Table 3.7 for further details regarding definitions

[b]See Table TS-4, Box TS-3.4 and Table 9.4

[c]Decreased frequency of cold days and nights (coldest 10%)

[d]Increased frequency of hot days and nights (hottest 10%)

[e]Warming of the most extreme days and nights each year

[f]Magnitude of anthropogenic contributions not assessed. Attribution for these phenomena based on expert judgment rather than formal attribution studies.

[g]Extreme high sea level depends on mean sea level and on regional weather systems. It is defined here as the highest 1% of hourly values of observed sea level at a station for a given reference period.

[h]Changes in observed extreme high sea level closely follow the changes in mean sea level (5.5, 5.6). It is very likely that anthropogenic activity contributed to a rise in mean sea level (9.5.2)

[i]In all scenarios, the projected global mean sea level at 2100 is higher than the reference period (10.6). The effect of changes in regional weather systems on sea level extremes has not been assessed.

Highlight 'Measuring CO_2 Exchange Between Ocean and Atmosphere' International Geosphere-Biosphere Program). Scholes and Noble (2001) pointed out that storing carbon in terrestrial sinks (Watson et al., 1992; Fan et al., 1998; Houghton et al., 1998; Herzog et al., 2000) will have little impact.

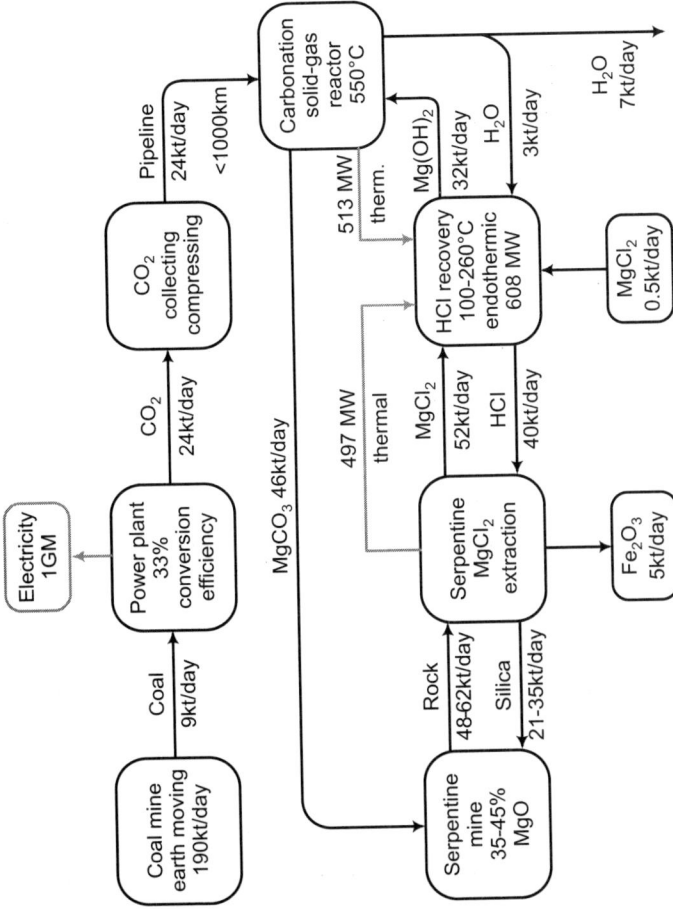

Fig. 3.12 Procedure for the removal of CO$_2$ by carbonation. "Mineral sequestration is the formation of carbonate salts by reaction of bicarbonate with dissolved calcium and magnesium. ..." Serpentinite reacts with carbon dioxide to form the carbon mineral magnesite, plus silica and iron oxide (magnetite). Serpentinite sequestration is favored because of the non-toxic and predictable nature of magnesium carbonate" (*http://co2.egi.utah.edu/mineralization/index.html*). "The numbers are chosen to reflect the overall mass flow for a gigawatt power plant with a conversion efficiency of 33%". From: *http://www.fas.org/sgp/otherg.../lib-.www/la-pubs/0032679.pdf.*

TABLE 3.11 Some Analytical Methods for Measuring Greenhouse Gases[a]

Method		Reference
Gas chromatography – isotope ratio mass spectrometry	Measure carbon isotopic composition of atmospheric methane	Miller, J.B. et al., 2002, J. Geophy. Res. 107:1-15
Thermo Environmental Instuments Model 17C Chemiluminescent NO_x/NH_3 Analyzer	Measure of concentrations of NO and NH_3	Walker, J.T. et al., 2002, Chemosphere 49:1389-1398
Fourier Transform Infrared Spectrometer	Detects CO, CO_2, CH_4 and N_2O	Gruther, M. 2003, Atmosfera 16:1-13
Sulphur hexafluoride tracer ratio methods	Determines methane fluxes from free ranging ruminants	Grieatorx, J.M. 2000; JTI-Institute for Jordbruks. Och miljotet-nik
Static chamber technique	Quality N_2O and CH_4	Grieatorx, J.M. 2000; JTI-Institute for Jordbruks. Och miljotet-nik
Micrometerological techniques	Measure the turbulent transfer of gases from ground surface to lower atmosphere	Grieatorx, J.M. 2000; JTI-Institute for Jordbruks. Och miljotet-nik
Gas chromatography	Quantify N_2O and CH_4	Grieatorx, J.M. 2000; JTI-Institute for Jordbruks. Och miljotet-nik
Infrared photoacoustic spectrometer	Detect N_2O, CH_4, CO_2 and NH_3 simultaneously	Grieatorx, J.M. 2000; JTI-Institute for Jordbruks. Och miljotet-nik
Tuneable diode laser absorption spectroscopy	Determination of CO, CO_2, N_2O, NO, NO_2, NH_3 and CH_4 concentrations	Grieatorx, J.M. 2000; JTI-Institute for Jordbruks. Och miljotet-nik
Ocean cycle models and model of global terrestrial carbon cycling	Explore the effects of changing terrestrial biosphere on atmospheric CO_2 residence time	Moore and Braswell (1994)
FACE Free – Ai-CO_2 Enrichment	CO_2 effects on tall vegetation	Hendrey (1993)

[a]Baldocchi et al. (1996) presented strategies for measuring CO_2 and water fluxes over terrestrial environments and Gravensten et al. (1995) reviewed gas monitoring in clinical practice.

TABLE 3.12 Summary of Some Research Projects Regarding Global Warming

Effort		Reference
National		
Evaluations of alternative refrigerants and technologies for ozone-depleting substances		C. Cage US EPA *http://www.epa.gov/appcdwww/apb/bio.htm*
Estimate forest and agricultural ecosystem fluxes and efluxes of radiatively important trace gases		C. Geron US EPA *http://www.epa.gov/appcdwww/apb/bio.htm*
Prevention and control technologies for protecting the stratospheric ozone layer, mitigation of global warming		W. J. Rhodes US EPA *http://www.epa.gov/appcdwww/apb/bio.htm*
Fuel cell application to waste methane gases		R. Spiegel US EPA *http://www.epa.gov/appcdwww/apb/bio.htm*
International		
RAINS Regional Air Pollution Information and Simulation	Strategies to reduce troposphere ozone formation in Europe	International Institute for Applied Systems Analysis, Austrialia

(Contd.)

(Contd.)

Infrared Remote Sensing Technology	Aerosol Remote Sensing	Japan Center for Climate System Research
Untied Nations Framework Convention on Climate Change, 1997 Bonn Cooperation with relevant international organizations Progress report on research systematic observation pp. 1-9.		
International Geosphere – Biosphere programs	Biotic Controls on selected trace gases	National Academy Press OIA 1992.

US-EPA. 1995. Proceedings of the 1995 symposium on Greenhouse Gas Emissions and Mitigation Research National Risk Management Research Laboratory, Cincinnati, OH, USA.

One of the chief treaties is the Kyoto Protocol which is designed to reduce CO_2 emissions. The United States has not signed the protocol even though it pumps more CO_2 into the atmosphere than any country in the world. A 2007 conference in Vienna proposed reducing CO_2 levels which was supported by 158 nations but not signed by the USA.

Attempts to reduce global warming (Yamin, 2005) involve the development of analytical methods for greenhouse gases (Table 3.11) and both natural and international research activities (Table 3.12). In Bangkok, Thailand, the IPCC (2007) proposed key mitigation technologies for ameliorating climate change. These involve improved energy efficiency in buildings, vehicles, electrical equipment use, crop and grazing land management, forest management and landfill recovery. Finally, the student should consult global warming-actions website (Table 3.13) for ways individuals can reduce CO_2 emissions. Of a controversial nature is the economics of controlling global warming. Tretenberg (1997) asserts that there are three main economic parameters, i.e., 1. damage, costs and optimal control strategy, 2. policy instrument choice and 3. ethics and discounting. These involve global politics (Paterson et al., 2001), ecological tax reform, role of developing countries and evaluation of international agreements (Mabey et al., 1997).

TABLE 3.13 Some Websites Relating to Climate Change

Global Climate Change	Student Information Guide
www.doc.mmu.oc.uk/aric/gccsg/ Working Group II: Imports, adaptation and vulnerability. United Nations Environmental Programming Panel on Climate Change (IPCC). www.usgcrp.gov/ipcc	Assessing scientific, technical aspects of climate change.
NASA's Global Change Master Directory gcmd.gsfc.nasa.gov/	Comprehensive directory of descriptions of data sets of relevance to global research change
NCDC: Global Climate Change. www.ncdc.noaa.gov/ol/climate/climateextremes.html	Global climate change
Climate change – Office of International Information Programs usinf.state.gov/topical/global/environ/climate/climate.htm	USA National Academy of Sciences report on global warming

(Contd.)

(Contd.)

Climate change and Human Health Integrated assessment Web. *wwwljhu.edu/~climate/*	Information on research conducted between 1998-2000
U.S. Global Change Research Program *www.usgcrp.gov/*	Changes in Ecosystems Global Carbon Cycle
Global Climate Change Digest Home Page *www.globalchange.org/digest.htm*	A guide to information on greenhouse gases and ozone depletion
Climate Change Research Center *www.grg.sr.unh.edu/ccrc/*	University of New Hampshire Climate Change Research Center.
Pacific Institute – Bibliography – climate change *www.pacinst.org/ccbib.html*	Comprehensive bibliography of the peer reviewed literature dealing with climate change
C H AMMP Home Page *www.epm.ovnl.gov/chammp/*	U.S. Dep't. of Energy Program to rapidly advance the science of decade – and longer-scale climate change.
Paleoclimate and Climate Change Home University of Santa Cruz Research Areas *www.ess.ucsc.edu*	Research on climate variability, warm climate
Intergovernmental Panel on Climate Change *www.ipcc.ch/*	IPCC assesses the scientific, technical and socio-economic information
United Nations Framework Convention on Climate Change *www.unfccc.de/*	Official site for U.N. Climate Secretariat: Kyoto Protocol
Common Questions about Climate Change *www.gcrio.org/ipcc/qa/cover.html*	United Nations Environment Program World Meteorological
EPA Global Warming Site *www.epa.gov/globalwarming/*	Focuses on the science and reports of global warming

TABLE 3.13 Websites and On Line Academic Information – Global Warming and Climate Change.

Climate Change Science: An Analysis of Some Key Questions committee on the Science of Climate Change, 2001.

Global Warming Central – Pace Energy Project

Carbon Cycle Science Program

U.S. Global Change Research Program

Climate Diagnostic Center (CDC)

EPA's Global Warming Website

Global Climate Change Digest Archives

Global Warming: and the Third World

Global Warming: Early Warning Signs

Global Warming: Focus on the Future

Global Warming: Information Page

Global Warming: News Zone

Global Warming: Understanding the Forecast

Global Warming: Update

Greenhouse Gases and the Kyoto Protocol

Intergovernmental Panel on Climate Change

Climate Change 2001: The Scientific Basis

(Contd.)

(Contd.)

National Academy Press
  Climate Change Science: An Analysis of Some Key Questions
  Reconciling Observations of Global Temperature Change
  Ozone Depletion, Greenhouse Gases and Climate Change
National Climatic Data Center (NOAA)
  Climate 2000, annual revisions
New Scientist – Global Warming Report
Ocean Remote Sensing Division (NOAA)
  Satellite Research Group
Pacific Institute for Studies in Development, Environment and Security
Reporting a Climate Change: Understanding the Science
U.S. Geological Survey
U.S. Global Change Research Information Office (GCRIO)
U.S. Global Change Research Program (USGCRP)

References

Allen, L.H. 1991. Evapotranspiration responses of plants and crops to carbon dioxide and temperature. *http://www.nal.usda.gov/ttic/tektran/data/000010/61/0000106193. html*.

Allen, M., Raper, S. and Mitchell, J. 2001. Uncertainty in the IPCC's third assessment report. *Science* 293: 430-433.

Amthor, J.S. 1995. Terrestrial higher-plant response to increasing atmospheric $[CO_2]$ in relation to the global carbon cycle. *Global Change Biol.* 1: 243-2774.

Anderson, S and Newell, R. 2004. Prospects for carbon capture and storage technologies. *Ann. Rev. Env. Resour.* 29: 109-142.

Arnell, N. 1996. *Global Warming: River Flows and Water Resources*. Wiley, Chichester, England.

Arp, W.J. and Drake, B.G. 1991. Increased photosynthetic capacity of *Scirpus olneyi* after 4 years of exposure to elevated CO_2. *Plant Cell Environ.* 14: 1003-1006.

Arp, W.J., Drake,B.G., Pockman, W.T., Curtis, P.S. and Whigham, D.F. 1993. Interactions between C_3 and C_4 salt marsh plant species during four years of exposure to elevated CO_2. *Vegetation* 104: 133-143.

Badiani, M., Raschi, A., Paolacci. A.R. and Miglietta, F. 1999. Plant responses to elevated CO_2: A perspective from natural CO_2 Springs. In: (Agrawal, S.B. and M. Agrawal, eds.). *Environmental Pollution and Plant Responses*. Lewis Publishers, Boca Raton, FL. pp. 45-81.

Baldocchi, D., Valentini,R., Runnig,S., Oechel, W. and Dahlman, R. 1996. Strategies for measuring and modeling carbon dioxide and water vapour fluxes over terrestrial ecosystems. *Global Change Biol.* 2: 159-168.

Balinus, S. and Soon, W. 2001. *Global Warming A Guide to the Science*. Fraser Institute, Vancouver BC, Canada.

Barnola, J.M., Raynaud, D., Korotkevitch, Y.S. and Lorius, C. 1987. Vostok ice core: A 160,000-year record of atmospheric CO_2. *Nature* 329: 408-414.

Bazzaz, F.A. 1990. The response of natural ecosystems to the rising CO_2 levels. *Ann. Rev. Ecol. Syst.* 21: 167-196.

Bazzaz, F.A., Bassow, S.L., Berntson, F.M. and Thomas, S.S. 1996. Elevated CO_2 and terrestrial vegetation: implications far and beyond global carbon budget. In: (Walker, B.H. and W.L. Steffen, eds.) *Global Change and Terrestrial Ecosystems*. Cambridge Univ. Press, Cambridge, UK.

Benestad, R.E. 2002. *Solar Activity and Earth's Climate*. Springer, New York, NY, USA.

Berger, A. and Lautre, M.F. 2002. Climate: An exceptionally long interglacial ahead. *Science* 297: 1287-1288.

Boden, T.A., Kancirak,P., Farell, M.P. 1990. *Trends '90. A. Compendium of Data on Global Change*. Oak Ridge, TN, USA.

Boese, S.R., Wolfe, D.W. and Melkonian, S.J. 1997. Elevated CO_2 mitigates chilling-induced water stress and photosynthetic reduction during chilling stress. *Plant Cell Environ*. 20: 625-632.

Bonnie, R., Schwartzman, S., Oppenheimer, M. and Bloomfield, J. 2000. Counting the cost of deforestation. *Science* 228: 1763-1764.

Bousquet, P., Peylin, P., Cials, P., LeQuere, C., Friedlingstein, P. and Tans, P.P. 2002. Regional changes in carbon dioxide fluxes of land and oceans since 1980. *Science* 290: 1342-1346.

Brackett, N.C., Jr., Wingo, O., Muren and Solano, J.T. 1969. Acid-base response to chronic hypercapnia in man. *New Engl. J. Med*. 280: 124-130.

Bruce, K.D., Cannon, P.F., Hall, G.S., Hartley, S.E., Howson, G., Jones, C.G., Kampichler, C., Kandeler, E. and Ritchie D.A. 1998. Impacts of using atmospheric carbon dioxide on model terrestrial ecosystems. *Science* 280: 441-443.

Burrough, W.J. 2007. *Climate Change. A Multi disciplinary Approach*. Cambridge Univ. Press, Cambridge, U.K.

Carpenter, S.R., Fisher, S.G., Grimm, N.B. and Kitchell, J.F. 1992. Global change and freshwater ecosystems. *Ann. Rev. Ecol. Syst*. 23: 119-139.

Carton, J.A., Grese, B.S. and Gvodsky, S.A. 2005. Sea level rise and the warming of the oceans in the simple ocean data assimilation (SODA) ocean reanalysis. *J. Geophys. Res*. 110: 9006-9008.

Cav, M. and Woodward, F.I. 1998a. Dynamic responses of terrestrial ecosystem carbon cycling to global climate change. *Nature* 393: 249-252.

Cav, M. and Woodward, F.I. 1998b. Net primary and ecosystem production and carbon stocks of terrestrial ecosystems and their responses to climate change. *Global Change Biol*. 4: 185-198.

Cazenave, A. and Nerem, R.S. 2004. Present-day sea level change: Observations and causes *Rev. Geophys*. 42: RG3001.

Chanton, J. 2002. Global warming and rising oceans. *http://www.actionbioscience.org/environment/chanton.html*.

Chaves, M.M. and Pereira, J.S. 1992. Water stress, CO_2 and climate change. *J. Exp. Bot*. 43: 1131-1139.

Chon, W.R. 2007. *Globol Warming and Agriculture Impact Estimated by Country* Peterson Institute, Jamestown, NY, USA.

Cowie, J. 2007. *Climate Change: Biological and Human Aspects*. Cambridge Univ. Press, Cambridge, UK.

Ciceione, R.J. and Oremland, R.S. 1988. Biogeochemical aspects of atmospheric methane. *Global Biogeochem. Cy*. 2:299-328.

Conroy, J.P., Milham, P.J., Reed, M.L. and Barlow, E.W.R. 1990. Increase in phosphorous requirements for CO_2-enriched pine species. *Pl. Physiol*. 92, 977-982.

Cook, G.C. 1992. Effects of global warming in the distribution of parasitic and other infectious diseases: A Review. *Journal of the Royal Society of Medicine*. 85: 688-691.

Cossins, A.R. and Raymond, R.S. 1987. Adaptive responses of animal cell membranes to temperature. In: *Temperature and Animal Cells*. Society for Experimental Biology 41 (K. Bowles and B.J. Fuller, eds.). Cambridge Univ. Press, Cambridge, England.

Crowley, T.J. 2000. Causes of climate change over the past 1,000 years. *Science* 289: 270-283.

Crowley, T.J. and Bener, R.A. 2001. CO_2 and climate change. *Science* 292: 870-872.

Cure, J.D. and Acock, B. 1986. Crop responses to carbon dioxide doubling: A literature survey. *Agr. Forest Meterol*. 38: 127-145.

Cushing, D.H.and Dickson, R.R. 1976. The biological response in the sea to climatic changes. *Adv. Mar. Biol*. 14: 1-22.

Dale, V.H. 1994. *Effects of Land Use Changes (LUC) on Atmospheric CO_2 Concentrations: Southeast Asia as a Case Study.* Springer-Verlag, New York, NY, USA.

Darden, T.R. 1972. Respiratory adaptations of a fossorial mammal, the pocket gopher (*Thomomys bottae*). *J. Comp. Physiol.* 78: 121-137.

Decker, W.L., Jones, V.K. and Achutuni, R. 1986. Impact of climate change from increased carbon dioxide on American agriculture. *U.S. Department of Energy Report TR-031*, Washington, DC, USA.

Dolsak, N. and Dunn, M. 2006. Investments in global warming mitigation: The case of activities implements jointly. *Policy Sci.* 39: 233-248.

Dow, K. and Downing, T.E. 2006. *The Atlas of Climate Change.* Earthscan, London. England.

Drake, B.G. et al. 1991. Does elevated CO_2 inhibit mitochondrial respiration in green plants? *Plant Cell Environ.* 22: 649-657.

Drake, F. 2000. *Global Warming.* Chapman and Hall, New York, NY,USA

Ehleringer, J., Cerling, T.E. and Dearing, M.D. 2005. *A History of Atmospheric CO_2 and the Effects on Plants, Animals and Ecosystems.* Springer, New York, NY, USA.

Falkowski, P. et al. 2000. The global carbon cycle: a test of our knowledge of earth as a system. *Science* 290: 291-296.

Fan, S., Gloor, M., Mahlman, J., Pocala, S., Sarmiento, J., Takahashi, T. and Tans, P. 1998. A large terrestrial sink in North America implied by atmospheric and oceanic carbon dioxide data and models. *Science* 282: 754-759.

Ferrey, S. and Cabraal, A.. 2005. *Renewable Power in Developing Countries: Winning the War on Global Warming.* Penn Well, Tulsa, OK, USA.

Firth, P. and Fisher, S.G. 1992. *Global Climate Change and Freshwater Ecosystems.* Springer-Verlag, New York, NY, USA.

Fisvold, G. and Kuhn, B. 1999. *Global Environment Change and Agriculture: Assessing the Impacts.* Edward Elgar, London, England.

Gates, D.M. 1993. *Climate Change and Its Biological Consequences.* Sinauer, Sunderland, MA, USA.

Gates, W.L., Mitchell, J.F.B., Boer, G.J., Cubasch, U. and Meleskko, V.P. 1992. Climate modeling, climate predictor and model validation. In: *Climate Change 1992: The Supplementary Report to the IPCC Scientific Assessment*, (Houghton, J.T., Callander, B.A. and Varney, S.K., eds.). Cambridge Univ. Press, Cambridge, England.

Gauthreaux, S.A., Jr. 1980. Long term and short term climate changes. In: *Animal Migration, Orientation and Navigation.* (Gauthreaux, S.A., ed.). Academic Press, New York, NY, USA.

Gifford, R.M. 1979. Growth and yield of CO_2-enriched wheat under water-limited conditions. *Aust. J. Plant Physiol.* 6: 367-378.

Grace, J. 1983. Plant-atmosphere relationships. In: *Online Studies in Ecology*, (Dunnet, G.M. and Gimmingham, C.H., eds.). Chapman and Hall, London, England.

Gravensten, J.S., Paulus, A. and Hayes, T.J. 1995. *Gas Monitoring in Clinical Practice.* Butterworth-Heinmann. Boston, MA, USA.

Graves, J. and Reavy, D. 1996. *Global Environmental Change. Plants, Animals and Communities.* Longman, Essex, England.

Grover, V. (ed.). 1995. *Climate Change. Perspective Five Years After Kyoto.* Science Publishers, Enfield, NH, USA.

Hansen, J., Sato, M., Riedy, R., Locis, A. and Clascoe, J. 1998. Global climate data and models: A reconciliation. *Science* 281: 930-932.

Hansen, J., Sato, M., Ruedy, R., Lo, K., Lea, D.W. and Medina-Elizade, M. 2006. Global temperature change. PNAS 103: 14288-14293.

Hansen, J., Sato, M., Ruedy, R., Lacis, A. and Vanis, V. 2000. Global warming in the twenty first century: An alternative scenario. PNAS 97: 9875-9880.

Hansen, J., Sato, M., Ready, R., Lo., D.W. and Medina-Elizade, M. 2006. Global temperature change. PNAS 103:14288-14293.

Harvell, D.D., Mitchell, C.E., Ward, J.R., Altizer, S., Dobson, P., Ostfeld, R.S. and Samuel, M.D. 2002. Climate warming and disease risks for terrestrial and marine biota. *Science* 296: 2158-2162.

Heath, J. and Mansfield, T.A. 1999. CO_2 enrichment of the atmosphere and the water economy of plants. In: (Agrawal, S.B. and M. Agrawal, eds.). *Environmental Pollution and Plant Responses*. Lewis Publishers, Boca Raton, FL, USA. pp. 33-43.

Hendrey, C.H. 1993. *FACE. Free-Air CO_2 Enrichment for Plant Research in the Field*. CRC Press, Boca Raton, FL, USA.

Herzog, H., Eliasson, B. and Kaarstad, O. 2000. Capturing greenhouse gases. *Sci. Am.* 282: 72-79.

Hocking, P.J. and Meyer, C.P. 1991. Effects of CO_2 enrichment and nitrogen stress on growth and partitioning of dry matter and nitrogen in wheat and maize. *Aust. J. Plant Physiol.* 18: 339-356.

Hophis, M. 2008. Ocean deserts are growing. Nature 10:/038/News. 2008. 795.

Houghton, J.T., Ding, Y., Griggs, D.J., Noguer, M., Van der Linden, P.J., Dai, Y., Maskell, K. and Johnson, C.A. 2001. *IPCC Climate Change 2001: The Scientific Basis*. Cambridge Univ. Press, Cambridge, UK.

Houghton, R.A., Davidson, E.A. and Woodwell, G.M. 1998. Missing sinks, feedbacks and understanding the role of terrestrial ecosystems in the global carbon balance. *Global Biogeochem. Cy.* 12: 25-34.

House of Lords Select Committee on Economic Affairs 2nd Report of Session 2005 – 2006. 2005. *The Stationary Office Environmental Studies*, London, Great Britain.

Idso, S.B. and Kimball, B.A. 1991. Downward regulation of photosynthesis and growth at high CO_2 levels. No evidence for either phenomenon in three-year study of sour orange trees. *Pl. Physiol.* 96: 990-992.

Idsu, K.E. and Idsu, S.B. 1994. Plant responses to atmospheric CO_2 enrichment in the face of environmental constraints: A review of the past 10 year's research. Agr. *Forest Meteorol.* 69: 153-203.

Intergovernmental Panel on Climate Change. 2001. *IPCC WG third assessment report: summary for policymakers*. Geneva, Switzerland. *http://www.ids.oc.uk/eldis/WGI.htm*.

Intergovernmental Panel on Climate Change. 2007. *IPCC WG fourth assessment report summary for policy makers*. Geneva, Switzerland.

Johansen, B.E. 2006. *Global Warming in the 21st Century*. Praeger, Westport,CT,USA.

Karl, T. and Trenberth, K. 1999. The human impact on climate. *Sci. Am.* 281: 100-145.

Keppler, F., Hamilton, J.T.G., Brass, M. and Rockman, T. 2006. Methane emissions from terrestrial plants under aerobic conditions. *Nature* 439: 187-191.

Kimball, B.A. 1982. Carbon dioxide and agricultural yield. An assemblage and analysis of 430 prior observations. *WCL Report* 11, U.S. Water Conservation Laboratory, Phoenix, AZ, USA.

King, D.A. 2004. Climate change science: Adapt, mitigate or ignore. *Science* 303: 176-177.

Kobayashi, F., Yamane, A. and Ikeda, T. 1984. The Japanese pine sawyer beetle as the vector of pine with disease. *Ann. Rev. Entomol.* 29: 115-135.

Koch, G.W. and Mooney, H.A. 1996. *Carbon Dioxide and Terrestrial Ecosystems*. Academic Press, San Diego, CA, USA.

Koerner, C. and Bazzaz, F.A. eds. 1996. *Carbon Dioxide, Populations and Communities*. Academic Press, San Diego, CA, USA.

Kruper, S.V. and Groth, J.V. 2000. Global climate change crop responses. Uncertainties associated with the current methodologies. In: *Environmental Pollution and Plant Responses*. (Agrawal, S.B. and Agrawal, M. eds.). CRC Press, Boca Raton, FL, USA.

Law, B.E., Ryan, M.G. and Anthoni, P.M. 1999. Seasonal and annual respiration of ponderosa pine ecosystem. *Global Change Biol.* 5: 69-182.

Law, B.E., Goldstein, A.H., Anthoni, P.M., Unsworth, M.H., Panek, J.A., Bauer, M.R., Frochebord, J.M. and Huttman, N. 2001. CO_2 and water vapor exchange by young and old ponderosa pine ecosystems during a drought year. *Tree Physiol.* 21: 299.

Lawrence, O.M., Slater, A.G., Thomas, R.A., Hollard, M.M. and Desev. 2008. Accelerated artic land warming and permafrost degradation during lapid sea use loss. Geophysical Research Letters. 35L11506, doc.10.1029/2008 GLO 33985.

Leaf, A. 1989. Potential health effects of global climate change and environmental change. *New Engl. J. Med.* 321: 1577.

Lemon, E.R. 1983. *CO$_2$ and Plants: The Response of Plants to Rising Levels of Atmospheric Carbon Dioxide.* Westview Press, Boulder, CO, USA.

Leon, A. Jr. 2002. *Evapotranspiration responses of plants and crops to carbon dioxide and temperature. http://www.nol.usda.gov/ttic/tektran/data/oooo10/61/0000106193.html.*

LeQuere, C., Aumont, O., Monfray, P. and Orr, J.C. 2003. Propagation of climatic events on ocean stratification, marine biology and CO$_2$ case studies over the 1979-1999 period. *J. Geophys. Res.* 108: 3375-3392.

Linacre, E. 1998. *Effect of carbon dioxide on plants. http://www-dos.anyo.edu/~geerts/cwx/notes/chap01/ CO$_2$-bio.html.*

Lincoln, D.E., Fajer, E.D. and Johnson, R.H. 1993. Plant-insect herbivore interactions in elevated CO$_2$ environments. *Trends in Ecol. Evol.* 8: 64-68.

Linden, E. 2007. *The Winds of Change: Climate, Weather and the Destruction of Civilization.* Simon and Schuster, London, UK.

Lindzen, R.S. 1997. Can increasing carbon dioxide cause climate change? *OMAS* 94: 8335-8342.

Lombardi, S., Altunia, L.K. and Beaubien, S.E. 2006. *Advances in the Geological Storage of Carbon Dioxide: Approaches to Reduce Anthropogenic Greenhouse Gas Emissions.* Springer, Dordrecht.

Long, S.P. 1991. Modification of the response of photosynthetic productivity to rising temperature by atmospheric CO$_2$ concentrations: has its importance been underestimated? *Plant Cell Environ.* 14: 729-739.

Long, S.P. and Drake, B.G. 1991. Effects of the long term elevation of CO$_2$ concentration in the field on quantum yield of photosynthesis of the C$_2$ sedge *Scirpus olneyi. Pl. Physiol.* 96: 221-226.

Long, S.P., Ainsworth, E.A., Leakey, A.D.B., Nösberger, J. and Ort, D.R. 2006. Food for thought: Lower than expected crop yield stimulation with rising CO$_2$ concentrations. Science 312: 1918-1921.

Lucht, W., Colin, I., Mynei, R.B., Sitch, S., Friedligstein, P., Cramer, W., Bosquet, P., Buermann, W. and Smith, B. 2002. Climatic control of the high latitude vegetation greening trend and Pinatubo effect. *Science* 296: 1687-1689.

Luft, U.C., Finkelstein, S. and Elliot, J.C. 1974. Repiratory gas exchange, acid-base balance and electrolytes during and after maximal work breathing 15 mm HG PICO2. In: *Carbon Dioxide and Metabolic Regulations.* (G. Nahas and K.E. Schaefer, eds.). Springer-Verlag, New York, NY, USA. pp. 273-281.

Luo, Y., Field, C.B. and Mooney, H.A. 1994. Predicting responses of photosynthesis and root function to elevated [CO$_2$] interactions among carbon, nitrogen and growth. *Plant Cell Environ.* 17: 1195-1204.

Luv, Y and Mooney, H.A. 1996. *Carbon Dioxide and Environmental Stress.* Academic Press, New York, NY, USA.

Mabey, N., Hall, S., Smith, C. and Gupta, S. 1997. *Argument in Greenhouse: The International Economics of Controlling Global Warming.* Routledge, New York, NY, USA.

Maslin, M. 2007. *Global Warming: Causes, Effects and The Future.* Voyageur Press, Osceola, WI, USA.

McCarthy, J.J. and McKenna, M.C. 2000. How earth's ice is changing. *Environ.* 42: 8-18.

McGranahan, G., Balk, D. and Anderson, B. 2007. The rising tide assessing the risks of climate change and human settlements in low elevation costal zones. *Environment and Urbanization* 19: 17-37.

Monnin, E., Indermuble, A., Dallenback, A., Fluckiger, J., Stauffer, B., Stocker, T.F., Raynaud, D. and Barnola, J.M. 2001. Atmospheric CO$_2$ concentrations over the last glacial termination. *Science* 291: 112-114.

Mooney, H.A., Fuentes, E.R. and Kranberg, B.I. 1993. *Earth System Responses to Global Change.* Academic Press, San Diego, CA, USA.

Moore, B. III and Braswell, B.H. 1994. The lifetime of excess atmospheric carbon dioxide. *Global Biogeochem. Cy.* 8: 23-38.

Moore, T.G. 1998. *Climate of Fear: Why We Shouldn't Worry About Global Warming*. Cato Institute, Washington, DC, USA.

Morrison, J.I.L., and Morecroft, M.D. 2006. *Plant Growth and Climate Change*. Blackwell Science Publications, Upper Saddle River, NJ, USA.

Newton, D.E. 1993. *Global Warming. A Reference Handbook*. ABC-CLIO, Santa Barbara, CA, USA.

Nicolas, G. and Sillans, D. 1989. Immediate and lateral effects of carbon dioxide on insects. *Ann. Rev. Entomol.* 34: 97-116.

Nijs, I, Impens, I. and Behaeghe, T. 1988. Effects of different CO_2 environments on the photosynthesis yield relationship and the carbon and water balance in white clover (*Trifolium repens L. cv. Blanca*) sward. *J. Exp. Bot.* 40: 353-359.

Nobel, P.S. 1991. Environmental productivity indices and productivity for *Opuntia ficus-indica* under current and elevated atmospheric CO_2 levels. *Plant Cell Environ.* 14: 637-646.

Norby, R.J., Kobayoski, K. and Kimball, B.A. 2001. Rising CO_2 - future ecosystems. *New Phytol.* 150: 215-221.

O'Brien, S.T., Hayden, B.P. and Shugait, H.H. 1992. (Historical Archive). *Global Climatic Change, Hurricanes and a Tropical Forest*. Kluwer Academic, New York, NY, USA.

Oreskes, N. 2004. Beyond the ivory tower: The scientific consensus on climate change. *Science* 306: 1686.

Pacala, S.W., Huitt, G.C. and Baker, D. 2001. Consistent land and atmosphere-based U.S. carbon sink estimates. *Science* 292: 2216-2320.

Parry, M. 1990. *Climate Change and World Agriculture*. Earthscan, London, England.

Paterson, M. 1996. *Global Warming and Global Politics*. Routledge, London, England.

Peters, R.L. and Lovejoy, T.B. 1992. *Global Warming and Biodiversity*. Yale University Press, New Haven, CT, USA.

Petit, J.R., et al. 1999. Climate and atmospheric history of the past 420,000 years form the Vostok ice core, Antarctica. *Nature* 399: 429-436.

Porter, J.H., Parry, M.L. and Carter, T.R. 1991. The potential effects of climatic change on agricultural insect pests. *Agric. For. Meteorol.* 57: 221-241.

Ravindranath, N.H. and Sathage, J.A. 2002. *Climate Change and Developing Countries*. Springer, New York, NY, USA.

Reiley, J., Stone, P.H., Forest, C.E., Webster, M.D., Jacoby, H.D. and Prinn, R.G. 2001. Uncertainty and climate change assessments. *Science* 293: 430-433.

Robertson, G.P., Paul, E.A. and Harwood, R.R. 2000. Greenhouse gases in intensive agriculture: Contributions of individual gases to the radiative forcing of the atmosphere. *Science* 289: 1922-1945.

Robinson, A.B., Balineis, S.L., Soon, W. and Robinson, L.W. 1996. Environmental effects of increased atmospheric carbon dioxide. Global Change *Biol.* 2: 429-440.

Rogers, H.H., Peterson, C.M., McCrimmon, J.N. and Cure, J.D. 1992. Response of plant roots to elevated atmospheric carbon dioxide. *Plant Cell Environ.* 15: 749-752.

Roleff, T.L. ed. 1997. *Global Warming: Opposing Viewpoints*. Greenhaven Press, San Diego, CA, USA.

Rosenzeweig, C. and Parey, M.L. 1994. Potential impact of climate change in world food supply. *Nature* 367: 133-138.

Rowland-Bamford, A.J., Baker, J.T., Allen, L.H. and Bowes, G. 1991. Acclimation of rice to atmospheric carbon dioxide concentrations. *Plant Cell Environ.* 14: 577-583.

Ryle, G.J.A. and Powell, D.E. 1992. The influence of elevated CO_2 and temperature on biomass accumulation of continuously defoliated white clover. *Plant Cell Environ.* 15: 593-599.

Sailor, S.F. 1989. *Global Climate Change Human and Natural Influence*. Paragon House, New York, NY, USA.

Sandalow, D.B. and Bowles, A. 2001. Fundamentals of treaty-making on climate change. *Science* 292: 1839-1840.

Schaefer, K.E. 1982. Effects of increased ambient CO_2 levels on human and animal health. *Experientia* 38: 1163-1168.

Schimel, D.S. 1995. Terrestrial ecosystems and the carbon cycle. *Global Change Biol.* 1: 77-91.

Schmidt, K. 1998. Coming to grips with the world's greenhouse gases. *Science* 281: 504-506.

Schmitdmann, E.T. and Miller, J.A. 1989. Effects of climatic warming in populations of the horn fly, with associated impact on weight gain and milk production in cattle. In: *The Potential Effects of Global Climate Change on the United States.* Appendix C: Agricluture. Volume 2 (Smith, J.B. and Tirpak, D.A., eds.). United States Environmental Protection Agency, Washington, DC, USA.

Scholes, R.J. and Noble, I.A. 2001. Climate change: Storing carbon on land. *Science* 294: 1012-1013.

Seager, R., *et al.* 2007. Model projections of an imminent transition to a more arid climate in southwestern north America. *Science* 316: 1181-1184.

Seely, T.D. 1974. Atmospheric carbon dioxide regulations in a honeybee ('Aphis mellifera') colonies. *J. Insect Physiol.* 20: 2301-2305.

Serreze, M.C., Holland, M.M. and Stroeve, J. 2007. Perspectives on the Arctic's shrinking sea-ice cover. *Science* 315: 1533-1536.

Shepherd, A. and Wingham, D. 2007. Recent sea-level contributions of the Antarctic and Greenland ice sheets. *Science* 315: 1529-1532.

Singer, S.F. 1999. Human contribution to climate change remains questionable. *EOS Transactions. American Geophysical Union* 80: 186-187.

Smith, J.B. and Tirpak, D. 1989. *Potential Effects of Global Climate Change on the United States.* U.S. Environmental Protection Agency, Washington DC, USA.

Socolow, R., Hotinski, R., Greenblott, J.B. and Pacala, W. 2004. Solving the climate problem. Technologies available to curb CO_2 emissions. *Environment* 46: 8-19.

Solomon, A.M. and Shugart, H.H. (eds.). 1993. *Vegetation Dynamics and Global Change,* Chapman & Hall, London, England.

Stanhill, G. and Cohen, S. 2001. Global dimming: a review of the evidence for a widespread and significant reduction in global radiation with discussion of its probable causes and possible agricultural consequences. *Agr. Forest Meterol.* 107: 255-278.

Stat, P.A., Tett, S.F.B., Jones, G.S., Allen, M.R., Mitchell, J.F.B. and Jenkins, G.J. 2001. External control of 20th century temperature by natural and anthropogenic forcings. *Science* 290: 2133-2137.

Stinner, B.R., Taylor, R.A.J., Hammond, R.B., Purington, F.R.F. and McCartney, D.A. 1989. *Potential Effects of Global Climate Change on the United States.* Appendix C: *Agriculture* Volume 2 (Smith, J.B. and D.A. Trrpak, eds.). United States Environmental Protection Agency, Washington, DC, USA.

Stitt, M. 1991. Rising CO_2 levels and their potential for carbon flow in photosynthetic cells. *Plant Cell Environ.* 14: 741-762.

Stulen, I. and Den Hertog, J. 1993. Root growth and functioning under atmospheric CO_2 enrichment. *Vegetation* 104/105: 99-115.

Sturges, W.T., *et al.* 2000. A potent greenhouse gas identified in the atmosphere: SF_5CF_3. *Science* 289: 611-613.

Sweeney, B.W., Jackson, J.K., Newbold, J.D. and Fink, D.H. 1992. Climate change and the life histories and biogeography of aquatic insects in Eastern North America. In: (Firth, P. and S.G. Fisher, eds.) *Global Climate Change and Freshwater Ecosystems.* Springer-Verlag, New York, NY, USA.

Tamburri, M.N., Peltzer, E.T., Friederich, G., Brewer, P.G., Aya, I. and Yamone, K. 2000. A field study of the effects CO_2 ocean disposal on mobile deep sea-animals. *Mar. Chem.* 72: 95-101.

Titus, J.G. and Richman, C. 2004. Maps of lands vulnerable to sea level rise. *http://yosemite. epa.gov/oar/globalwarming.nsf/content/ resourcecenterpublicationsslrm.*

Tolley, L.C. and Strain, B.R. 1985. Effects of CO_2 enrichment and water stress on gas exchange of *Liquidambar styraciflua* and *Pinus taeda* seedlings grown at different irradiance levels. *Oecologia* (Berlin) 65: 166-172.

Trenberth, K.E. 2000. Stronger evidence of human influences on climate. The 2001 IPCC assessment. *Environment* 43: 8-19.

Trenberth, K.E. 2005. Uncertainty in hurricanes and global warming. *Science* 17: 1753-1754.

Tretenberg, T. 1997. *The Economics of Global Warming*. E. Elgar Pub., Cheltenham, UK.

Tuba, Z. 2005. *Ecological Responses and Adaptations of Crops to Rising Atmospheric Carbon Dioxide*. Haworth Press, Binghampton, NY, USA.

Turino, G.M., Goldring, R.M. and Heinemann, H.O. 1974. The extracellular bicarbonate concentration and the regulation of ventilation in chronic hypercapnia in man. In: *Carbon Dioxide and Metabolic Regulations*. (G. Nahas and K.E. Schaefer, eds.). Springer-Verlag, New York, NY, USA. pp. 273-281.

Van Ypersele de Strihou, C. 1974. Acid-base equilibrium in chronic hypercapnia. In: *Carbon Dioxide and Metabolic Regulations*. (G. Nahas and K.E. Schaefer, eds.). Springer-Verlag, New York, NY, USA.

Yamin, F. 2005. *Climate Change and Carbon Markets. A Handbook of Emissions Reductions*. Earthscan London, England.

Vinnikov, K.Y. 1999. Global warming and northern hemisphere sea ice extent. *Science* 286: 1934-1937.

Vivin, P., Guehl, J.M., Clement, A. and Asseor, G. 1996. The effects of elevated CO_2 and water stress on whole plant CO_2 exchange, carbon allocation and osmoregulation in *Quercus vobur* seedlings. *Annales des Sciences Forestieres* 53: 447-449.

Vorosmarty, C.J., Green, P., Salisbury, J. and Lammens, R.B. 2000. Global water resources. Vulnerability from climate change and population growth. *Science* 289: 284-288.

Walker, B. and Steffen, W. 1996. *Global Change and Terrestrial Ecosystems*. Biodiversity Documentation Centre, Bangalore, India.

Watson, R.T., Meira, L.G., Sanjueza, E. and Janetos, A. 1992. *Sources and Sinks in Climate Change (1992) the Supplementary Report to the IPCC Scientific Assessment*. (Houghton, J.T., Callader, B.A. and Vaney, S.K., eds.). Cambridge Univ. Press, Cambridge pp. 25-46.

Weike, W.H. and Mertens, R. 1991. Human well-being, diseases and climate. In: *Climate Change: Science Impacts and Policy*, (Jager, J.

and Ferguson, H.L., eds.). Cambridge Univ. Press, Cambridge, England.

Wigley, T.M.L. and Raper, S.C.B. 2001. Interpretation of high proportion for global mean warming. Science 293: 451-454.

Wisniewski, J., Dixon, R.K., Kinsman, J.D., Sampson, R.N. and Logo, A.E. 1993. Carbon dioxide sequestration in terrestrial ecosystems. *Climate* 3: 1-5.

Wofsy, S.C. 2001. Where has all the carbon gone? Science 292: 2261-2263. Wolfe-Bellin, K.S., He, J-S, and Bazzaz, F.A. 2006. Low-level physiology, biomass, and reproduction of *Phytolacca americana* under conditions of elevated carbon dioxide and increased nocturnal temperature. *Int. J. Plant Sci.* 167: 1011-1020.

Wolfe, D.W., Schwartz, M.D., Lasko, A.N., Otsuki, Y., Pool, R.M. and Shaulis, N.J. 2005. Climate change and shifts in spring phenology of three horticultural woody perennials in northeastern USA. *Int. J. Biometerol.* 49: 303-309.

Wood, D.M. and McDonald, D.G. 1997. *Global Terrestrial Warming – Implications for Freshwater and Marine Fish*. Cambridge Univ. Press, Cambridge, England.

Woodward, F.I. 1987. Stomatal numbers are sensitive to increases in CO_2 from preindustrial levels. *Nature* 327: 617-618.

World Health Organization. 1990. Vector-borne disease Section. In: *Potential Health Effects of Climatic Change*. World Health Organization, Geneva, Switzerland.

Wright, D.G., Hendey, R.M., Loder, J.W. and Dobson, F.W. 1986. Oceanic changes associated with global sciences in atmospheric carbon dioxide: a preliminary report for the Atlantic Coast of Canada. *Canadian Technical Report of Fisheries and Aquatic Sciences* 1426.

Wyman, R.L. 1991. *Global Climate Change and Life on Earth*. Chapman and Hall, New York, NY, USA.

Xiao, X., Melillo, J.M., Kicklighter, D.W., McGuire, A.D., Prinn, R.G., Wang, C., Stone, P.H. and Sokolov, A. 1998. Transient climate change and net ecosystem production of the terrestrial biosphere. *Global Biogeochem. Cy.* 12: 345-360.

CHAPTER 4

Energy

TYPES OF ENERGY

The use of the world's energy is increasing at about 7% per year. If this rate continues, world energy utilization will double every 10 years. Given the world's fossil fuel and ore supplies, this is not sustainable over the long haul. Thus, there will eventually be a leveling off of energy consumption unless renewable forms of energy, e.g., solar and wind, are improved and others developed (Table 4.1). The USA energy consumption by source for 1960 – 1992 is shown in Table 4.2. Figure 4.1 and 4.3 compare world energy and North American consumptions. Maheshwari and Provorar (2004) have con-sidered energy usage in relation to both the environment and society.

NON-RENEWABLE SOURCES OF ENERGY

Coal – This fuel is in adequate supply until about 2030 in contrast to oil (Fig. 4.2). The latter should be depleted by 90% between 2020 and 2030. Coal can be converted to liq-uid fuel and methane. The latter is simple to transport and 'burns' cleanly (see discus-sion on natural gas). Coal can also be converted to kerosene, motor oil and other petroleum products. However, there are environmental drawbacks as high sulfur-

TABLE 4.1	Types of Energy
Renewable	*Non-renewable*
Biomass (Silverio, 2005)	Coal (Clean Coal Technology, 2001: US Dept. of Energy, 2001: Bailey et al., 2002)
Geothermal (Arnstead, 1983; Dickson and Fanelli, 1995)	Natural Gas (MacAvoy, 2000; Seddon, 2006; Taurus, 2001)
Hydroelectric (Spiro and Stigliani, 2003)	
Ocean and Tidal wave (Avery and Wu,1994; Falnes, 2002; Ross, 1995)	Nuclear (Nealey, 1990 ; Cockerham and Cockerham, 1994 ; Atkins, 2000)
Solar (Tiwari, 2002 ; Hanakawa, 2004)	
Wind (Ackerman, 2005 ; Ewing, 2003; Hau, 2005 ; Katzman, 1984 ; Koller et al., 2006 ; Larid, 2001 ; Mathew, 2006 ; Patel, 2006 ; Walker and Jenkins, 1997)	Oil (Abelson, 1997, 1994 ; Tickell et al., 2006)

TABLE 4.2   U.S. Energy Consumption by Energy Source, 2000-2004 (Quadrillion Btu)

Energy Source	2000	2001	2002	2003	2004[P]
Total[a]	98.961	96.464	97.952	98.714	100.278
Fossil Fuels	84.965	83.176	84.070	84.889	86.186
Coal	22.580	21.952	21.980	22.713	22.918
Coal Coke Net Imports	0.065	0.023	0.061	0.051	0.138
Natural Gas[b]	23.916	22.861	23.628	23.069	23.000
Petroleum[c]	38.404	38.333	38.401	39.047	40.130
Electricity Net Imports	0.115	0.075	0.078	0.022	0.039
Nuclear Electric Power	7.862	8.033	8.143	7.959	8.232
Renewable Energy	6.158	5.328	5.835	6.082	6.117
Conventional Hydroelectric	2.811	2.242	2.689	2.825	2.725
Geothermal Energy	0.317	0.311	0.328	0.339	0.340
Biomass[d]	2.907	2.640	2.648	2.740	2.845
Solar Energy	0.066	0.065	0.064	0.064	0.063
Wind Energy	0.057	0.070	0.105	0.115	0.143

[a]Ethanol blended into motor gasoline is included in both "Petroleum" and "Biomass", but is counted only once in total consumption.

[b]Includes supplemental gaseous fuels.

[c]Petroleum products supplied, including natural gas plant liquids and crude oil burned as fuel.

[d]Biomass includes: black liquor, wood/wood waste liquids, wood/wood waste solids, municipal solid waste (MWS), landfill gas, agriculture byproducts/crops, sludge waste, tires, alcohol fuels (primarily ethanol derived from corn and blended into motor gasoline) and other biomass solids, liquids and gases.

[P]= Preliminary

Note: Data revisions are discussed in highlighted sections. Totals may not equal sum of components due to independent rounding.

Sources: Non-renewable energy: Energy Information Administration (EIA), Monthly Energy Review (MER) March 2005, DOE/EIA-0035 (2005/03) (Washington, DC, March 2005,) Tables 1.3 and 1.4. Renewable Energy: Table 2 of this report. As a result totals in this table do not match the March MER.

From: *http://www.eia.doe.gov/cneaf/solar.renewables/page/trends/table1.html*

containing coals can emit sulfur dioxide when burned. Coal is mined either in tunnels or open pits (strip mines). While the former are dangerous, non-reclaimed strip mines are not pleasing to the eye.

Natural Gas-Natural gas consists of 75 – 100% methane (CH_4), 6 - 10% ethane, 5 - 8% propane and butane, 1 – 4% pentane and heavier hydrocarbons together with varying amounts of nitrogen, carbon dioxide, hydrogen sulfide and helium depending upon its location (Van Loon and Duffy, 2005). The greatest supplies of natural gas are in Eastern Europe, former Soviet Republics and the Middle East (Spriro and Stigliani, 2003). Natural gas usage pos-

sesses desirable environmental benefits, for example, it generates more heat and less CO_2 per unit of energy than either petroleum or coal. In addition, it contains less sulfur than other fuels and methane can be used for the synthesis of gasoline. Also, natural gas may enhance the transition to a hydrogen economy as a pipeline distribution system exists. The environment pitfalls of natural gas are leakage of natural gas at wells together with losses during processing and transport (Van Loon and Duffy, 2005). Atmospheric methane can serve as a greenhouse gas (see chapter 3).

Oil – Oil-based fuel became the dominant fuel in the 1950s. In the 1970s the USA,

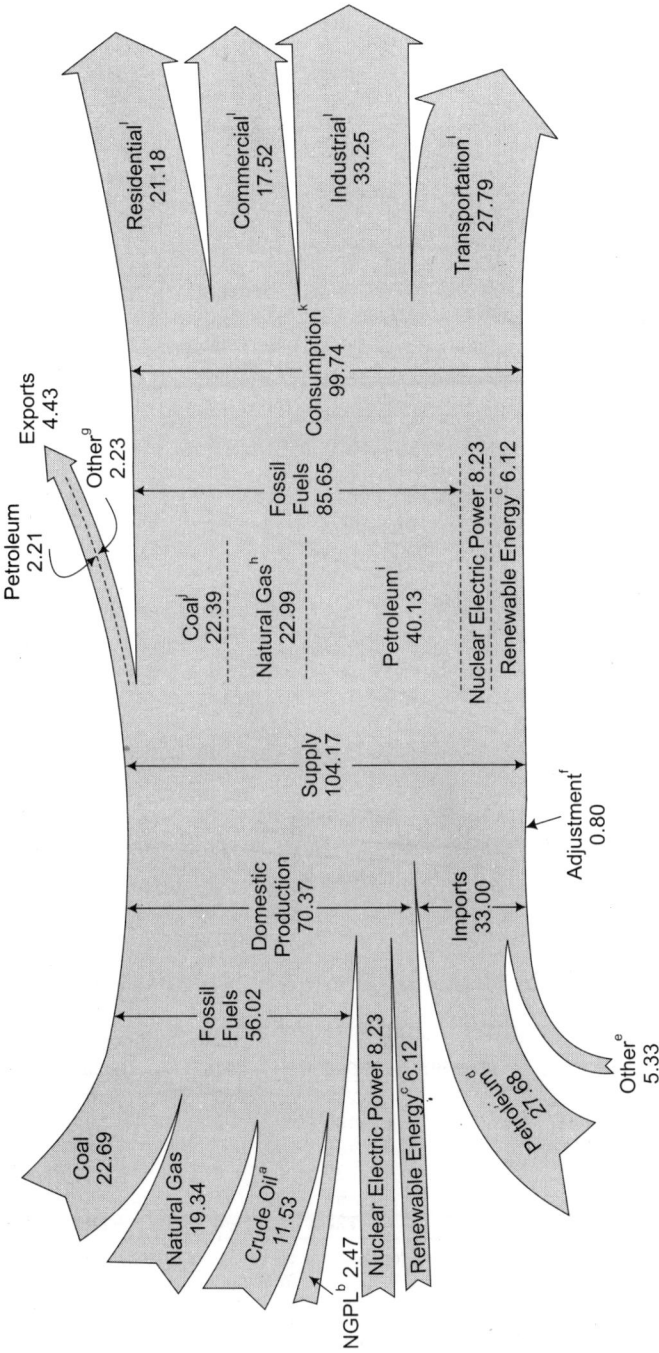

Fig. 4.1 Energy Flow, 2004. From: *http://www.eia.doe.gov/eneu/pdf/pages/secl_3.pdf*

a Includes lease condensate.
b Natural gas plant liquids.
c Conventional hydroelectric power, wood, waste, ethanol blended into motor gasoline, geothermal, solar, and wind.
d Crude oil and petroleum products. Includes imports into the Strategic petroleum Reserve.
e Natural gas, coal, coal coke, and electricity.
f Stock changes, losses, gains, miscellaneous blending components, and unaccounted-for supply.
g Coal, natural gas, coal coke, and electricity.
h Includes supplemental gaseous fuels.

i Petroleum products, including natural gas plant liquids.
j Includes 0.14 quadrillion Btu of coal coke net imports.
k Includes, in quadrillion Btu, 0.30 ethanol blended into motor gasoline, which is accounted for in both fossil fuels and renewable energy but counted only once in total consumption; and 0.04 electricity net imports.
l Primary consumption, electricity retail sales, and electrical system energy losses, which are allocated to the end-use sectors in proportion to each sector's share of total electricity retail sales. See Note, "Electrical Systems Energy Losses," at end of Section 2.

Notes: • Data are preliminary. • Totals may not equal sum of components due to independent rounding.

Sources: Tables 1.1, 1.2, 1.3, 1.4, 2.1a, and 10.1.

Energy consumption by Fuel, 1970-2020

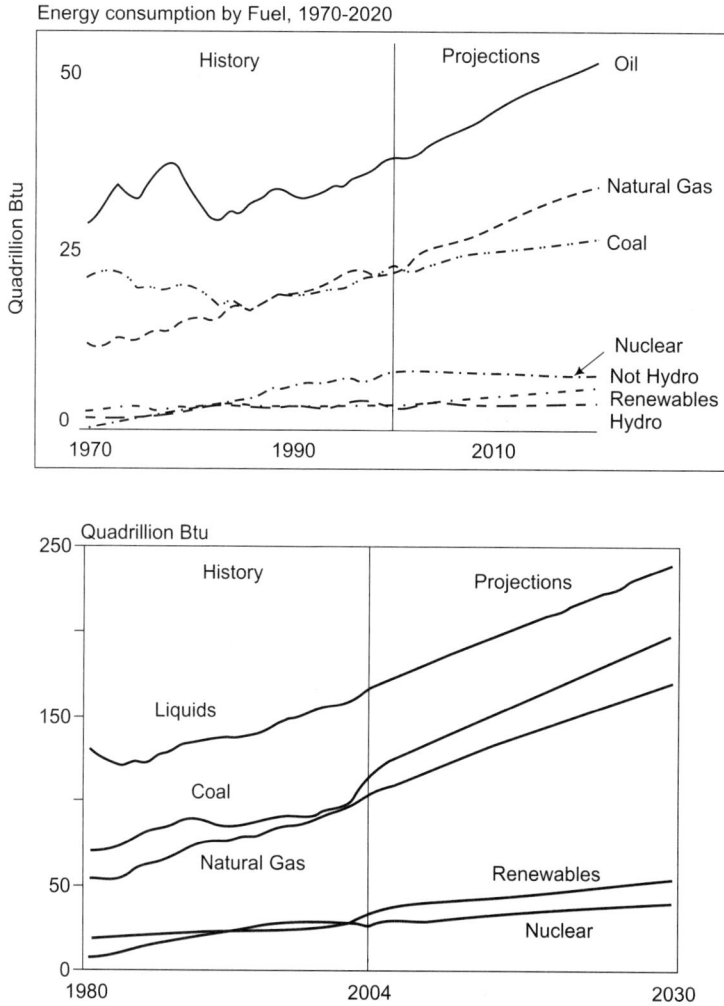

Comparison of *IFO*2007 and *IEO*2006 World Energy Consumption by Fuel, Reference Case, 2015 and 2030 (Quadrillion Btu)

Fuel	2015		2030		Change in *IEO*2007	
	*IEO*2007	*IEO*2006	*IEO*2007	*IEO*2006	2015	2030
Liquids..............................	198	199	239	239	−2	0
Natural Gas......................	134	140	170	190	−6	−19
Coal.................................	152	144	199	196	7	4
Nuclear............................	33	31	40	35	2	5
Renewable/Other..............	43	48	53	61	−5	−8
Total.................................	**559**	**563**	**702**	**722**	**−4**	**−20**

Sources: **IEO2007**: Energy Information Administration (EIA), System for the Analysis of Global Energy Markets (2007). **IEO2006** EIA, *International Energy Outlook* 2006, DOE/EIA-0484(2006) (Washington, DC, June 2006), Table A2, pp. 84-85.

Fig. 4.2 World-wide Time-Dependent Energy Consumption by Fuel (1980-2030).
From: Energy Information Administration International Energy Annual, 2004. *http://www.eia.doe.gov/oiaf/iev/world.html.*

Japan and European countries became dependent upon foreign oil, e.g., OPEC (Organization of Petroleum Exporting Countries)

What are the World's Estimated Oil Reserves?

There are educated guesses and proven reserves.

Educated Guesses

Geologists make these regarding where oil and/or natural gas may be expected to be found.

Proven Reserves

These are derived from exploratory drilling to ascertain the extent and depth of oil deposits.

Thus, educated guesses and proven reserves are added for the estimate. Unfortunately, the world is using fossil fuels faster than the reserves can be regenerated. The environmental drawbacks of employing oil involve its transport. Large tankers have experienced accidents resulting in massive oil spills. Oil from slicks has washed up on beaches destroying wildlife. Perhaps, the most well-known of these spills is the 1989 Exxon Valdez accident. The toxicity of oil has been reviewed by Overton et al. (1994).

Electricity

Electricity is a secondary energy source, i.e., requires a primary energy source to turn the generators. Most of our electricity is produced in steam turbines. However, electricity can be generated by water turbines and gas turbines.

Power Source (Coal, Oil or Nuclear Fuel)
↓
Hot Water in Boiler to Yield Hot High Pressure Steam
↓
Steam Expands against Blades of Turbine
↓
Spinning Turbine Operates a Generator (Rotating Coil of Wire in a Magnetic Field)
↓
Electricity
↓
Spent Steam Flows into a Condenser Where it is Cooled and Liquefied. Water is Returned to the Boiler.
From: (*http://www.tva.gov/power/coalart.htm*)

Transmission of Electricity

Electricity is carried by low voltage service lines (telephone poles) and low voltage trunk lines across metal towers.

The disadvantages of electrical power are:

1. Land destruction
2. Pollution or hazardous waste products
3. Thermal pollution.

An electrical power plant employs water as a coolant. Heated water when added to aquatic systems can often result in fish kills and dirty water. Turk and Turk (1997) discuss ways of controlling thermal pollution, e.g. magneto-hydrodynamics.

Finally, Abelson (2000) discussed the future supplies of electricity.

Nuclear Power There are both advantages and disadvantages to nuclear power (Nuttall, 2007).

Advantages vs. Disadvantages

Advantages

1. Electricity generated by a nuclear power plant is inexpensive and plentiful.

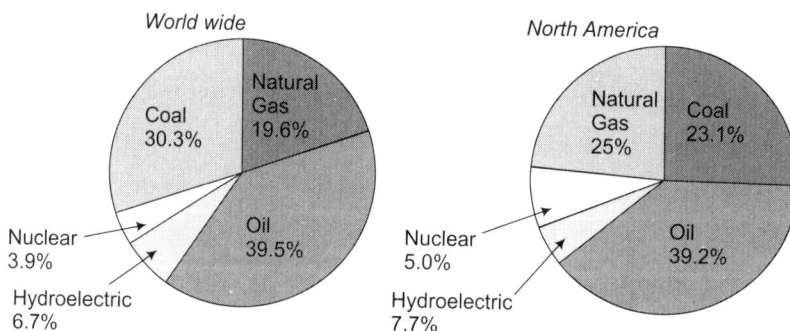

Fig. 4.3 Comparison of Global and North American Energy Use. From: *http://www.globalchange.umich.edu/globalchange2/current/lectures/energy1.html.*

2. When operating normally, nuclear power plants emit less radioactivity into the atmosphere than coal-fired power plants (Fig. 4.5a).

3. The generated electricity can be used directly.

Disadvantages

1. Costly to operate
2. Dangers of nuclear accident, i.e., a melt-down.
3. Waste disposal

Reactors

How they Work

A schematic of an operating nuclear power station is shown in Fig. 4.5b. The station produces heat which is employed to generate electricity. The following references can be consulted for an in depth view of nuclear energy (Nealey, 1990; Kruschke, 1990; Hoheremser, 1992; Wolfson, 1993; Beacham, 1993; Bauer, 1995; Williams, 1998; Makhijani et al, 2000).

Types of Reactors

Fission – When nuclei of U-235 are bombarded with neutrons (Fig. 4.4), they split resulting in copious energy (note: Student should review neutrons, electrons and protons as well as orbital theory). Fission occurs when heavier elements are broken into lighter ones. An example of this is the following reaction:

$$1_0n + 235_{92}U \rightarrow 141_{56}Ba + {}^{92}_{36}Kr + 3{}^1_0n$$

Fission creates radioactive products together with energy which can be utilized to heat water which can drive a steam turbine to produce electricity.

Radioactive waste can be released if the fission reaction is not controlled, i.e. excess U235 yields neutrons spontaneously releasing additional neutrons and so on. 235U + neutron $\rightarrow$ fission products + 2 to 3 neutrons + energy. The chain reaction is controlled by moderators (gases and control rods) such that energy is given off as heat over a lengthy time period. The rods are composed of graphite which absorb neutrons slowing the chain reaction. The control rods are lowered between the fuel rods. Thus, nuclear reactors differ from a nuclear bomb in that the concentration of fissionable material is diluted and there are moderators.

Breeder Fission Reactors

Breeder fission reactors are similar to fission reactors except that breeder fission

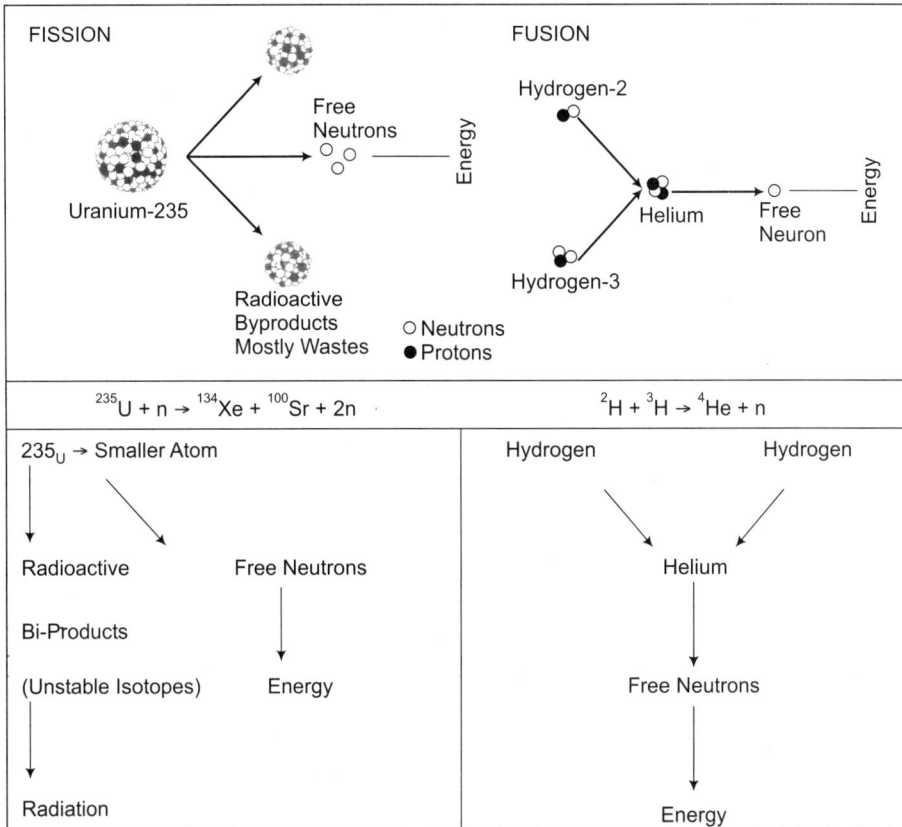

$$^{235}U + n \rightarrow \,^{134}Xe + \,^{100}Sr + 2n$$

$$^{2}H + \,^{3}H \rightarrow \,^{4}He + n$$

$235_U \rightarrow$ Smaller Atom

Radioactive   Free Neutrons

Bi-Products

(Unstable Isotopes)   Energy

Radiation

Hydrogen   Hydrogen

Helium

Free Neutrons

Energy

Fig. 4.4 Comparison of Fission and Fusion Processes

reactors can employ both U235 and U238. The latter contaminates U235 during mining. The U238 can be converted to plutonium 239 providing additional fuel. 238U + neutron → plutonium 239. 239 + 1 neutron → fission products + 2 + 3 neutrons + energy. Another way of releasing nuclear energy is through fusion, i.e., the "melting" together of small atoms to form a larger atom. The small atoms are often hydrogens.

$$^{1}_{1}H + \,^{1}_{1}H \rightarrow \,^{2}_{1}H + \,^{0}_{1}e + v$$

Nuclear Safety

Nuclear reactors possess steel and concrete shells (Avenhaus, 1986). Breeder reactors appear to be safer than other fission reactors. However, electricity produced from breeder reactors is more expensive to produce than fission reactors. In addition, breeder reactors are more dangerous as the breeder core lacks a moderator to temper the neutrons' energies. Other problems include a compactly spaced fuel which is richer in fissionable isotopes. Thus, it is closer to the critical mass, i.e., the maximum mass, required to support a self-sustaining chain reactor. Additional problems to breeder reactors include short neutron half-lives and liquid sodium which could result in hydrogen gas if contact with water occurs. This could breach the containment

Schematic Diagram of a Coal-fired Power Plant

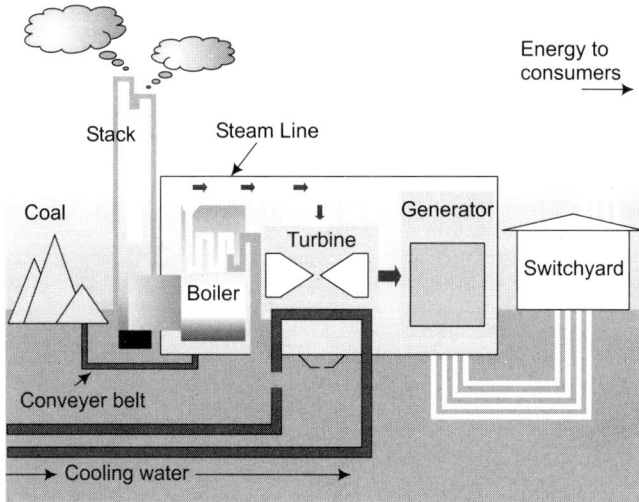

a

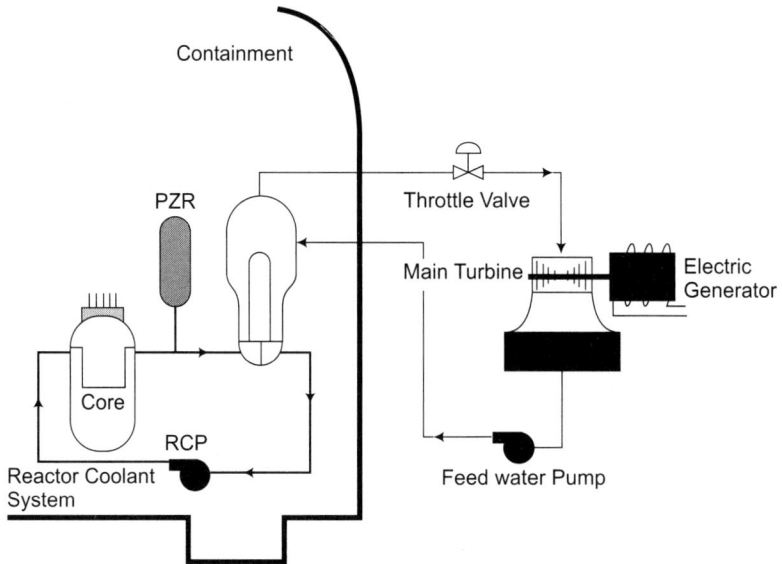

b

Fig. 4.5a Coal-fired Power Plant. From: *http://www.tva.gov/power/coalart.htm.* **b** Pressurized Water Reactor. From: *http://www.nvc.gov/reactors/prois.html*

NUCLEAR REACTOR SCHEMATIC

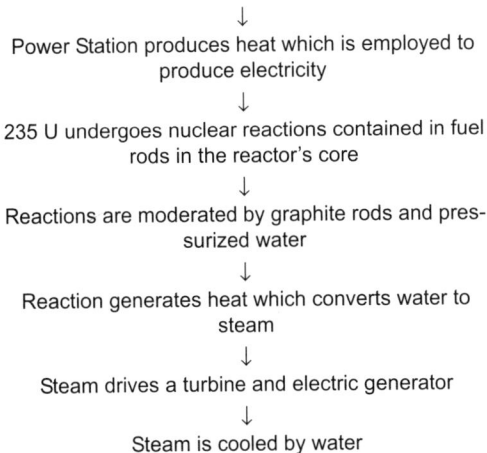

↓

Power Station produces heat which is employed to produce electricity

↓

235 U undergoes nuclear reactions contained in fuel rods in the reactor's core

↓

Reactions are moderated by graphite rods and pressurized water

↓

Reaction generates heat which converts water to steam

↓

Steam drives a turbine and electric generator

↓

Steam is cooled by water

vessels. Radioactive wastes have been buried in salt mines via an access shaft and a transporter as well as sea (Ringus, 2000).

Renewable Energies

Given the time-dependent loss of oil reserves and the cost of oil, there has been marked interest in developing alternative forms of energy (Tester et al., 2005; Kruger, 2006). Table 4.3 presents renewable forms of energy (Bernstein, 2001; Knops, 2003; Boyle, 2004; Fauet and Simnoes, 2006; Quaschning, 2005) whose environmental impacts have been considered (OECD Compass Project, 1988).

Biomass- Ethanol can be generated from fermenting biomass (Karp and Shield, 2008). Figure 4.6 illustrates the generation utilizing a continuous cascade fermentation system (*http://www.eere. energy.gov/inventions/pdfs/bioproc.pdf.*) The system can employ waste sugars from the food processing industry. Ethanol can be blended with gasoline reducing petroleum usage. The blending reduces CO_2 emissions.

Fuel Cells

The cells (Hoffmann, 2001; Sorenson, 2005) convert energy derived from electrochemical reactions directly to electricity without a combustion process (Manahan, 2005).

Fuel cells provide water and heat besides to electricity. In addition, fuel cells possess several other advantages, i.e., they generate smaller amounts of greenhouse gases and less of the air pollutants which create smog and health issues. Figure 4.7 illustrates how a fuel cell operates.

Geothermal Energy

The energy (Fig. 4.8a,b) derives from geothermal steam as the Earth generates heat from its molten core and from the decay of naturally occurring radioisotopes (Duffield and Sass, 2003; Spiro and Stigliani, 2003; Gupta and Roy, 2006). However geother-

TABLE 4.3	Other Sources of Energy	
Source	*Advantage*	*Disadvantage*
Biomass	Reduces CO_2 emissions	Cost
Solar	Every 15 min. of solar energy is equivalent to man's power needs for a year at 1970's consumption	Large cost and amount of land needed
Hydroelectric	A form of solar energy	Provides only 4% of power used in US; not enough dams available
Geothermal	Considerable future energy by drilling through the Earth's crust and its molten core	Drilling might cause slippage of rock layers and thus earthquakes
Tides	Generate electricity as tides turn turbines	Output from a single generator would be small
Wind	Small windmills could produce electricity	Wind doesn't blow constantly

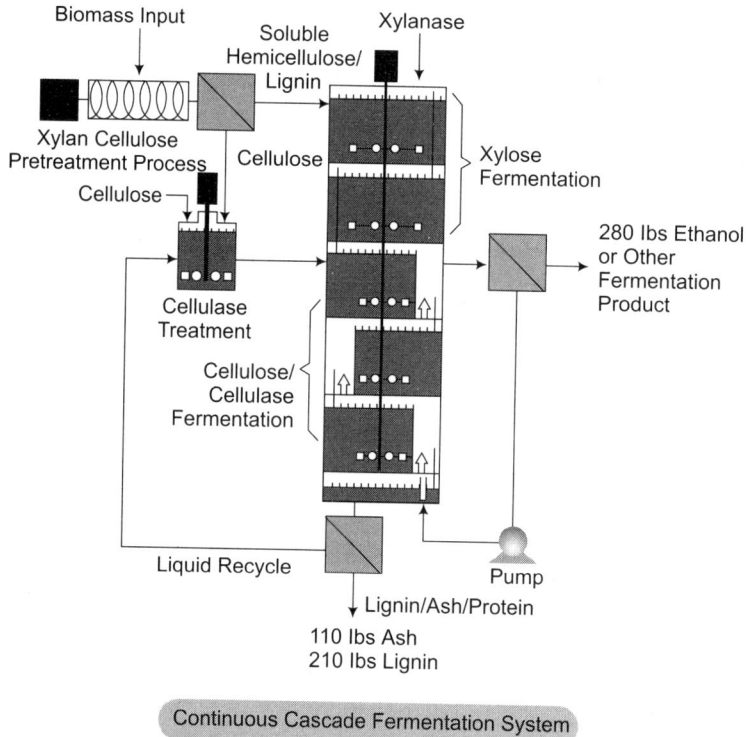

Fig. 4.6 Low-Energy Continuous System for Converting Waste Biomass to Ethanol. From: US Department of Energy, Office of Industrial Technologies, Energy Efficiency and Renewable Energy A schematic for ethanol productions from corn can be viewed at *http://www.freemarketnews.com/analysis/86/4618/2006-94-24.asp?wid=86nid=4618*

mal steam is not worldwide and thus geothermal steam plants are not ubiquitous. Geothermal energy comprises three categories (*http://www.las.asu.edu/~holbert/eee463/hydrotheraml.html*), high (>150°C), moderate (90 - 150°C) and low (<90°C). Whereas the high category can be employed for electricity generated by geothermal steam plants, the moderate form can be utilized for space heating, industrial process steam, greenhouse and agriculture. The low category is used in conjunction with heat pumps for home heating and cooling (Spiro and Stigliani, 2003). Geothermal energy development may be limited by land subsidence and siesmic activities (Manahan, 2005). However, development is encouraged as geothermal energy can generate power continuously (Yen, 1999).

Hydroelectricity

Hydroelectric plants use a portion of the energy derived from the solar–driven hydrological cycle (Spiro and Stigliani, 2003). Excessive rain water (water not lost by evapotranspiration) 'runs-off' rivers into the oceans. The runoff can be employed to turn a turbine and generate electricity. Hydroelectric plants (Fig. 4.10) utilize water directly or rely on dams to enhance water

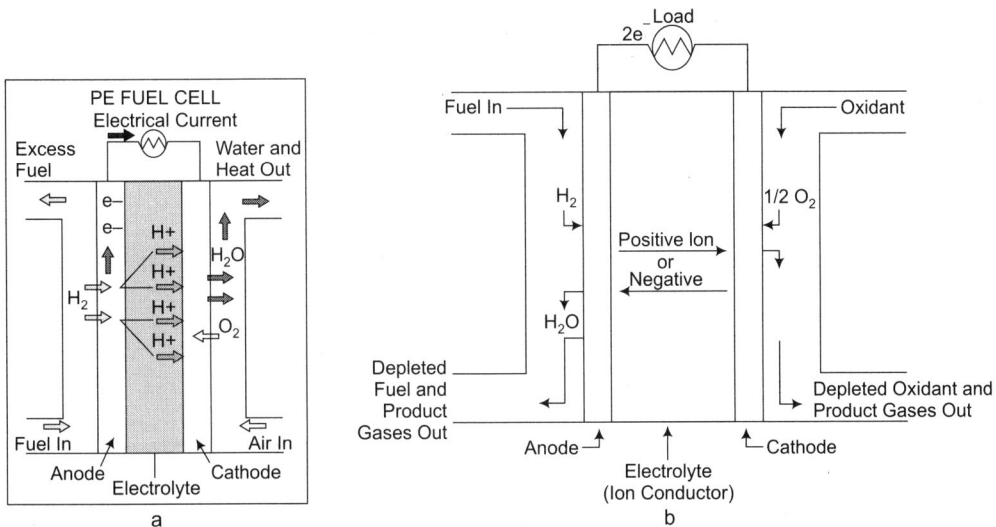

Fig. 4.7 a, b Fuel Cell Schematics. A. PEM Fuel Cell, B. Individual Fuel Cell
a. From: *http://www.eere.energy.gov/hydrogenandfuelcells/pdrs/review04/f*
b. From: *http://www.its.ucdavis.edu/education/classes/pathwaysclass/fchon*

Single fuel cell comprising an electrolyte and two catalyst-coated electrodes, i.e., an anode and a cathode

Hydrogen or a hydrogen-rich fuel is supplied to the anode (negative electrode)

↓

Catalyst separated hydrogens electrons from protons
Yielding H_2O and (OH)

Polymer electrolyte membrane and phosphoric acid fuel cells	Alkaline, molten carbonate and solid oxide fuel cells
Protons migrate through the electrolyte to the cathode to combine with O_2 electrons producing water and heat	Negative ions move through the electrolyte to the anode where they combine with H_2 to yield H_2O and electrons

Electrons from the anode connect pass through the electrolyte to the cathode (positive electrodes) the electrons travel via an electrical circuit.

Adapted from: *http://www.eere.energy.gov/hydrogenandfuelcells/fuelcells/basics.html?pint.*

pressure and 'even out' the flow permitting continuous electricity production (Spiro and Stigliani, 2003). Although hydroelectricity does not emit CO_2 or other emissions to the atmosphere, water backing up behind dams (Fig. 4.9) can flood the shoreline ecosystems and impair migratory fish.

Ocean Energy

The world's oceans store energy as tides, waves (Ross, 1995) and gradients of temperature and salt concentrations (Spiro and Stigliani, 2003). These energy forms are highly dispersed and the number of tidal

a

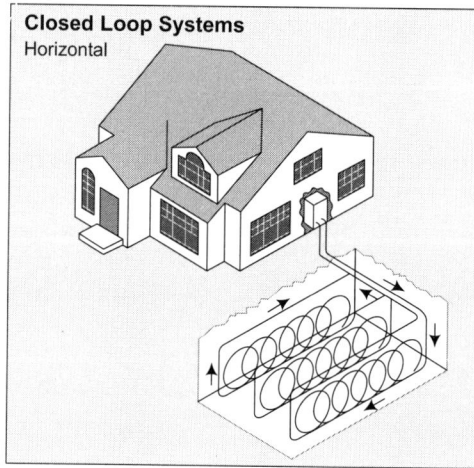

b

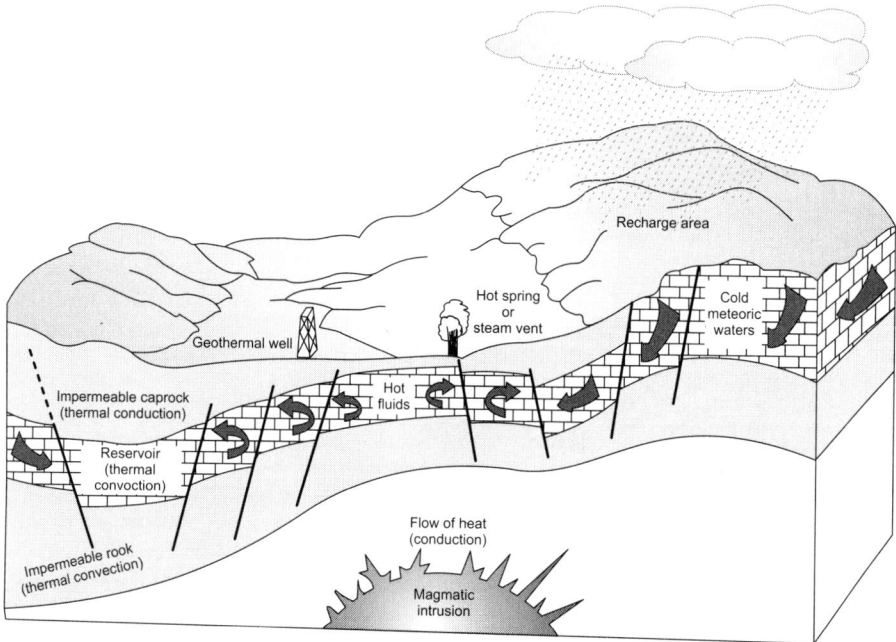

Fig. 4.8a Diagram of a home and a geothermal heat pump system. From: *http//activerain.com/blogsview/303434/ -Dreaming-A-Little* who cite the US Dept. of Energy as the source. Another useful diagram can be found at *geoheat@oit.edu*. **b** Schematic representation of an ideal geothermal system. From: *http://iga.igg.cnr.it/ geoenergy.php.*

sites is limited rendering tidal energy (Boyle, 2004) an insignificant global energy supply. Although waves can carry significant quantities of energy, mechanical devices (Fig. 4.11a) for extracting wave energy are as, of yet, not practical. An oscillating water column (Fig. 4.11b) originates electricity in two steps. During the first step, a wave enters the column and forces air into the column past a turbine. The result is an enhanced column pressure. When the wave retreats, the air is drawn

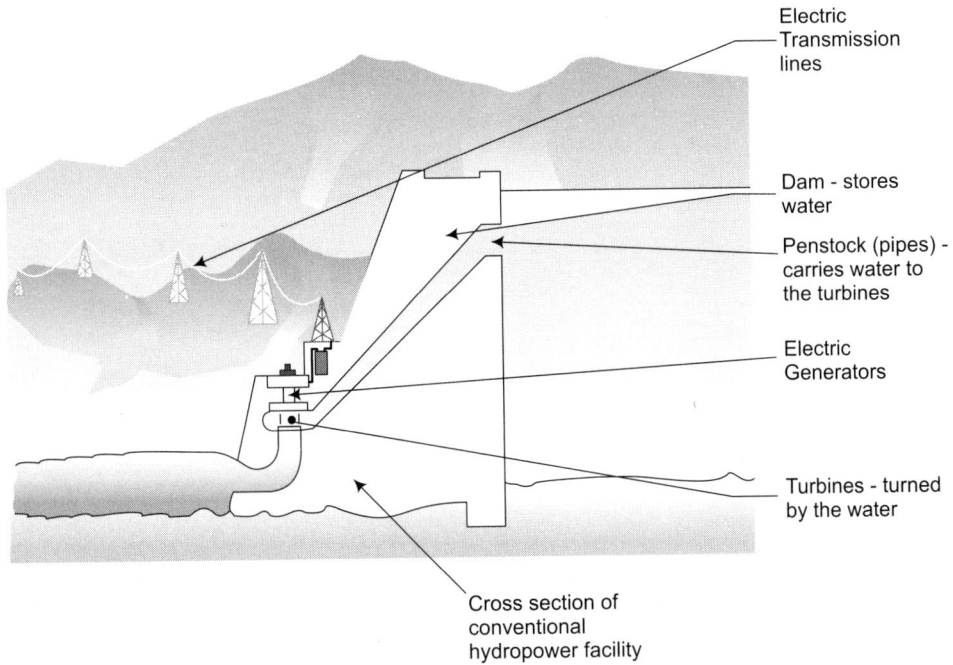

Electric Transmission lines

Dam - stores water

Penstock (pipes) - carries water to the turbines

Electric Generators

Turbines - turned by the water

Cross section of conventional hydropower facility

Fig. 4.9 Hydroelectric Dam. US Dept. of Energy. From: *http://www.tva.gov/power/hydro.htm.*

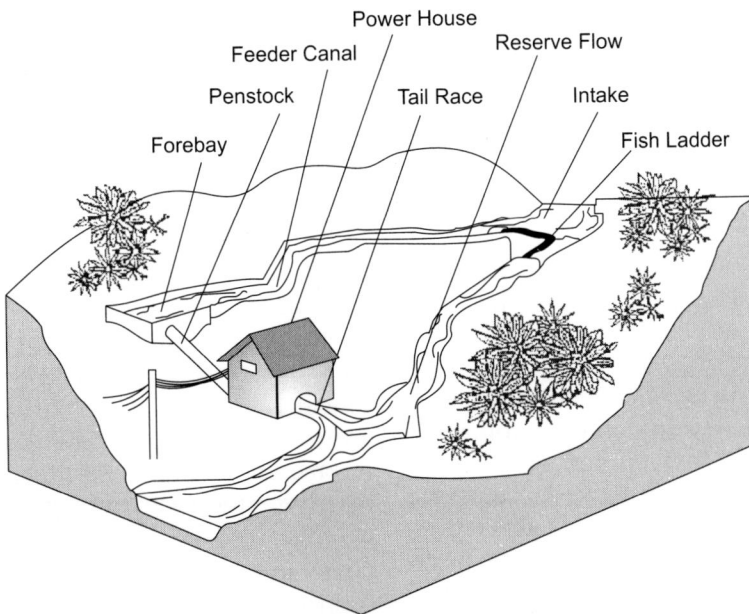

Power House

Feeder Canal   Reserve Flow

Penstock   Tail Race   Intake

Forebay   Fish Ladder

Fig. 4.10a Schematic of a Typical 100 KW Micro-Electric Plant. From: *http://www.p2pays.org/uef/32/31342.pdf.*

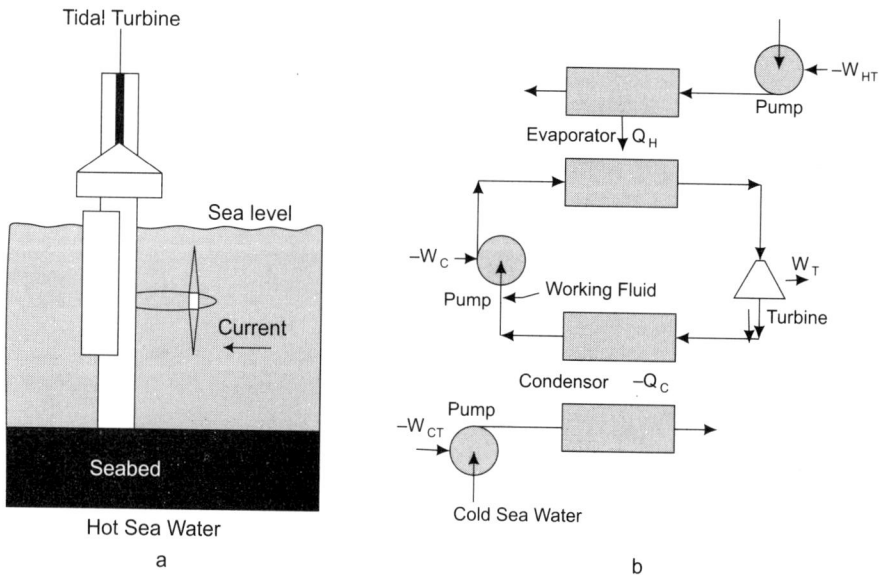

Fig. 4.11 Diagram of OTEC Cycle. From: *http://digitallibrary.okstate.edu/oas/oas_image_files/v56/p114_120fig1.*

back past the turbine because of reduced air pressure on the turbine's ocean side *http://rise.org.air/reslab/resfiles/wave/text.html).* Finally, ocean thermal energy conversions (OTEC) can be employed to generate electricity by exploiting the energy stored in ocean thermal gradients, i.e., the temperature difference between surface and deep water in the tropics.

Solar Energy

Photovoltaic is the most direct means of converting solar radiation into electricity (Goetzberger and Hoffmann, 2005; Chubb, 2007). It is the emergence of an electrical voltage between two electrodes attached to either a solid or liquid surface following shining light onto a photovoltaic system (Hamakawa, 2004). Solar cells contain a semiconductor water treated to generate an electric field, positive on one side and negative on the other. Electrons are released from the atoms in the semiconductor material when light energy impinges upon the solar cell. Electricity can be obtained by attaching electrical conductors to the positive and negative sides (*http://science.nasa.gov/headlines/y2002/solarcells.htm*). Photovoltaic modules (Fig. 4.12) can be connected in both serial and parallel electrical arrangements to yield any required voltage and current combination (*http://science.nasa.gov/headlines/y2002/solarcells.htm*).

Wind Energy

This type of renewable energy is restricted to geographical areas where the wind is relatively constant. The components of a wind turbine (Fig. 4.13) are:

Tower – Reaches from its base on the ground into the air and supports the turbine's rotor blades; the higher the tower, the greater the electricity generated because of longer blades.

Generator – At the tower's top a rotor is connected to a generator via a shaft; the generator produces energy by turning the rotor's shaft via wind.

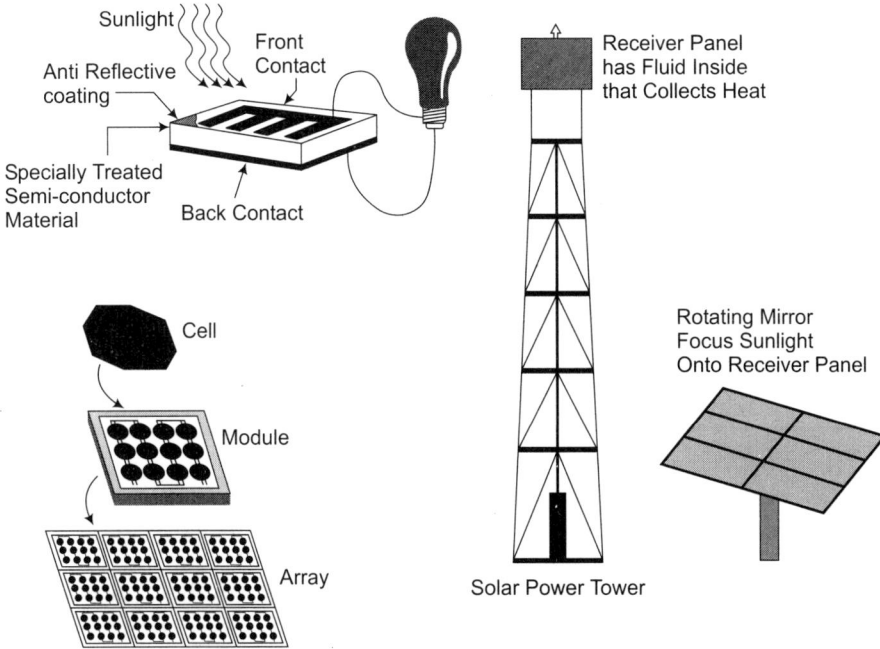

Fig. 4.12 Photovoltaic Cells. Left From: *http://science.nasa.gov/headlines/y2002/solarcell.htm*
Right Solar Power Tower From: *www.mms.gov/offshore/assets/photos/solar_power288.gif.*

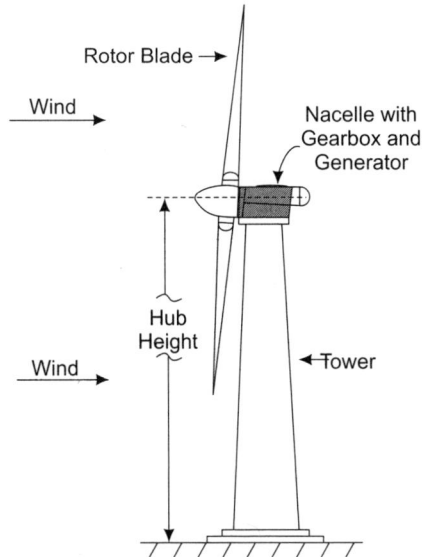

Fig. 4.13 Diagram of a Wind Power Machine.
From: *http://www.web-one.gc.ca/energyreports/emaemergingtechnal.*

TABLE 4.4 Summary of Select Energy Research Activities

Research Effort	Reference
United States	
"Coal evaluation for coke making designing coal blends for better blast furnace operation; predicative modeling of coal/coke properties"	http://www.nrcan.gc.ca/es/etb/cetcol/htmldocs/services/indes_e.htm
"Provides project management, evaluation, design, modification and retrofit of energy systems"	As above
Develop a science-based solution that harnesses fusion energy to power USA homes and industries	http://www.science.doe.gov/sub/mission.strategic_plan/fev-2004-5
Research and Development in renewable energies (solar, wind, biomass, geothermal, hydrogen and fuel cells)	US Nation Renewable Energy Laboratory (NRER) http://www.nrel.gov/ Battelle Pacific Northwest National Laboratory, US Department of Energy conducts research on fuel cells.

Cables – These carry the generated electricity down the turbine tower to the ground where the turbine is connected to a utility grid.

Nacelle – Streamlined casing that encloses the rotor and generator.

Energy of the Future

Nuclear risks must be balanced against air pollution caused by the burning of gas, oil and/or coal (Winteringham, 1992). Turk and Turk (1997) propose judgments that must be rendered regarding the future of nuclear energy. In this connection, the monographs by Kruschke (1990), Nealey (1990) and Wolfson (1993) are relevant.

Table 4.4 offers some research efforts concerned with improving existing fossil energy sources and renewable energies. The incentives for protecting the environment are presented in Table 4.5.

TABLE 4.5 Summary of International Incentives for Protecting the Environment[a]

Incentive	Detail
Direct Fees and Taxes	Most used market mechanism intenationally
Deposit-refund systems	
Trading Regimes Nordhaus, W. 2008. *A Question of Balance: Weighing the Options on Global Warming Policies.* Yale University Press, New Haven, CT, USA	Shift into capped allowance systems from more open-ended mechanisms
Greenhouse Gas Emission Control	Carbon taxes and greenhouse gas trading regimes.
Reductions In Environmentally Harmful Subsidies	Lenders often make the elimination of environmentally harmful subsidies a condition for lending
Liability	Harm caused to the environment is increasingly being used as a tool to limit pollution and environmentally damaging activities
Information	Used in many new applications including product labeling, categorizing firms according to their environmental performance and disclosure of pollution releases.
Voluntary programs	Encourage firms to improve their environmental practices

[a] From: http://yosemite.epa.gov/ee/epa/eed.nsf/wevpages/internationalincentivesreport.html.

References

Abelson, P. 2000. Future supplies of electricity. *Science* 287: 973.

Abelson, P.H. 1994. US petroleum: past and future. *Science* 263: 303.

Abelson, P.H. 1997. Improved fossil energy technology. *Science* 276: 511.

Ackerman, T. 2005. *Wind Power in Power Systems*. Wiley, Chichester, West Essex, England.

Arking, A. 1996. Absorption of solar energy in the atmosphere: Discrepancy between model and observations. *Science* 273: 779.

Arnstead, H.C.H. 1983. *Geothermal Energy Its Past, Present and Future Contributions to the Energy Needs of Man*. E and FN Span, London, England.

Atkins, S.E. 2000. *Historical Encyclopedia of Atomic Energy*. Greenwood Press, Westport, CT, USA.

Avenhaus, R. 1986. *Safeguards System Analysis With Applications to Nuclear Safeguards and Other Inspection Problems*. Plenum Press, New York, NY, USA.

Avery, W.H. and Wu, C. 1994. *Renewable Energy From the Ocean*. Oxford Uni. Press, New York, NY, USA.

Bailey, R.A., Clark, H.M., Ferris, J.P., Krause, S. and Strong, R.L. 2002. *Chemistry of the Environment*. Academic Press, London, UK.

Bauer, N. 1995. *Resistance to New Technology: Nuclear Power, Information Technology and Biotechnology*. Cambridge Univ. Press, New York, NY, USA.

Beacham, W. 1993. *Beacham's Guide to Environmental Issues and Sources*. Beacham Publishing Inc., Washington, DC, USA.

Bernstein, P. 2001. *Alternative Energy Facts, Statistics and Issues*. Greenwood Publishing Group, Westport, CT, USA.

Boyle, G. 2004. *Renewable Energy*. Oxford Univ. Press, Oxford, UK.

Clean Coal Technology. 2001. *Environment Benefits of Clean Coal Technologies*. Clean Coal Technology, Washington, DC, USA.

Cockerham, L.G. and Cockerham, M.B. 1994. Environmental ionizing radiation. In: *Basic Environmental Toxicity*. (Cockerham, L.G. and Shane, B.S., eds.). CRC Press, Boca Raton, FL, USA. pp. 231-261.

Committee on R and D Opportunities for Advanced Fossil-Fueled Energy Complexes. 2000. Vision 2, Chapter file 1, National Academy Press, Washington, DC, USA.

Chubb, D.L. 2007. *Fundamentals of Thermophotovoltaic Energy Conversion* Oxford, Amsterdam, The Netherlands.

Dickson, M.H. and Fanelli, M. 1995. *Geothermal Energy*. Wiley, New York, NY, USA.

Duffield, W.A. and Sass, J.H. 2003. *Geothermal Energy: Clean Power From the Earth's Heat*. U.S. Geological Survey, Denver, CO, USA.

Ehleringer, J.R., Cerling, T.E. and Dearing, M.D. 2005. *A History of Atmospheric CO_2 and Its Effects on Plants, Animals and Ecosystems*. Springer, New York, NY, USA.

Eisler, R. 1994. *Radiation hazards to fish, wildlife and invertebrates: A synoptic review*. US Dept. of Interior National Biological Services, Washington, DC. USA.

El-Hawary, M.E. 2000. *Electrical Energy Systems* (Electronic Resource), CRC Press, Boca Raton, FL, USA.

Elliott, D. 2003. *Energy, Society and Environment*, 2nd Ed. Routledge, New York, NY, USA.

Ewing, R.A. 2003. 'Power With Nature. Solar and Wind Energy Demplified.' Pixyjach Press, Masonville, Co, USA.

Falnes, J. 2002. *Ocean Waves and Oscillating Systems: Linear Interactions Including Wave-Energy Extraction*. Cambridge Univ. Press, Cambridge, UK.

Fauet, F.A. and Simnoes, M.G. 2006. *Integration of Alternative Sources of Energy*. Wiley-IEEE Press, New York, NY, USA.

Franchi, J.R. 2005. *Energy in the 21st Century*. World Scientific, Hackensack, NJ, USA.

Goetzberger, A. and Hoffman, V.U. 2005. *Photovoltaic Solar Energy Generation*. Springer, New York, NY, USA.

Gupta, H.K. and Roy, S. 2006. *Geothermal Energy: An Alternative Resource for the 21st Century*. Elsevier, New York, NY, USA.

Hanakawa, Y. 2004. *Thin-film Solar Cells. Next Generation Photovoltaic and its Application*. Springer, Berlin, Germany.

Hau, E. 2005. 'Wind Turbines: Fundamentals, Technologies, Applications, Economics.' Springer, Berlin, Germany.

Heyes, A. 2001. *The Law and Economics of the Environment*. E. Elgar, Cheltenham, UK.

Hoffman, P. 2001. *Tomorrow's Energy: Hydrogen, Fuel Cells And the Prospects for a Cleaner Planet*. MIT Press, Cambridge, MA, USA.

Hoheremser, C. 1992. *Nuclear Power In The Energy Environment Connection*. (Hollander, J.M., ed.). Island Press, Washington, DC, USA.

Hollander, J.M. (ed.). 1992. *The Energy-Environment Connection*. Island Press, Washington, DC., USA.

Karp, A and Shield, I. 2008. Bioenergy from plants and the sustainable yield challenges New Phytologists 179: 15-32.

Katzman, M.T. 1984. *Solar and Wind Energy: An Economic Evaluation of Current and Future Technologies*. Rowman and Allanhead, Totowa, NJ, USA.

Knops, H.P.A. 2003. Promoting renewables which support scheme gets the green light in the internal market? In *Energy and the Environment*. (Brebbia, C.A. and Sakellaus, I., eds.). WIT Press, Southampton, UK. pp. 133-142.

Koller, J., Koppel, J. and Peters, W. 2006. '*Offshore Wind Energy. Research on Environmental Impacts.*' Springer, Berlin, Germany.

Kruger, P. 2006. '*Alternative Energy Sources: The Energy Quest for Sustainable Energy.*' Wiley, Hoboken, NJ, USA.

Kruschke, E.R. 1990. *Nuclear Energy Policy: A Reference Handbook*. ABC – CLIO, Santa Barbara, CA, USA.

Larid, F.N. 2001. *Solar Energy, Technology Policy and Institutional Values*. Cambridge Univ. Press, Cambridge, UK.

Liu, P.I.H-Fei. 1993. *Introduction to Energy and the Environment*. Van Nostrand, New York, NY, USA.

MacAvoy, P.W. 2000. *The Natural Gas Market Sixty Years of Regulation and Deregulation*. Yale Univ. Press, New Haven, CT, USA.

Maheshwari, A. and Parmar, G. 2004. *A Textbook of Energy, Environment and Society*. Anmol Publications Pvt. Ltd., New Delhi, India.

Makhijani, A.H., Hu, H. and Yih, K. 2000. *Nuclear Wasteland: A Global Guide to Nuclear Weapons Production and Its Health and Environmental Effects*. MIT Press, Cambridge, MA, USA.

Manahan, S.E. 2005. *Environmental Chemistry*. CRC Press, Boca Raton, FL, USA.

Manella, D.W. 2005. *Environmental Health*. Harvard Univ. Press, Cambridge, MA, USA.

Mathew, S. 2006. *Wind Energy: Fundamentals, Resource Analysis and Economics*. Springer, Berlin, Germany.

National Renewable Energy Laboratory Technical Report. 2005. 21st Century Power Technologies Data Book. National Renewable Energy Laboratory Technical Profiles of Biomass, Geothermal, Hydrogen, Hydropower, Solar and Wind Energy, Superconductivity, Fuel Cells, Batteries, Advanced Storage, Electricity Restructuring, Electricity Demands, Progressive Management. Washington, DC, USA.

National Technical Information Services. 1999. *Wind Power Today*. 1998 Wind Energy Program Highlights, Washington, DC, USA.

Nealey, S.M. 1990. *Nuclear Power Development: Prospects in the 1990's*. Batelle Press, Columbus, OH, USA.

Nebel, B.J. 1990. *Environmental Science: The Way the World Works*. Prentice Hall,Englewood Cliffs, NJ, USA.

Nurse, P. 2000. The incredible life and times of biological cells. Science 289: 1711-1716.

Nutall, W.J. 2007. Nuclear renaissance requires nuclear enlightenment. In *Nuclear or Not: Does Nuclear Power Have A Place in a Sustainable Energy Future?* (Elliott, D., ed.) Palgrave MacMillan, USA. pp. 221-238.

OECD 1988. Emission Controls in Electricity Generation Energy and Industry. OECD Publications and Information Center. Paris, France.

Overton, E.B., Sharp, W.D. and Roberts, P. 1994. *Toxicity of Petroleum in Basic Environmental Toxicology*. Boca Raton, FL, USA, pp. 133-136.

Patel, M.R. 2006. *Wind and Solar Power Systems: Design, Analysis and Operation*. Taylor and Francis, Boca Raton, FL, USA.

Puttaswamaiah, K. 2002. *Cost-benefit Analysis Environmental and Ecological Perspectives*. Transaction Publishers, New Brunswick, NJ, USA.

Quaschning, V. 2005. *Understanding Renewable Energy Systems*. Earthscan, London, England.

Rhodes, R. 1987. *The Making of the Atomic Bomb*. Simon and Schuster, New York, NY, USA.

Rhodes, R. 1995. *Dark Sin: The Making of the Hydrogen Bomb*. Simon and Schuster, New York, NY, USA.

Ringus, L. 2000. *Radioactive Waste Disposal at Sea. Public Ideas, Transnational Policy Entrepreneurs and Environmental Regimes*. MIT Press, Cambridge, MA, USA.

Ross, D. 1995. *Power From the Waves*. Oxford Univ. Press, Oxford, UK.

Seddon, D. 2006. *Gas Usage and Value; The Technology and Economics of Natural Gas Use in the Process Industries*. Pen Well, Tulsa, OK, USA.

Silverio, S. 2005. *Bioenergy: Realizing the Potential*. Oxford: Elsevier, Amsterdam, The Netherlands.

Silyok, V.A. 2001. *Environmental Laws Summaries of Statutes Administered by the Environmental Protection Agency*. Nova Science Publishers, Inc., Huntington, NY, USA.

Sorenson, B. 2004. *Renewable Energy*; Third Edition. Academic Press, New York, NY, USA.

Sorenson, B. 2005. *Hydrogen and Fuel Cells: Emerging Technologies and Applications*. Elsevier Academic Press, Amsterdam, The Netherlands.

Speiv, J., Denito, B. and Theodore, L. 2000. *Regulatory Chemicals Handbook* (Electron Resource). Marcel Dekker, New York, NY, USA.

Spiro, T.G. and Stigliani. W.M. 2003. *Chemistry of the Environment*. Prentice Hall, Upper Saddle River, NJ, USA.

Szabo, A.S. 1993. *Radioecology and Environmental Protection*. Ellis Harwood, New York, NY, USA.

Taurus, I.B. 2001. *The Future of Natural Gas in the World Energy Market*. I.B. Taurus and Co., United Kingdom.

Tester, J.W., Drabe, E.M., Driscoll, M.J., Golay, M.W. and Peters, W.A. 2005. *Sustainable Energy Choosing Among Options*. MIT Press, Pensacola, FL, USA.

The OECD Compass Project. 1988. *Environmental Impacts of Renewable Energy*. OECD Publications, Paris, France.

The Research Revolution. 2002. *The Atomic Age*. National Video Resources, New York, NY, USA.

Tickell, J. 2006. *Biodiesel Amercia. How to Achieve Energy Security, Free America From Middle-East Oil Dependence and Make Money Growing Fuel*. Tickell Energy Consulting. Tallahasee, FL, USA.

Tiwari, G.N. 2002. *Solar Energy Fundamentals, Design, Modeling and Applications*. CRC Press, Boca Raton, FL, USA.

Turk, J. and Turk, A. 1997. *Environmental Science*. Harcourt Brace, Orlardo, FL., USA.

Turner, J.E. 1986. *Atoms, Radiation and Radiation Protection*. Pergamon Press, New York, NY, USA.

US Dept of Energy, Office of Fossil Energy 2001. *Coal and Power Systems: Ensuring Clean Affordable Energy*. US Dept. of Energy, Washington, DC, USA.

US Dept. of Energy. Energy Efficiency and Renewable Energy. (*www.eeree.energy.gov/*).

US Dept. of Energy. 2000. *The Changing Structure of the Electric Power Industry*. Energy Information Administration, Washington, DC, USA.

Van Loon, G.W. and Duffy, S.J. 2005. *Environmental Chemistry. A Global Perspective*. Oxford Univ. Press, Oxford, UK.

Walker, J.F. and Jenkins, N. 1997. *Wind Energy Technology.* John Wiley, Chichester, England.

Williams, D.R. 1998. What is safe?: The risks of living in a nuclear age. Royal Society of Chemistry Information Services, Cambridge, England.

Winteringham, F.P.W. 1992. *Energy Use and the Environment.* Lewis Publishers, Boca Raton, FL, USA.

Wolfson, R. 1993. *Nuclear Choices: A Citizen's Guide to Nuclear Technology.* MIT Press, Cambridge, MA, USA.

Wooley, D., Pugh-Smith,J., Langham,R. and Upton, W. 2000. *Environmental Law* (electronic resource). Oxford Univ. Press, Oxford, England. World Health Organization. 1994. *Nuclear Power and Health: The Implications for Health of Nuclear Power Productions.* World Health Organization, Copenhagen, Denmark.

Yen, T.F. 1999. *Environmental Chemistry Essentials of Chemistry for Engineering Practice.* Prentice Hall, Upper Saddle River, NJ, USA.

CHAPTER 5

Air Pollution

What is Air Pollution and What are its Sources?

An air pollutant is any substance in ambient air that is either detrimental to human health or threatening to other organisms (Austin et al., 2002; Chhatwal, 2003; Cheremisinoff, 2006). Certain of these pollutants can cause property damage, e.g., degradation of stone by acid rain.

There are two forms of air pollutants, primary and secondary (Hill, 2004). Whereas the former have an immediate effect, the latter are harmful when converted to another form. There appear to be three categories of air pollution sources, area, mobile and point (Table 5.1). A summary of the six major air pollutants is presented in Table 5.2. While their trends in the USA over the years are depicted in Figs. 5.1 – 5.7, global emissions of SO_2 and NO_2 can be observed in Fig. 5.8. In the UK, the two have decreased 68% and 36% from 1990 to 2001 (*http:www.defra.gov.uk/news/2001/011210a. html*).

Chemical Reactions and Transport of Air Pollutants

Gases are injected into a region of the atmosphere known as the boundary layer of the troposphere (Fig. 5.9). This is a mixing layer in which gases are exchanged between the atmosphere and terrestrial surfaces. The boundary layer varies in depth and time of day. It can reach from the ground to the temperature inversion layer at about 1 km (Mead et al., 1997). However, under convective conditions, the upper boundary layer can be several kilometers. Boundary layer winds have been discussed in depth at *http:www.2010.atmos.uiuic.edu(Ch)/ gurdes/mtr/fw/bndy.rml*. The volume of air into which a pollutant is injected varies with time and meteorological conditions.

Processes which can occur in the troposphere include precipitation scavenging, advection and convection. Precipitation scavenging is the removal of pollutants from the air via snow or rain.

There are two forms of scavenging, rainout and washout. Whereas rainout involves the in-cloud capture of particles as condensation nuclei, washout is the below-cloud capture of gaseous pollutants and particles by raindrops (*http: amsglossary. allenpress.com/glossary/seact?id=scavenging-by-precipitation.*). The efficiency of precipitation scavenging of air pollutants can be quantified by a weather radar.

Convection and advection are wind processes caused by differences in temperature. Convection transfers heat energy from

TABLE 5.1 Categorization of Air Pollutants[a]

Category	Characteristics
Area (Dry cleaners, gas stations and auto part shops).	Small pollution sources emit < 10 tons/yr of criteria or hazardous air pollutants or < 25 tons/yr of combined pollutants.
Mobile (Cars, trucks, buses, ships, airplanes, agricultural and constructional equipment).	On road vehicles and off-road equipment contribute significantly to air pollution.
Point (Chemical plants, fertilizer manufactures, steel mills, oil refineries, paper mills, power plants and hazardous waste incinerators).	Major industrial facilities emit 10 tons/yr of any of the criteria pollutants or 25 tons/yr of a mixture of air toxics.

[a]Adapted form: *http://scorecard.org/env-releases/def/air-source.html.*

ATMOSPHERIC AIR POLLUTION[a,b,c]

Acid Rain NO_x SO_x	Acid Rain SO_x NO_x and Ozone	Greenhouse Gases CFC, CH_4 CO_2, N_2O, O_3, H_2O Vapour
Local Air Pollution NO_x = Nitrogen oxides SO_x = Sulfur oxides	*Regional Air Pollution*	*Global Air Pollution* CFC = Chlorofluorocarbon Compounds from aerosols CH_4 = Methane CO_2 = Carbon dioxide O_3 = Ozone

[a] See section on chemical reactions and atmospheric conditions for the formation of acid rain from NO_x and SO_x and the generation of ozone.

[b] While all of the pollutants can be anthropogenic in origin (Table 5.2), there are natural stratospheric O_3 levels also.

[c] **Note** transport of air pollutants between regional, local and global areas can occur. In addition, the areas depend upon the number of vehicles, coal-fire power plants and the like.

TABLE 5.2 Summary of Air Pollutants and Their Sources[a]

Air Pollutant	Type	Source
Carbon monoxide (CO)	Primary – a colorless, odorless, poisonous gas	Incomplete burning of carbon-based fuels; component of motor vehicles exhaust
Ozone (O_3) Ground level	Secondary – a gas which is similar to O_2 but contains three atoms of oxygen, primary constituent of smog.	Occurs in nature in the stratosphere and is produced by lightning; produced during the combustion of coal, gasoline and other fuels in the presence of sunlight.
Lead (Pb)	Primary – an element	Large furnaces, incinerators and battery plants; lead-based paints.
Sulfur dioxide (SO_2)	Primary/Secondary – a gas, can be converted to sulfuric acid	Produced by burning coal in power plants and industrial processes, paper production and metal smelting

(Contd.)

(Contd.)

| Nitrogen dioxide (NO$_2$) | Secondary – a reddish brown, highly reactive gas; NO$_x$ is the sum of NO, NO$_2$ and other oxides of nitrogen | Produced from burning fuels, major sources of NO$_2$ are power plants and vehicles that use gasoline or diesel |
| Particulate matter | Mixture of solid particles and liquid droplets; Dust and soot particles < 10 PM$_{10}$ or < 2.5 PM 2.5 | Produced by burning of diesel fuels, incineration of garbage, mixing and application of fertilizers and pesticides, root construction, fuel places, etc. |

[a] These were classified as the six major air pollutants in the US. Clean Air Act of 1970. The Clean Air Act amendments of 1990 list 189 air pollutants which can cause serious health problems.

[b] The National Emission Inventory (NEI) of EPA database lists sources that emit air pollutants and their precursors.

warm areas close to the Earth's surface to more elevated regions of the atmosphere. Advection involves the horizontal movement of air and heat energy. In contrast, convection is the vertical movement of air.

The reactions which can occur in the boundary layer are: Oxidation of NO$_2$ and SO$_2$ in the gas phase (Lastovicka et al., 2006; Seinfeld and Pandis, 1998; Shipman et al., 2006).

$$OH + SO_2 \rightarrow HSO_{3\_}$$

Products are presented in particles

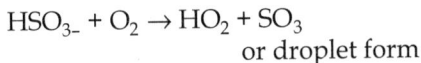

$$HSO_{3\_} + O_2 \rightarrow HO_2 + SO_3$$

or droplet form

$$NO_2 + OH \rightarrow HNO_3$$

Major gas phase oxidation pathway

$$NO_2 + O_3 \rightarrow NO_3 + O_2$$

NO$_3$ may be present in particles or droplets as NO$_{3\_}$

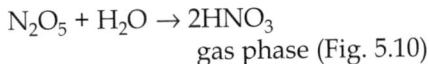

$$NO_3 + NO_2 \rightarrow N_2O_5$$

$$N_2O_5 + H_2O \rightarrow 2HNO_3$$

gas phase (Fig. 5.10)

In another major pathway for atmospheric transformation, gases are oxidized within the air on the surfaces of particles or clouds in fog droplets.

Clouds: SO$_2$ dissolves at typical cloudwater pH's mainly to bisulphite

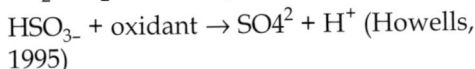

$$SO_2 + H_2O \rightarrow HSO_{3\_} + H^+$$

$$HSO_{3\_} + oxidant \rightarrow SO4^2 + H^+ \text{ (Howells, 1995)}$$

Droplet phase: Most important oxidants are

H$_2$O$_2$ and O$_3$ (Guderian, 1989)

H$_2$O$_2$ is very soluble and reacts with HSO$_3$

O$_3$ is much less soluble than H$_2$O$_2$

Ozone (O$_3$) is not emitted directly into the air but is produced by the reactions of volatile organics (VOCs) and NOx under the proper amount of sunlight.

NH$_3$ emitted from agricultural land after application of animal waste or fertilizer is readily incorporated into acidic atmospheric aerosols and cloud droplets. The deposition of aerosols on either snow or ice may decrease surface albedo (Law and Stohl, 2007).

Transport

Primary gases at the proximity of sources 0 – 100 km can be converted to secondary pollutants by the above reactions and transported 500 – 2,000 km over 1 – 3 days (Zeeduk and Velds, 1973).

Wet deposition (Fig. 5.11) Acidic *pollutants transported* ≥ 1,000 km are in the form of rain or snow (Rosenfield, 2000) or deposited as particles or cloud droplets (Carson and Munton, 2000). Pollutant deposition 0 – 100 km downwind from source areas is dominated by dry depositions.

Trends in Carbon Monoxide Emissions

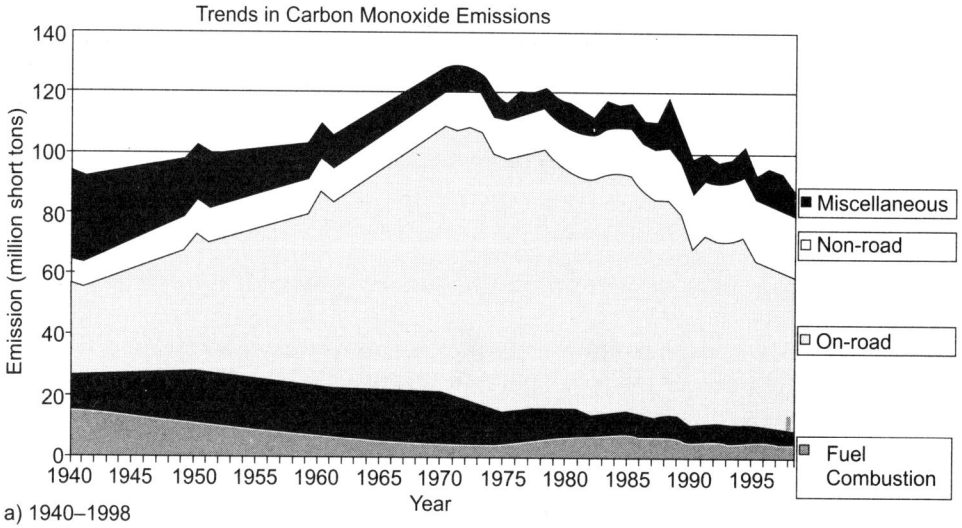

a) 1940–1998

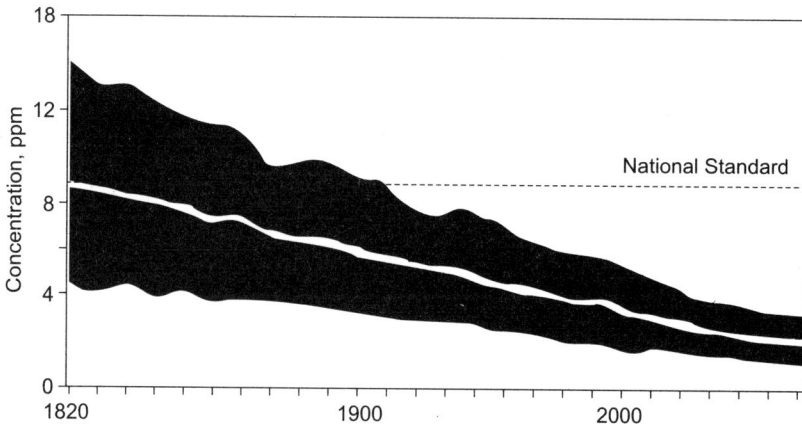

b) 1820–2007

Fig. 5.1

Trends in Nitrogen Oxide Emissions, 1940 to 1998

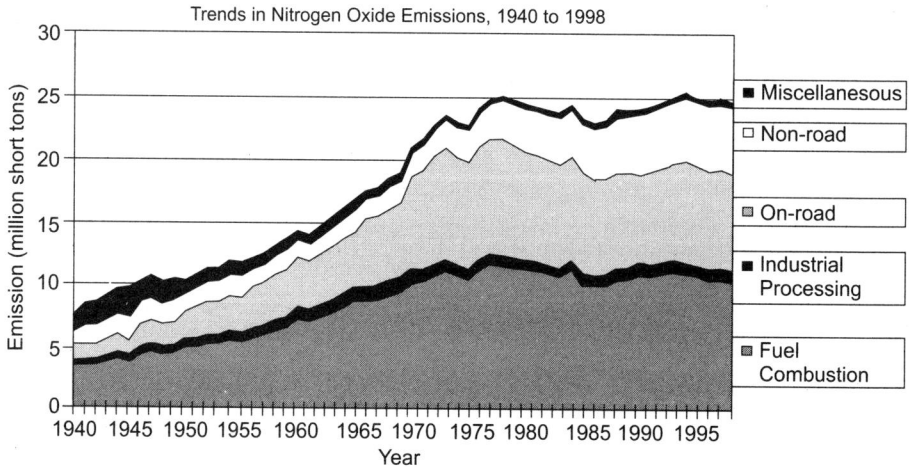

Fig. 5.2

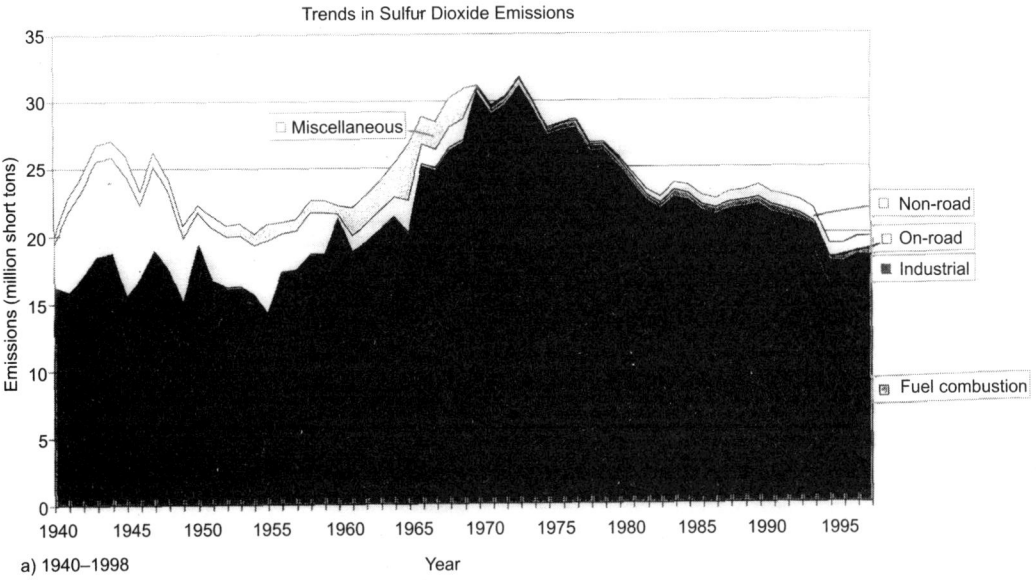

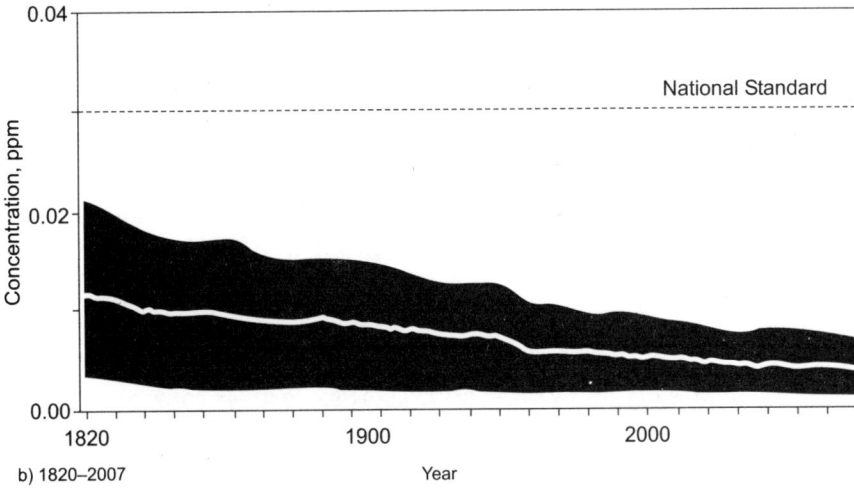

Trends in Sulfur Dioxide Emissions

a) 1940–1998

b) 1820–2007

Fig. 5.3

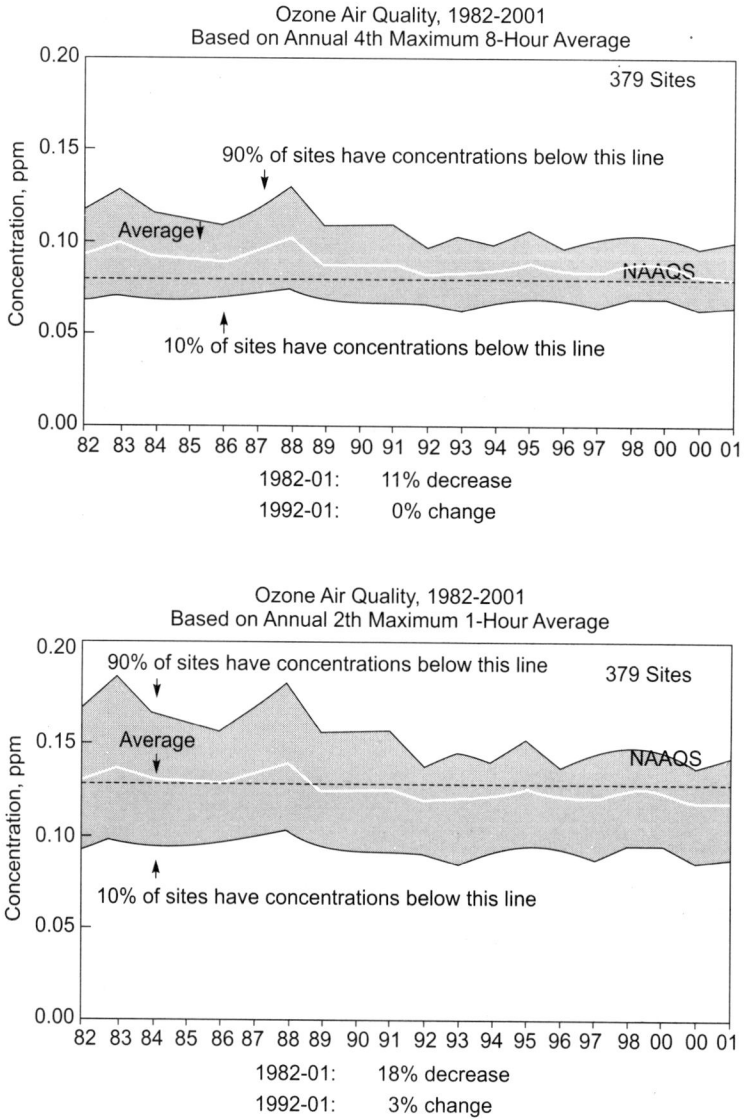

Ozone Air Quality, 1982-2001
Based on Annual 4th Maximum 8-Hour Average

379 Sites

90% of sites have concentrations below this line

Average

NAAQS

10% of sites have concentrations below this line

82 83 84 85 86 87 88 89 90 91 92 93 94 95 96 97 98 00 00 01

Concentration, ppm

1982-01:   11% decrease
1992-01:   0% change

Ozone Air Quality, 1982-2001
Based on Annual 2th Maximum 1-Hour Average

90% of sites have concentrations below this line   379 Sites

Average

NAAQS

10% of sites have concentrations below this line

82 83 84 85 86 87 88 89 90 91 92 93 94 95 96 97 98 00 00 01

Concentration, ppm

1982-01:   18% decrease
1992-01:   3% change

Fig. 5.4

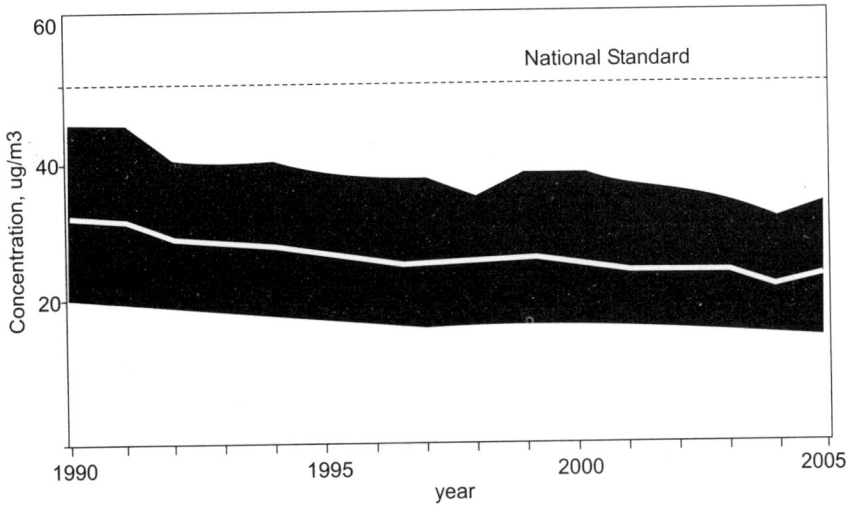

Fig. 5.5

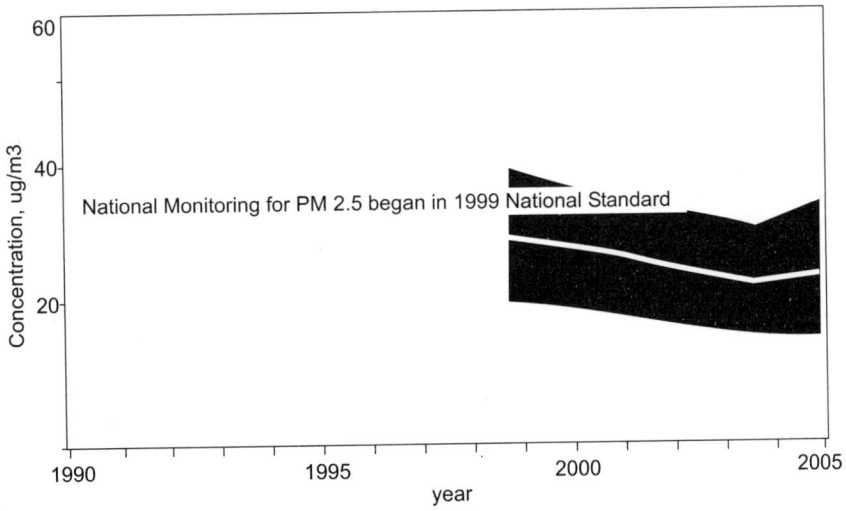

Fig. 5.6

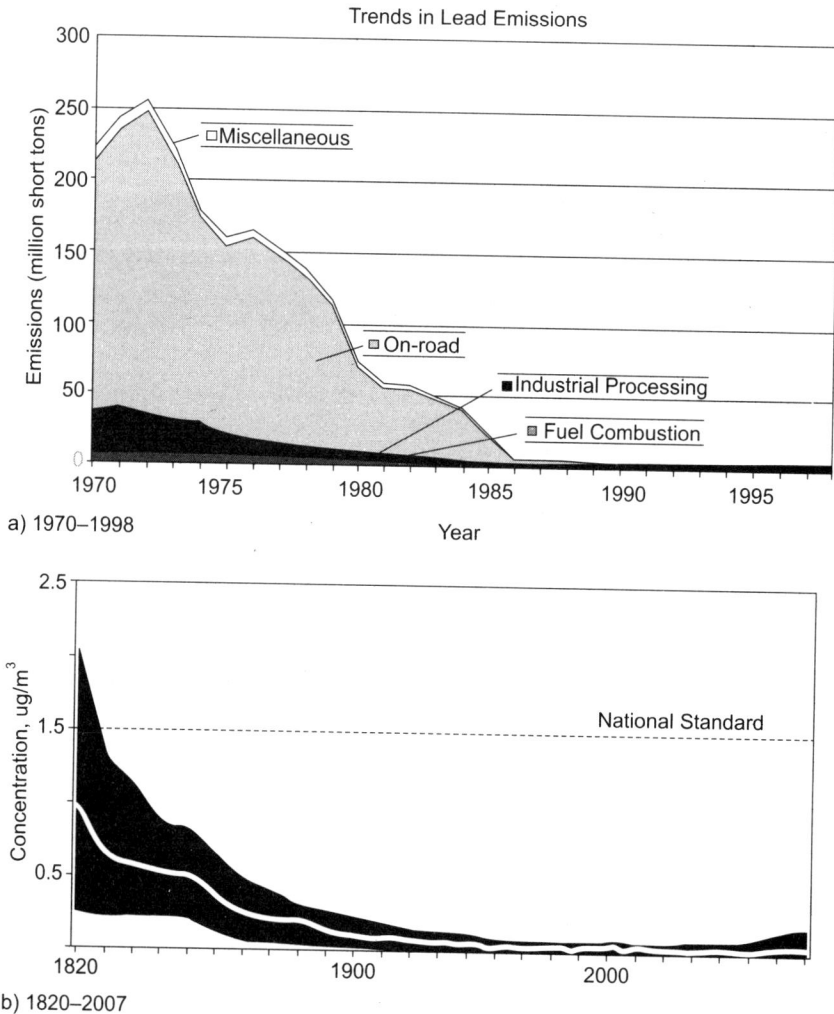

Trends in Lead Emissions

a) 1970–1998

b) 1820–2007

Fig. 5.7

Fig. 5.1 – 5.7 Trends in major US atmospheric pollutants.
5.1. Carbon Monoxide, **5.2.** Nitrogen Oxide, **5.3.** Sulfur Oxide, **5.4.** Ozone, **5.5.** PM$_{2.5}$, **5.6.** PM$_{10}$ and **5.7.** Lead.
From: *http://www.epa.gov/airtrends/sulfur.html.*
Data for 1970 – 2003 are presented in table form
http://www.epa/airtrends/aqtrn04/econ-emissions.html.

Effects of Air Pollutants on Organisms

The amount of research literature regarding air pollutants and plants is very extensive (see reviews of Kozlowski, 1980; Smith, 1990; Baker and Tingey, 1992; Leer, 1999).

Table 5.3 presents a generalized summary of the effects on plant systems. As with animals, certain plants can serve as bioindicators of air pollution (Grant, 1998).

There is a copious amount of information regarding the effects of air pollutants

TABLE 5.3 Generalized Summary of the Effects of Air Pollutants on Plant Systems

Plant System	Effect
Algae, Blue-Green	Susceptible to a whole range of air pollutants
Lichens	Very responsive to air pollution
Bryophytes mosses especially sensitive	Highly sensitive to many air pollutants; tree-living and bog
Pteridiophytes	Some fern and many club mosses decline in polluted air
Gymnosperms	Many tree species decline in polluted environments
Angiosperms	Increasing evidence that herbaceous flowering plants decline
Acid rain on forests Acid rain = pH < 5.0 Smith (1990); Godbold and Houttmann (1994)	1983, Germany- 34% of country's forest damaged, including the Black Forest, from acid rain and or acid cloud droplets that fell on needles of trees, leach of nutrients (Karnosky,2003; http://www.elmhurstedu/~chm/vchembovk/195lakeeffects.html.)
Acid rain, Acid fog and Acid vapor Bresser and Salomons (1989); White and Terninko (2003)	Damages surfaces of leaves and needles, Reduces a tree's ability to withstand cold, Inhibits plant germination and reproduction, Loss of ability to withstand drought, disease, insect pests and cold weather (Schulze et al., 2000; http://www.ec,gl.ca/acidrain/acidforest.html).

Adapted from: http://www.acidrain.org/biodiv.htm. The student is referred to the following: Agrawal and Agrawal (1999), Bell and Marshall (1999), Bell and Treshow (2002), Darall (1989), Dlamini et al. (1994), Godbold and Houttmann (1994), Koziol and Whatley (1984), Krupa, (1971), Olszyk et al. (2001), Schulze et al. (1989), Taylor (1984) and Winner (1994).

and acidified environments on animals (see literature cited in Beacham, 1993; Chhatwal, 2003). Instead of reviewing the results of these research investigations, Table 5.4 presents some generalizations. These observed effects have prompted the use of certain animal species, sentinel species, as indicators of adverse effects associated with air pollution (Saldiva and Bohn, 1998). These effects have been associated with bioconcentration of trace elements, alterations of physiological parameters and genotoxicity.

TABLE 5.4 Generalised Effects of Air Pollutants on Animals[a]

Pollutant	Effect
Amphibians	Many species decline in acidified waters primarily because of reproductive failure.
Arthropods	Many crustaceans and insects decline in acid waters; air pollution is likely to cause many species to decline on land.
Birds	Some species have adapted to acidified waters; others have declined because of food chain losses.
Fish	Some species decline in slightly acid waters; others withstand fairly severe acidification.
Mammals	Some suffer from disturbances in the food chain.
Reptiles	Little information
Soft-bodied Invertebrates	Almost all lower invertebrates decline in acid waters; mollusks can be affected by air pollution on land.

[a]Adapted from: http://www.acidrain.org/biodiv.htm. The student is referred to Dockery and Pope (1994); Dudley and Stolton (1996) and Albers et al. (1998) for additional information.

TABLE 5.5 Summary of Acid Rain Effects on Aquatic Life[a]

Acidification	Effects
Lakes	Reduction in acid-sensitive species of zooplankton, phytoplankton and fish, frogs, snails, mayflies depending upon the pH
Fish	pH limits
Perch (yellow) and Trout (lake)	4.5 – 5.0
Trout (Brown)	5.0 – 5.5
Bass (Small Mouth)	6.0 – 6.5
Zooplankton	4.0 – 5.0
Frogs	4.0 – 4.5
Mayfly	5.5 – 6.0

[a]Adapted from *http://www.sft.no/publikasjovierintermasjonalt 1860/tal1860.pff* and *http://www.elmhurst.edu/~chm/vchembrook/195/akeffects.html.*

TABLE 5.6 Effects of Six Major Air Pollutants on Human Health

Pollutants	Health Effects
Carbon monoxide	Enters the bloodstream through the lungs (Fig. 5.12) and reduces O_2 delivery to organs and tissues; elevated levels can reduce visual impairment, work capacity, and manual dexterity and result in poor learning ability and difficulty in performing complex tasks.
Ground level ozone	Decrease in lung function and increased respiratory sub-symptoms such as chest pain and cough, increased susceptibility to respiratory infection, lung inflammation and aggravation of asthma.
Lead	Accumulates in the blood, bone, soft tissues; can adversely affect kidneys, liver, nervous system and may cause neurological impairments.
Sulfur dioxide	High concentrations can result in temporary breathing impairment for asthmatic children and adults active outdoors; in conjunction with high PM levels can aggravate existing cardiovascular disease, respiratory illness and alterations in lung's defenses
Nitrogen dioxide	Short-term exposure (< 3hr) to low levels of NO_2 may lead to changes in airway responsiveness and lung function in individuals with pre-existing respiratory illness. Long-term exposure may lead to increased susceptibility to respiratory infection and may cause irreversible alteration in lung structure.
Particlulate matter	Particles < 10 PM (Fig. 5.13) can get into the lungs and cause numerous health problems and possible heart and lung disease; particles can aggravate asthma and bronchitis and have been associated with cardio-arrhythmias and heart attacks (Lipmann, 2003; Nel, 2005). Ultrafine particles can damage mitochondria (Li et al., 2003).

Adapted from *http://www.epa.gov/airtrends/pm.html.* The student is referred to Ostro (1993), Schwartz (1994) and Krzyanowski et al. (2005).

Air Pollution Research Efforts. There is extensive air pollution research both nationally and internationally (Table 5.8). Much of this research centers about developing analytical methods for detecting and quantifying air pollutants (Table 5.7). Of particular interest is the continued search for both animal and plant species which can serve as bioindicators.

Whereas Table 5.5 depicts some effects of acid rain on aquatic organisms, Table 5.6 illustrates the human health effects of six major pollutants. With regard to the latter, the reader is referred to Ostro (1993).

Emissions Tg(10^{12})/year	SO$_2$/IPCC	NOx/Graedel	Volatile Organic Comp.
Biomass burning	2.2	6	45
Volcanoes	9.3	-	-
Lightning	-	5	-
Biogenic emissions from land areas	1.0	15	350 (isoprenes) + 480 (terpenres)
Biogenic emissions from oceans	24	-	27
Industrial and utility activities	76	22	45
Solvents	-	-	15
Total natural emissions	34	21	855
Total anthropogenic emissions	78	27	105
Total emissions	113	48	960

Fig. 5.8   Global, natural and anthropogenic emissions of SO$_2$ and NO$_2$
From: *http://www.eoearth.org/article/air-pollution-emissions*.

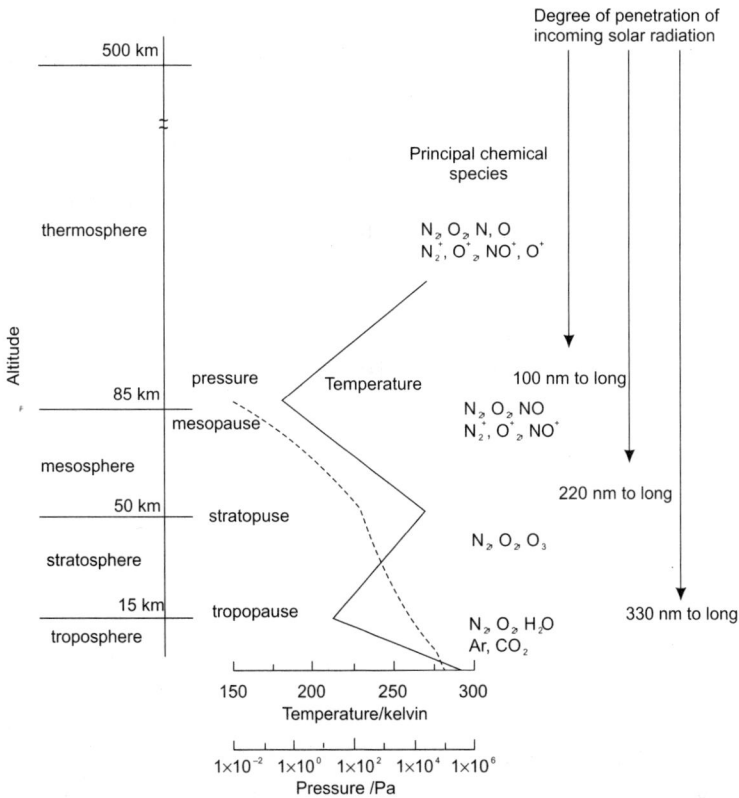

Fig. 5.9   Layers of the atmosphere. Note the troposphere
From: Van Loon, G.W. and Duffy, S.J. (2005).

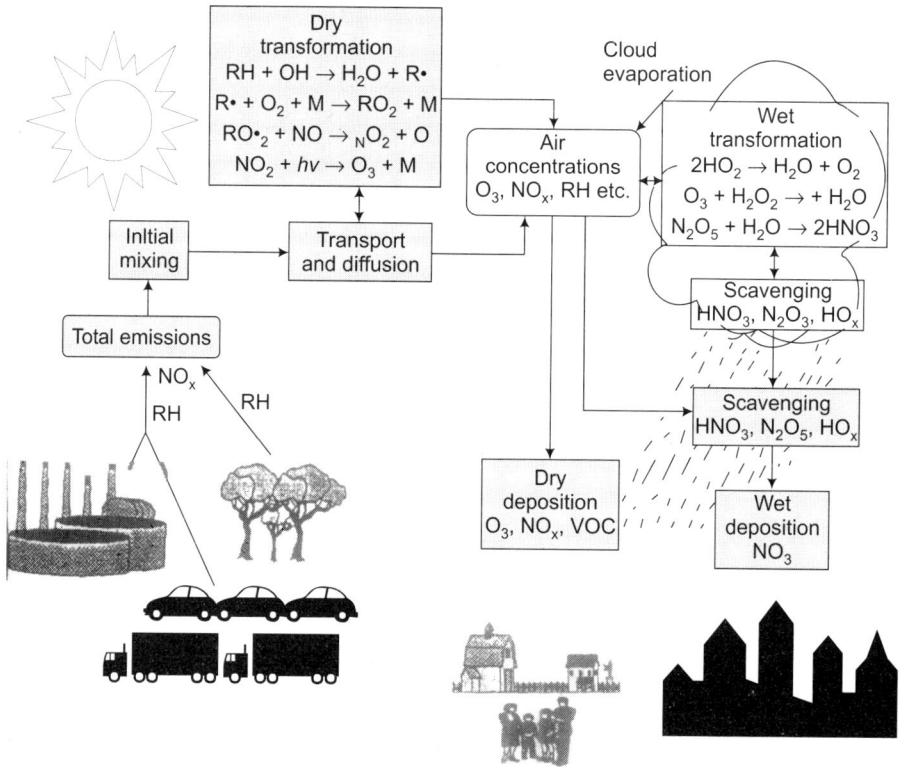

Fig. 5.10 Schematic diagram of photochemical air pollution from emission to depositon.
From: *http://www.EPA.gov/ttnamtii/files/ambient/criteria/velfocis/r-98-002.pdf.*

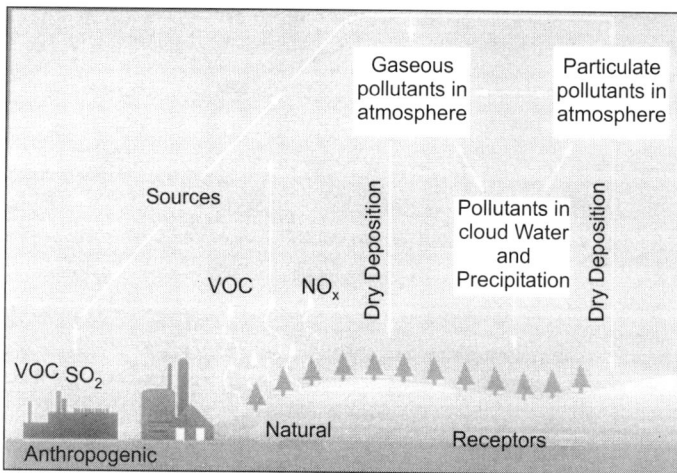

Fig. 5.11 Pollutant wet and dry depositions.
From: *http://www.epa.gov/airmarkets/acidrain/*

Tongue

Mandible

Epiglottis

Hyoid bone

Glottis

Tyroid cartilage

Cricoid cartilage

Wind pipe

Upper section of the pharynx

Middle section of the pharynx

Lower section of the pharynx

Esophagus

Muscle strands

Terminal bronchioles

Alveolus

Lungs

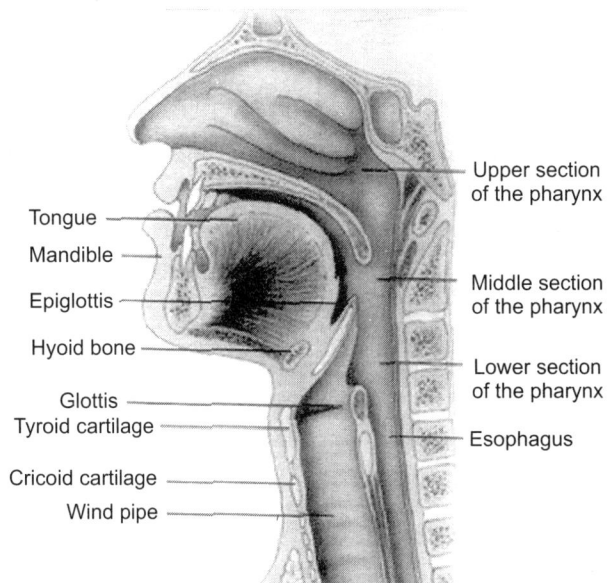

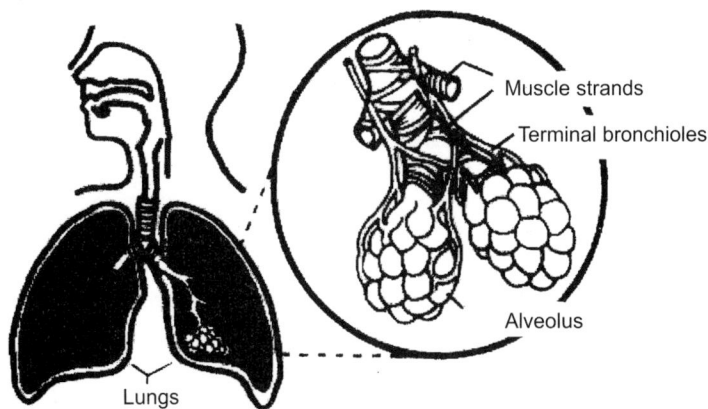

Human Breathing
(Involuntary Process Controlled by Respiration Center)

Phases	Activity
Inhalation or inspiration (Active process supplying oxygen)	Chest cavity expands and air enters
Exhalation or expiration (Passive process elimination carbon dioxide)	Air is expelled

Fig. 5.12 Diagram of respiratory anatomy. A more detailed diagram can be observed at: *http://www.nb.lung.ca/images/flast/respiratory.swf*. Top. From: *http://www.contmedrarisa*

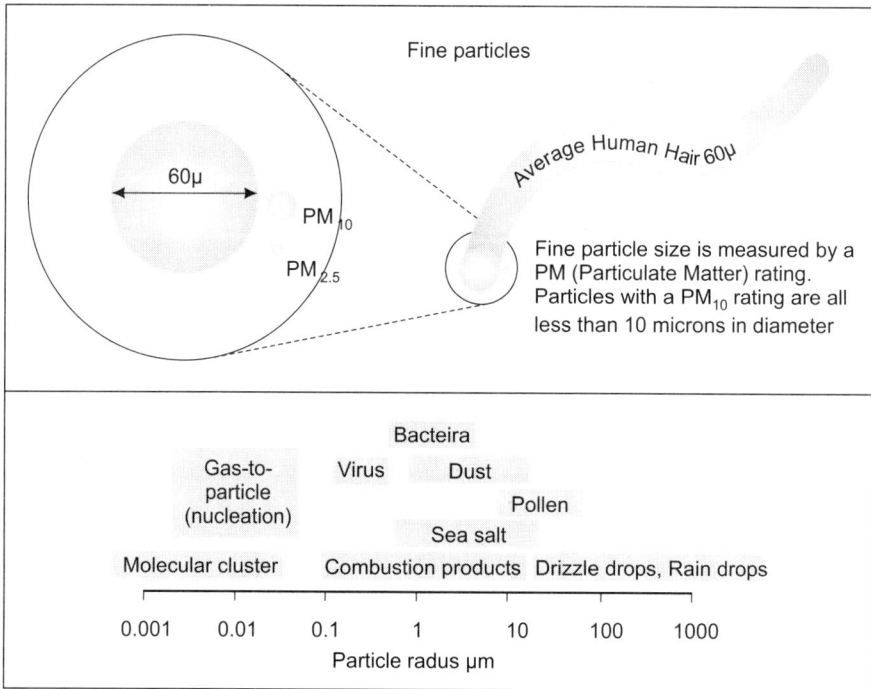

Fig. 5.13 Comparative sizes of PM$_{2.5}$ and PM$_{10}$ with human hair and sand. From: *http://www.epa.gov/aor/pareticlepollution/humanhair.html.*

Pollution Control and Economic Impact

The costs resulting from air pollution (Reis, 2004) derive from monitoring and controlling pollutants (Coglianese, 2000; Jorgensen, 2000; *http:www.epa.gov/air/ criteria.html*). There are many air pollutant monitoring stations located throughout various regions of the world. In the USA, monitoring sites are operated by local, state and federal agencies. A prominent example of the latter is the USA Environmental Protection Agency (EPA). Obtained air pollutant data are entered upon spreadsheets and then graphed presenting air pollution (e.g., ozone levels) over time. For ozone, its levels are given a color code to render the public aware of daily conditions relative to health.

The costs concerning air pollution (Reis, 2004) are those for generating, passing and enforcing legislation.. This involves installing scrubbers in coal-fire power plants. A critical US legislation is the Clean Air Act and its amendments (see chapter 14). The EPA's responsibilities under the Act (*http:// www.epa.gov/airtrends/sixpoll.html.*) are:

- Setting NAAQS for the six principal pollutants that are considered harmful to public health and the environment.

- Ensuring that air quality standards are met (in cooperation with state, tribal, and local governments) through national standards and strategies to control air pollutant emissions from vehicles, factories and other sources.

TABLE 5.7 Summary of Methods for Detecting Air Pollutants[a]

Method	Procedure	Reference
Porolyticol Chromatographic Methods	Air Supply & Analysis	OSHA Salt Lake City, UT, USA OSHA US Department of Labor
Spectroscopic Methods	Air Supply and Analysis	OSHA Salt Lake Technical Center Sandy, UT, USA. www.osha.gov/sltc/ sampling analysis/analysis.html.
Tunable Diode Laser Absorption Spectroscopy	Ozone Detection	
Animal Indicators Sentinel Species rats, mice, rabbits and dogs (emphasis rats) Invertebrate Metallothionein Levels	Bioaccumulators of environmental trace elements	Saldiva and Bohn (1998)
Plant Indicators Tradescantia	Mutatation Chromosome abnormalities	Grant (1998) Ma et al. (1994)
Allium and Vicia root tips	Chromosome breaks for genotoxicity	Ma et al. (2005)
Arabidopsis	Mutation bioassay for genotoxicity	Ma et al. (2005)
Glycine max Hordeum vulgare	Soluble proline in SO_2 fumigated seedlings	Erickson and Dashek (1982)

[a]A manual for sampling and analysis of contaminates in workplace air has been published by NIOSH (http://www.cdu.gov/aviosh/ Nnan). Keith and Walker (1995) have published an extensive handbook for sampling and analysis of air toxics. In 1999, the USA EPA published: A compendium of methods for determination of inorganic compounds in ambient air. EPA/625/SR-96/010a. Cincinatti, OH, USA.

TABLE 5.8 Some Air Pollution Research Effects[a]

Research Project	Effect	Reference
International European Urban Air Quality Project	Scientific breakthroughs in relation to: 1. Improvement in understanding of sources of pollutants such as particles. 2. Processes which test influence pollutant dispersion in street and urban scale situations.	http://www.nilu.no/clear/proj2.htm
Australia CSIRO Atmospheric Research	Investigate emissions of organic compounds from Australian eucalyptus trees and gases which contribute to photochemical smog formation.	http://www.au/ar.climate change.
University of Aberdeen	Extent to which it is possible to value the benefits of UK ecosystems resulting from air pollution reduction.	http://www.abn.ac.uk
International study of long-range air transport using instruments aboard aircraft		http://earthobservatorynasa.gov/ newsroom/nasanews/2006/20006 031221818.html
National USA Air Pollution Prevention and Control Division of EPA	Characterization of major sources of air pollution and verification of the performance of innovative technologies	http://www.epa.gov/apped www/index.html.

(Contd.)

(Contd.)

	Studies concerned with particulate matter exposure to elevated levels of small solid and liquid particles in the air damaging health effects.	http://www.epa.gov/pm-research/
University of Georgia, USA	Develop biological markers for water-borne pollutants from rainfall and validate markers in the field	http://www.uga.edu/publichealth/ ehs/research/areas.html.
US-EPA Atmospheric protection	Risk management research that will make a difference in the Earth's environment and its inhabitants' health.	http://www.epa.gov/appcdiow/ apb/indext.htm.
US-EPA Emissions Characterization and Prevention	Research and develop new and more accurate methods for estimating emissions from sources posing a risk to human health; develop and demonstrate technological solutions to these emissions	http://www.epa.gov/appcdwww/ lepb/index.htm.

[a]Regional air pollution in developing countries has been addressed by RAPIDC (http://www.york.ac.uk/inst/sei/rapidc2/book.htm)

TABLE 5.9 Fuel economy and greenhouse gas (GHG) emissions on vehicle propulsion technologies

	Feed-stock	Fuel economy		Zero to	GHG emissions	
		miles per gallon	performance index	60 miles per hour (seconds)	grams per gallon	performance index
Conventional gasoline spark ignition (SI)	Fossil	20.2	100	7.9	544	100
Conventional diesel compression ignitions (CI)	Fossil	23.8	85	9.2	472	87
Gasoline SI hybrid electric vehicle (HEV)	Fossil	24.4	83	6.3	454	83
Diesel HEV	Fossil	29.4	69	7.2	384	71
Gasoline fuel cell	Fossil	27.2	74	10.0	408	75
Methanol fuel cell	Fossil	30.3	6	9.4	371	68
Ethanol (100%) fuel cell	Renewable	28.6	71	10.0	35	6
Hydrogen fuel cell	Fossil	43.2	47	8.4	330	61
Hydrogen fuel cell	Fossil	48.1	42	10.0	296	54

Source: General Motors Corporation/Argonne National Library, Well-to-Wheel Energy Use and Greehouse Gas Emissions of Advanced Fuel/Vehicle Systems—North American Analysis (Argonne, Ill.; Argonne National Library, 2001), Table 2.1 and Appendix 3C.

- Reducing emissions of SO_2 and NO_x that cause acid rain.
- Reducing air pollutants such as PM, SO_2 and NO_x that can cause visibility impairment across large regional areas, including many of the nation's most treasured parks and wilderness areas.

- Ensuring that toxic air pollutant sources which may cause cancer and other adverse human health and environmental effects are well controlled and that risks to public health and the environment are substantially diminished.

- Limiting the use of chemicals that damage the stratospheric ozone layer in order to prevent increased levels of harmful ultraviolet radiation.

The long-term cost associated with SO_2 pollution and resulting aquatic (White and Terninko, 2003) and terrestrial (Albers et al., 1998) acidification will only become apparent with time (Commission on Life Sciences, 1989).

Perhaps, the most contentious issue concerning economic measures and protection of the environment centers about the automobile industry which yields noxious pollutants (Harrington and McConell, 2003). Shifting from petroleum-based fuels (Jensen and Ross, 2000) would involve: (a.) a lengthy change over to parts for electrical controls, (b.) replacement of battery sales and rentals to electrical repairs, (c.) alterations in highway designs as well as (d.) taxation. Table 5.9 compares fuel economy and greenhouse gas emissions and vehicle propulsion technologies.

References

Albers, P.H., Heinz, G. and Ohlendorf, H.M 1998. *Environmental Contaminates and Terrestrial Vertebrates Effects on Population, Communities and Ecosystems*. SETAC Special Publication Series. Pensacola, FL, USA.

Argawal, S.B. and Argawal, M. 1999. *Environmental Pollution and Plant Responses*. Lewis Publishers, Boca Raton, FL, USA.

Austin, J., Brimblecambe, P. and Sturges, W. 2002. *Air Pollution Science for the 21st Century*. Pergamon, Burlington , MA, USA.

Barker, J.R. and Tingey, D.T. 1992. *Air Pollution Effects and Biodiversity*. Van Nostrand, Reinhold, New York, NY, USA.

Beacham, W. 1993. *Beacham's Guide to Environmental Issues and Sources*. Volume 3 Beacham Publishing Inc., Washington, DC, USA.

Bell, J. and Treshow, M. 2002. *Air Pollution and Plant Life*. Wiley, New York, NY, USA.

Bell, J.M.B. and Marshall, F.M. 1999. Field studies on impacts of air pollution on agricultural crops. In: *Environmental Pollution and Plant Responses*. (Agrawal, S.B. and Agrawal,M., eds.) Lewis Publishers, Boca Raton, FL, USA pp.99-110..

Bresser, A.H.M. and Salomons, W. 1989. 'Acidic Precipitation'. Vol. 5. *International Overview and Assessment*. Springer Verlag, New York, NY, USA.

Carson, N. and Munton, D. 2000. Flows in the conventional wisdom on acid deposition. Environment 2: 33-35.

Chapman, D. 1992. *Water Quality Assessments*. Chapman and Hall, London, England.

Cheremisinoff, N.P. 2006. *Handbook of Air Pollution Prevention and Control*. Butterworth-Heinemann, Oxford, UK.

Chhatwal, G.R. 2003. *Encyclopedia of Environmental Air Pollution*. Vadams Books, New Delhi, India.

Coglianese, C. 2000. The constitution and the costs of clean air. *Environment* 42: 32-37.

Commission on Life Sciences. 1989. *Biologic Markers of Air-Pollution Stress and Damage in Forests*. National Academy of Sciences, Washington, DC, USA.

Darall, N.M. 1989. The effects of air pollutants on physiological processes in plants. *Plant Cell Environ* 12: 1 - 30.

Dlamini, E.T., Williams, A.L., Dashek, W.V., Swank, W.T. and Vose, J.M. 1994. Protein contents of ozone- and air-fumigated 'Pinus strobus' needles. In: *Biodeterioration Research 4*. (Llewellyn, G.C., Dashek, W.V. and O'Rear, C.E., eds.), Plenum Press. New York, NY, USA. pp 379-390.

Dockery, D.W. and Pope, C.A. 1994. Acute respiratory effects of particle air pollution. *Annu. Rev. Public Health* 15: 107-132.

Dudley, N. and Stolton, S. 1996. *Air Pollution and Biodiversity*. WWF International, Glard, Switzerland.

Erickson, S. and Dashek, W.V. 1982. Accumulation of foliar soluble proline following fumigation of *Glycine max*, cv. "Essex" and *Hordeum vulgare*, cvs. 'Proctor' and 'Excelsior' seedlings with sulfur dioxide. *Environ. Pollut.* 28: 89-108.

Forster, B.A. 1993. *The Acid Rain Debate Science and Special Interests in Policy Formation.* Iowa State University Press, Ames, Iowa, USA.

Godbold, D.L. and Houttmann, A. 1994. *Effects of Acid Rain and Forest Processes.* Wiley-Liss, New York, NY, USA.

Grant, W.F. 1998. Higher plant assays for the detection of geotoxicity in air polluted environments. Ecosyst. Health 4: 210.

Guderian, R. 1989. *Pollution by Photochemical Oxidants.* Springer Verlag, Berlin, Germany.

Harrington,W. and McConell, V. 2003. A lighter trend. Policy and technology options for motor vehicles. *Environment* 45: 10-22.

Hill, M. 2004. *Understanding Environmental Pollution.* Cambridge Univ. Press, Cambridge, England.

Howells, G. 1995. *Acid Rain and Acid Waters.* Ellis Horwood, New York, NY, USA.

Jaegle, L., Steninberger, J., Martin, R.V. and Chance, K. 2005. Global portioning of NOx sources using satellite observations. Relative roles of fossil fuel combustion, biomass and soil emissions. *Faraday Discuss.* 130: 407-423.

Jensen, M.W. and Ross, M. 2000. The ultimate challenge developing an engine for fuel cell vehicles. *Environment* 42: 10-22.

Jorgensen, S.E. 2000. *Principles of Pollution Abatement. Pollution Abatement for the 21st Century.* Elsevier, Amsterdam, The Netherlands.

Kabel, R.J. and Heinsohn, R.L. 1998. *Sources and Control of Air Pollution.* Prentice Hall, Upper Saddle River, NJ, USA.

Karnosky, D.F., Percy, K.E., Chappelaw, A.H., Simpson, C., and Pikkarinen, J. 2003. *Air Pollution Global Change and Forests in the New Millennium.* Elsevier, Burlington, MA, USA.

Keith, L.H. and Walker, M.M. 1995. *Handbook of Air Toxics. Sampling, Analysis and Properties.* Lewis, Boca Raton, FL, USA.

Kosugl, T., Meredith, C.P. and Hollander, A. 1983. *Perspective Basic Life Sciences.* Plenum, New York, NY, USA.

Koziol, M.J. and Whatley, F.R. 1984. *Air Pollutants and Plant Metabolism.* Butterworth, London, England.

Kozlowski, T.T. 1980. *Impacts of Air Pollution on Forest Ecosystems.* American Institute of Biological Sciences. Madison, WI, USA.

Krupa, S.V. 1971. *Air Pollution and Plants.* American Phytopathological Society Press, St. Paul, MN, USA.

Krzyzanowski, M., Kuna-dibbert, B., and Schneider, J. 2005. *Health Effects of Transport-Related Air Pollution.* World Health Organization, Geneva, Switzerland.

Lastovicka, J., Akmaev, R.A., Bremer, J. and Emmert, J.T. 2006. Global change in the upper atmosphere. Science 314:1253-1254.

Law, K.S. and Stohl, A. 2007. Artic air pollution. Origins and impacts. *Science* 315: 1537-1540.

Laurence, J.A., Retzlaff, W.A., Kern, J.S., Lee, E.H., Hogseth, W.E. and Weinstein, D.A. 2001. Predicting the regional impact of ozone and precipitation on the growth of loblolly pine and yellow-poplar using linked TREGO and ZELIG models. *Forest Ecol. Manag.* 146: 247-263.

Leer, H.R. 1999. *Plant Responses to Environmental Stress.* Marcel Dekker, New York, NY, USA.

Lerdau, M.T., Munger, J.W. and Jacob, D.J. 2000. The NO_2 flux conundrum. *Science* 289: 2291-2293.

Levingtion, J.S. 2001. *Marine Biology: Function, Biodiversity, Ecology.* Oxford Univ. Press, New York, NY, USA.

Li, N., Sioutas, C., Cho, A., Schmitz,D., Way, C.M., Oberley, Froines, J. and Nel, A. 2003. Ultrafine particulate pollutants induce oxidative stress mitochondrial damage. Environ. Health Perspect. 111: 455-460.

Lipmann, M. 2003. Human health: Effects of ambient air particulate matter. In: *Acid Rain: Are the Problems Solved?* (White, J.C. and Terninko, J., eds.). American Fisheries Society, Bethesda, MD, USA. pp. 83-92.

Lisowski, M. and Williams, R.A. 1997. *Wetlands.* Franklin Watts Library, New York, NY, USA.

Ma, T.H., Labvera, G.L., Chen, R., Gill, B.S., Sandhu, S.S., Van Der Berg, A.L. and Salamore, M.F. 1994. *Tradescantia* micronucleus bioassay. *Mutat. Res.* 310: 221-230.

Ma, T.H., Xu, Z., Xu, C. McConnell, H., Rabogo, E.V., Arreola, G.A. and Zhang, H. 2005. The improved *Allium/Vicia* root tip. Micronucleus assay for clastogenicity of environmental pollutants. *Mutat. Res.* 334: 185-195.

Mead, J.B., Hopercraft, G., Frasier, S.J., Pollard, B.D., Cherry, C.D., Schar, D.H. and McIntosh, R.E. 1997. A volume-imaging radar wind profiler for atmomspheric transboundary turbulence studies. *J. Atmos. Ocean Tech.* 15:849-859.

Nel, A. 2005. Air pollution related illness: Effect of particles. *Science* 308: 804-806.

Novarty, V. and Olens, H. 1994. *Water Quality: Prevention, Identification and Rearrangement of Diffuse Pollution.* Van Nostrand Reinhold, New York, NY, USA.

Olszyk, D.M., Johnson, M.G., Phillips, D.L., Seidler, R.J., Tingey, D.T. and Watrad, L.S. 2001. Interactive effects of CO_2 and O_3 on a ponderosa pine plant/litter/soil mesocosm. *Environ. Pollut.* 115: 447-462.

Ongley, E.D. 1994. Global water pollution: Challenges and opportunities. *Proceedings: Integrated Measures to Overcome Barriers to Minimizing Harmful Fluxes from Land to Water.* Publication No. 3, Stockholm, Sweden, pp. 23-30.

Ongley, E.D., Krishnappan, B.G., Droppo, I.G., Rao, S.S. and Maguire, R.J. 1992. Cohesive sediment transport: Emerging issues for toxic chemical management. *Hydrobiologia* 235/236:177-187.

Ostro, B. 1993. The association of air pollution and mortality: examining the case for inference. *Arch. Environ. Health* 48: 336.

Ostry, R.C. 1982. Relationship of water quality and pollutant loads to land uses in adjoining watersheds. *Water Resources Bull.* 18:99-104.

Pepper, S.L. 1996. *Pollution Science.* Academic Press, New York, NY, USA.

Philip, R.B. 2001. *Ecosystems and Human Health. Toxicology and Environmental Standards.* Lewis, Boca Raton, FL, USA.

Reis, S. 2004. *Costs of Air Pollution Control.* Springer, New York, NY, USA.

Reynolds, T.B. and Metcalfe-Smith, J.L. 1992. An overview of the assessment of aquatic ecosystem health using benthic invertebrates. *J. Aquatic Ecosystem Health* 1: 295-308.

Rosenfeld, D. 2000. Suppression of rain and snow by urban and industrial air pollution. *Science* 287: 1793-1796.

Saldiva, P.H.S. and Bohn, G.M. 1998. Animal indicators of adverse effects associated with air pollution. *Ecosyst. Health* 4: 1526-0992.

Schulze, C.D., Large, O.L. and Over, R. 1989. *Forest Decline and Air Pollution.* Springer-Verlag, New York, NY, USA.

Schwartz, J. 1994. Air pollution and daily mortality: a review and metanalysis. *Environ. Res:* 64: 36-52.

Seinfeld, J.H. and Pandis, S.N. 1998. *Atmospheric Chemistry and Physics: From Air Pollution to Climate Change.* Wiley, New York, NY, USA.

Shipman, J.T., Wilson, J.D. and Toddy, A.W. 2006. *An Introduction to Physical Science. Atmospheric Effects.* Houghton Mifflin, Boston, MA, USA.

Shugar, G.J., Drum, D.A., Lauber, J. and Bauman, S.L. 2000. *Environmental Field Testing and Analysis Ready. Reference Handbook.* McGraw Hill, New York, NY, USA.

Smith, W.H. 1990. *Air Pollution and Forests. Interaction Between Air Contaminants and Forest Ecosystems.* 2nd Edition. Springer-Verlag, New York, NY, USA.

Sutherland, W.J. 2000. *The Conservation Handbook Research, Management and Policy.* Blackwell. Oxford, England.

Taylor,C. 1984. Organismal responses of higher plants to atmospheric pollutants: Photochemical and other. In: *Air Pollution and Plant Life.* Treshow, M. (ed.). Wiley and Sons, New York, NY, USA pp.215-223.

Theodore, L. 1999. *Pollution Prevention: The Waste Management Approach to the 21st Century.* CRC Press, Boca Raton, FL, USA.

Van Loon, G.W. and Duffy, S.J. 2005. *Environmental Chemistry. A Global Perspective.* Oxford University Press, Oxford Univ. Press, Oxford, England.

White, J.C. and Terninko, J. 2003. *Acid Rain: Are the Problems Solved?* American Fisheries Society, Bethesda, MD, USA.

Winner, W.E. 1994. Mechanistic analysis of plant responses to air pollution. *Ecol. Appl.* 4: 651-661.

Yu, Ming-Ho. 2005. *Environmental Toxicology Biological and Health Effects of Pollutants.* CRC Press, Boca Raton, FL, USA.

Zeeduk, H. and Velds, C.A. 1973. The transport of sulfur dioxide over a long distance. *Atoms. Environ.* 7: 849-862.

CHAPTER 6

Metals and Volatile Organics Pollution

TOXIC METALS AND THEIR SOURCES

There are two overall classifications of pollutant sources, i.e., point and non-point. Whereas the former originates from a concentrated point (e.g., industrial and sewage treatment plants), the latter comes from a wide range of sources. "Non-point pollution arises by rainfall or snowmelt moving over and through the ground" (*http: www. epa.gov/owow/nps/qa.html*).

This chapter is concerned with toxic metals (Bloemen, 1993; Harrison, 1995; Market and Friese, 2000; Hill, 2004) and organic volatiles (Wang et al., 1996). Both plants (Shkolink, 1984) and animals (Castillo-Duran and Cassorla, 1999) require metals (Table 6.1) for growth and development. However, major environmental questions are at what levels do these metals become toxic and where do environmentally harmful levels originate from?

The 'light' metals are Na, Mg, K and Cu and the heavy metals are V, Ch, Mn, Fe, Co, Cu, Zn, Mo and Ti. With regard to the anthropogenic origins of metal pollutants, fossil coal combustion appears to be a major source of Cr, Hg, Mn, Sb, Se, Sn and Ti and Ni and for V oil combustion yields (Pacyna and Pacyna, 2001). According to these authors, the major source of atmospheric Pb is the combustion of low-level leaded and unleaded gasoline. The most significant source of atmospheric As, Cd, Cu and Zn is non-ferrous metal production. Adriano (2001) has published a thorough description of both the natural and anthropogenic sources of Ag, As, Be, Cd, Cr, Cu, Hg, Ni, Pb, Sb, Se, Ti and, Zn. Another comprehensive listing of toxic metals in the environment is that in Ross (2001).

Certain fertilizers can be sources of harmful toxic metals (Shaffer, 2001). The California Public Interest Research Group Charitable Trust and Washington's Safe Food and Fertilizer found 22 toxic metals (Table 6.2) in 29 fertilizers sold in the US (Shaffer, 2001). The environmental problem is that fertilizers are subject to agricultural runoff.

Another source of toxic metals is sewage sludge (Table 6.3) which is sometimes applied to farmlands (Logan and Chaney, 1983). The USA production of sewage sludge is about 7 million metric tons/year. Only about 36% of the sludge is applied to the land for beneficial purposes, e.g., agricultural, turf grass production or reclamation of surface mining areas. Sludge

| TABLE 6.1 | Metal Locations and Functions in Animals and Plants[a] |

Animals

Metal (Chemical Symbol)	Location	Function
"Healthy" Metals		
Iron (Fe)	Binds to molecules throughout the body (e.g., hemoglobin, nitric oxide synthase)	Helps body transport oxygen and certain chemical messengers
Copper (Cu)	Binds to enzymes throughout the body (e.g., superoxide dismutase)	Defends body against damage from free radicals
Zinc (Zn)	Binds to enzymes throughout the body, to DNA and to some hormones (e.g., insulin)	Plays a role in sexual maturation and regulation of hormones; helps some proteins bind tightly to DNA
Sodium (Na) Potassium (K)	Throughout the body (Na outside cells, K inside cells)	Helps communicate electrical signals in nerves, heart
Calcium (Ca)	Bones, muscle	Muscle and nerve function; blood clotting
Cobalt (Co)	Forms the core of vitamin B_{12}	Necessary ingredient for making red blood cells
"Unhealthy" Metals		
Arsenic (As)	Rocks, soil	Can cause cancer, death
Lead (Pb)	Old paint (before 1973)	Can cause cancer, neurological damage, death
Mercury (Hg)	Contaminated fish (especially from the Great Lakes region of the United States)	Binds to sulfur-containing molecules in organelles; can cause neurological damage, death

Plants[a]

Element	Concentration in Matter (mmol/kg)
Macronutrients	
Hydrogen	60,000
Carbon	40,000
Oxygen	30,000
Nitrogen	1,000
Potassium	250
Calcium	125
Magnesium	80
Phosphorous	60
Sulfur	30
Micronutrients	
Chlorine	3.0
Boron	2.0
Iron	2.0
Manganese	1.0
Zinc	0.3
Copper	0.1
Nickel	0.05
Molybdenum	0.001

[a] http://publications.nigms.nih.gov/chemhealth/chapter2.html and Shkolink (1984)

TABLE 6.1 (Continued) The Biogeochemistry of Life – Essential Elements

Essential to all animals and plants	Essential to several classes of animals and plants	Essential to a wide variety of species in one class	Essential to one or two species only	Recent work indicates essentiality, but of unknown function
Hydrogen (H)	Silicon (Si)	Boron (B)	Lithium (Li)	Rubidium (Rb)
Carbon (C)	Vanadium (V)	Fluorine (F)	Aluminum (Al)	**Tin (Sn)**
Nitrogen (N)	Cobalt (Co)	Chromium (Cr)	Nickel (Ni)	
Oxygen (O)	**Molybdenum (Mo)**	**Bromine (Br)**	**Strontium (Sr)**	
Sodium (Na)			**Barium (Ba)**	
Magnesium (Mg)	**Iodine (I)**			
Phosphorus (P)				
Sulfur (S)				
Chlorine (Cl)				
Potassium (K)				
Calcium (Ca)				
Manganese (Mn)				
Iron (Fe)				
Copper (Cu)				
Zinc (Zn)				
Selenium (Se)				

Elements in bold type are generally considered to be trace elements.
From: *http://interactive2.usgs.gov/faq/list_faq_by_category/get_answer.asp?id=248* and
http://geology.er.usgs.gov/eastern/environment/environ.html. A thorough list of the major and minor elements and their deficiency symptoms for plants can be found in Brewer (2006) and Malmstroem (2006). (see appended IUPAC periodic table of elements)

TABLE 6.2 Summary of Harmful Metals Found in Twenty-Nine Fertilizers Used in the US[a]

Metal	Metal
Aluminum (Al)	Manganese (Mn)
Antimony (Sb)	Mercury (Hg)
Arsenic (As)	Molybdenum (Mo)
Barium (Ba)	Nickel (Ni)
Boron (B)	Selenium (Se)
Cadmium (Cd)	Silver (Ag)
Chromium (Cr)	Thallium (Ti)
Cobalt (Co)	Vanadium (V)
Copper (Cu)	Uramium (U)
Iron (Fe)	Zinc (Zn)
Lead (Pb)	

[a]Adapted from: *http://www.pirg.org/toxics/reports/wastelands/*

appears to possess very little agricultural impact from national farmers' perspectives because of environmental issues involving transportation and the varying composition of nutrients and trace elements (Table 6.3) due to differences among types of sludge (Smith, 1996). Tables 6.3, 6.4 compare the metal element contents of soil (Kabata-

IUPAC Periodic Table of the Elements

Key:

atomic number
Symbol
name
standard atomic weight

1	2	3	4	5	6	7	8	9	10	11	12	13	14	15	16	17	18
1 H Hydrogen 1.00794(7)																	2 He Helium 4.002602(2)
3 Li Lithium 6.941(2)	4 Be Beryllium 9.012182(3)											5 B Boron 10.811(7)	6 C Carbon 12.0107(8)	7 N Nitrogen 14.0067(2)	8 O Oxygen 15.9994(3)	9 F Fluorine 18.9984032(5)	10 Ne Neon 20.1797(6)
11 Na Sodium 22.989770(2)	12 Mg Magnesium 24.3050(6)											13 Al Aluminum 26.981538(2)	14 Si Silicon 28.0855(3)	15 P Phosphorus 30.973761(2)	16 S Sulphur 32.065(5)	17 Cl Chlorine 35.453(2)	18 Ar Argon 39.948(1)
19 K Potassium 39.0983(1)	20 Ca Calcium 40.078(4)	21 Sc Scandium 44.955910(8)	22 Ti Titanium 47.867(1)	23 V Vanadium 50.9415(1)	24 Cr Chromium 51.9961(6)	25 Mn Manganese 54.938049(9)	26 Fe Iron 55.845(2)	27 Co Cobalt 58.933200(9)	28 Ni Nickel 58.6934(2)	29 Cu Copper 63.546(3)	30 Zn Zinc 65.409(4)	31 Ga Gallium 69.723(1)	32 Ge Germanium 72.64(1)	33 As Arsenic 74.92160(2)	34 Se Selenium 78.96(3)	35 Br Bromine 79.904(1)	36 Kr Krypton 83.798(2)
37 Rb Rubidium 85.4678(3)	38 Sr Strontium 87.62(1)	39 Y Yttrium 88.90585(2)	40 Zr Zirconium 91.224(2)	41 Nb Niobium 92.90638(2)	42 Mo Molybdenum 95.94(2)	43 Tc Technetium (97.9072)	44 Ru Ruthenium 101.07(2)	45 Rh Rhodium 102.90550(2)	46 Pd Palladium 106.42(1)	47 Ag Silver 107.8682(2)	48 Cd Cadmium 112.411(8)	49 In Indium 114.818(3)	50 Sn Tin 118.710(7)	51 Sb Antimony 121.760(1)	52 Te Tellurium 127.60(3)	53 I Iodine 126.90447(3)	54 Xe Xenon 131.293(6)
55 Cs Caesium 132.90545(2)	56 Ba Barium 137.327(7)	57-71 Lanthanoids	72 Hf Hafnium 178.49(2)	73 Ta Tantalum 180.9479(1)	74 W Tungsten 183.84(1)	75 Re Rhenium 186.207(1)	76 Os Osmium 190.23(3)	77 Ir Iridium 192.217(3)	78 Pt Platinum 195.078(2)	79 Au Gold 196.96655(2)	80 Hg Mercury 200.59(2)	81 Tl Thallium 204.3833(2)	82 Pb Lead 207.2(1)	83 Bi Bismuth 208.98038(2)	84 Po Polonium (208.9824)	85 At Astatine (209.9871)	86 Rn Radon (222.0176)
87 Fr Francium (223.0197)	88 Ra Radium (226.0254)	89-103 Actinoids	104 Rf rutherfordium (261.1088)	105 Db Dubnium (262.1141)	106 Sg Seaborgium (266.1219)	107 Bh Bohrium (264.12)	108 Hs Hassium (277)	109 Mt Meitnerium (268.1388)	110 Ds Darmstadtium (271)	111 Rg Roentgenium (272)							

Lanthanoids

57 La Lanthanum 138.9055(2)	58 Ce Cerium 140.116(1)	59 Pr Praseodymium 140.90765(2)	60 Nd Neodymium 144.24(3)	61 Pm Promethium (144.9127)	62 Sm Samarium 150.36(3)	63 Eu Europium 151.964(1)	64 Gd Gadolinium 157.25(3)	65 Tb Terbium 158.92534(2)	66 Dy Dysprosium 162.500(1)	67 Ho Holmium 164.93032(2)	68 Er Erbium 167.259(3)	69 Tm Thulium 168.93421(2)	70 Yb Ytterbium 173.04(3)	71 Lu Lutetium 174.967(1)

Actinoids

89 Ac Actinium (227.0277)	90 Th Thorium 232.0381(1)	91 Pa Protactinium 231.03588(2)	92 U Uranium 238.02891(3)	93 Np Neptunium (237.0482)	94 Pu Plutonium (244.0642)	95 Am Americium (243.0614)	96 Cm Curium (247.0704)	97 Bk Berkelium (247.0703)	98 Cf Californium (251.0796)	99 Es Einsteinium (252.0830)	100 Fm Fermium (252.0951)	101 Md Mendelevium (258.0984)	102 No Nobelium (259.1010)	103 Lr Lawrencium (262.1097)

Notes

- 'Aluminum' and 'cesium' are commonly used alternative spellings for 'aluminium' and 'caesium'.
- IUPAC 2001 standard atomic weights (mean relative atomic masses) are listed with uncertainties in the last figure in parentheses [R. D. Loss, *Pure Appl. Chem.* **75**, 1107-1122 (2003)].
  These values correspond to current best knowledge of the elements in natural terrestrial sources. For elements with no IUPAC assigned standard value, the atomic mass (in unified atomic mass units) or the mass number of the nuclide with the longest known half-life is listed between square brackets.
- Elements with atomic numbers 112, 113, 114, 115, 116 have been reported but not fully authenticated.

Copyright @ 2004 IUPAC, the International Union of Pure and Applied Chemistry. For updates to this table, see *http://www.iupac.org/reports/periodic_table/.* **This version is dated 1 November 2004.**

TABLE 6.3 Metal Element Contents of Soil and Sewage Sludge and Crops[a]

| | mg kg −1 | | |
Metal Elements	Sewage Sludge	Sewage Treated Soil	Crops
Arsenic	850 – 150,000	6,000 – 1,375,000	8 – 1,000
Copper	74,000 – 3,650,000	1,000 – 240,000	100 – 13,700
Lead	10,000 – 3,500,000	30 – 710,000	30 – 3,200
Manganese	120,000 – 1,540,000	188,000 – 2,750,000	500 – 173,000
Mercury	1,400 – 26,000	30 – 3,300	150
Nickel	26,000 – 940,000	3,000 – 74,000	10 – 7,500

[a]Adapted with modification from Crompton (1998).

TABLE 6.4 Chemical Compositions of Sewage Sludge and Animal Manure

Constituent (Unit)	Animal Manure[a] Range	Sewage Sludge Range	Typical
Nitrogen (% dry weight)	1.7 – 7.8	<0.1 – 17.6[b]	3.0
Total phosphorus (% dry weight)	0.3 -2.3	<0.1 -14.3[b]	1.5
Total sulfur (% dry weight)	0.26 – 0.68	0.6 – 1.5[b]	1.0
Calcium (% dry weight)	0.3 -8.1	0.1 -25[b]	4.0
Magnesium (% dry weight)	0.29 – 0.63	0.03 – 2.0[b]	0.4
Potassium (% dry weight)	0.8 - 4.8	0.02 – 2.6[b]	0.3
Sodium (% dry weight)	0.07 – 0.85	0.01 – 3.1[b]	-
Aluminum (% dry weight)	0.03 – 0.09	0.1 – 13.5[b]	0.5
Iron (% dry weight)	0.02 – 0.13	<0.1 – 15.3[b]	1.7
Zinc (mg/kg dry weight)	56 – 215	101 – 27,800[b]	1200
Copper (mg/kg dry weight)	16 – 105	6.8 – 3120[c]	750
Manganese (mg/kg dry weight)	23 – 333	18 – 7,100[b]	250
Boron (mg/kg dry weight)	20 -143	4 – 757[b]	25
Molybdenum (mg/kg dry weight)	2 – 14	2 – 976[b]	10
Cobalt (mg/kg dry weight)	1	1 – 18[b]	10
Arsenic (mg/kg dry weight)	12 – 31	0.3 – 316[c]	10
Barium (mg/kg dry weight)	26	21 – 8,980[b]	-

[a]Data summarized from Azevado and Stout (1974), California Agricultural Exp. Station V
[b]Data summarized from Dowdy et al. (1976) , Soil Conservation Society of America
[c]Data summarized from Kuchenrither and Carr (1991), Water Environment Federation
Full references in a, b and c are in Municipal Wastewater, Sewer Sludge and Agriculture (www.epa.gov/owm/pipes/sludmis/mstr.ch2.pdf).

Pendias, 2000; Basta et al., 2005), sewage sludge and crops. It is apparent that toxic copper, lead, mercury and nickel levels in sewage sludge exceed those in soil.

Mining of copper, gold, iron ore, lead, silver, titanium and zinc can be a significant source of toxic metal pollution. This pollution can result from acidic drainage of mine openings, waste tailings (pools of mining wastes), groundwater seepage and surface water runoff. Finally, a thorough summary of the natural and anthropogenic metal sources has been published by Adriano (2001). WHO provided the common forms of metals in wastes (Table 6.5).

TABLE 6.5	Maximum Permissible Concentrations of Various Metals in Natural Waste[a]	
Metal	Chemical Symbol	$mg\ m^{-3}$
Mercury	Hg	0.144
Lead	Pb	5
Cadmium	Cd	10
Selenium	Se	10
Thallium	Ti	13
Nickel	Ni	13.4
Silver	Ag	50
Manganese	Mn	50
Chromium	Cr	50
Iron	Fe	300
Barium	Ba	1000

[a]Source: US EPA (1987); Federal Register 56 (110): 26460 – 26564 (1991).

What are the Biological Effects of Pollutant Levels of Metals?

Contaminants such as metals can be biomagnified as the food chain accelerates from producers to tertiary consumers (*http://www.uaf.edu/seagrant/nosb/papers/2004/seeuonline-septicsystems.html.*). The effects of toxic metals on humans are shown in Table 6.6.

Attempts have been made to understand the toxicity of several elements at the cellular level (Hall, 2002; *http://www.portfolio.mvm.ed.oc.uk/studentwebs/session2/group29/diatox2.htm*). These attempts have centered around free radical generation, co-ordination complexes, competition with physiological elements and mimicking substrates (see toxicity figure for several elements at the above website). The biology involves mitochondria and the nucleus at the cellular level and the renal tubule in the kidney at the tissue level.

A thorough treatment of the biochemical/molecular aspect of heavy metal intoxication can be found at *http://www.unc.edu/courses/2005/spring/envv/132/001/toxic%20effec.* Two examples, methylmercury and lead, provide some insights into the actions of metals in humans. Figure 6.1 illustrates that methylmercury can enter the cell by combining with cysteine. The hematological effects of lead involve inhibition of two enzymes in the synthesis of heme (Fig. 6.2). Leslie (2005) also presents information regarding treatment for metal intoxication. The treatment employs common chelators, e.g., EDTA, pentetic acid, dimercaprol, succimer, penicillamine, trietine and deferoxamine to form metal ion complexes which can be excreted. According to Leslie (2005) the drawbacks for chelation treatment are: 1. not amenable to all metals, 2. can enhance toxicity of certain metals, 3. may deplete essential metals and, 4. chelators can have toxic side effects.

As expected, there is copious literature regarding the effects of toxic metals on fish (Ditoro et al., 2001; Taylor, 1996). Of particular concern is that metals form a biotic liquid that affects calcium channel proteins in the gill surface which regulates the ionic composition of the blood (Ditoro et al., 2001). The adverse effects of certain organometals on birds have been described (Bryan and Langston, 1992). Table 6.6 offers the effects that toxic levels of metals exert on animals and plants.

What are the Ways That Environmentally Toxic Metals Can be "Cleaned-Up"?

The US EPA reported that precipitation (Fig. 6.3) can separate heavy metals from contaminated groundwater (Allen et al., 1993). Precipitation alters dissolved heavy metal contaminants into a solid form which can be removed from water. The decontaminated water can be subsequently pumped back into the ground coupled with

| TABLE 6.6 | Biological Effects of Heavy Metals as Pollutants |

Metal	Effect
Aluminum	Asthmatic conditions, Bronchitis, Encephalopathy, Genital abnormalities, Pneumonia, Laryngitis, Neuritis, Neuro-fibrillary tangles, Pulmonary fibrosis, Rashes
Arsenic	Abnormal EEG's, Alteration in peripheral nerves, Anemia, Abdominal pain, Birth defects, Various cancers, Cardiovascular disease, Decreased white blood cell count, Encephalopathy, Hepatotxicity, Headache, Hypertension, Kidney failure, Muscle pain, Nausea, Neuritis, Numbness in extremities, Peripheral neuropathy, Pneumonia, Laryngitis, Reproductive dysfunction, Restrictive airway disorders, Skin effects
Cadmium	Adverse Effects on the Kidney, Reproductive Dysfunctions
Copper	Chloreiform Movements; Difficulty Swallowing, Talking and Walking, Disturbances in Menstrual Cycles; Headache, Muscle Pain
Lead	Birth defects; Convulsions, Encephalopahty; Immune Suppression; Inhibits Enzymes, Muscle Pain, Poor Concentration and Memory; Protein Coagulation in Cells; Vascular Disease
Manganese	Toxic to Nervous System
Mercury	Abnormal Gait/Posture/Walking; Birth Defects; Blurred Vision; Bronchitis; Chloreiform Movements; Coma; Death; Difficulty in Swallowing, Talking and Walking; Male Fertility Problems; Muscle Wasting; Pneumonia, Laryngitis; Poor Memory, Disturbances in Menstrual Cycle; Hearing Loss.

[a]Adapted from a variety of sources including: *http://www.u-ok.net/metal-effects.html* and other medical chelation websites which modified information from Alternative and Complimentary Therapies and Towesend's letters; Skukla and Singhal, 1984; Clarkson, 1995; Hu, 2002; earthwatch.unep.net/emergingissues/toxichen/heavy metals, 2006; Nriagu, J.O., 1988. Environ. Pollut. 50:139 has estimated the magnitude of the extent of trace element poisonings.

Plants		
Metal	Effect	References
Aluminum	Can cause stubby roots, leaf scorch, molting	Antonovics et al. (1971)
	May increase lipid peroxidation	Cakmark and Horst (1991)
Cadmium	Possible increase in lipid peroxidation	Pereira et al. (1999) Cakmark and Horst (1991)
Chromium	Can promote yellow leaves with green veins	Shah et al. (2001) Antonovics et al. (1971)
Copper	May result in lipid peroxidation, can yield dead patches in lower leaves, purple stems, stunted roots, chlorotic leaves containing green veins	Antonovics et al. (1971)
Nickel	Can promote white dead patches on leaves, may affect acid phosphatase activity	Antonovics et al. (1971) Gabbrielli et al. (1989)
Lead	Possible decreases in chlorophyll content, carotenoids, proteins, nitrate reductase activity, leaf structure, and phenol levels	Ewias (1997) Xiang (1997) Fargasova (2001) Kevresan et al. (2001) Singh et al. (1997) Kovacevic et al. (1999) Lavid et al. (2001a, b)
Zinc	Condensation of nuclear chromatin, dilution of nuclear membrane, disruption of cortical cells	Rout and Das (2003)
	Can result in chlorotic leaves with green veins, neurotic regions on leaf tips, stunted roots	Antonovics et al. (1971)

The physiological effects of heavy metals on plant growth and function have been reviewed by Liphadzi and Krirkham (2006). In: 'Plant Environment Interactions'. (Huang,B., ed.). CRC Press, Boca Raton, FL, USA.

Aquatic Organisms
(Sadiq, 1992; Tessier and Turner, 1995; Langston, 1986; Langston and Bebianno, 1998)

Metal	Effect	References
Mercury	Mercury is accumulated by fish, invertebrates and aquatic plants; Methyl mercury (one of three forms of mercury) bioaccumulates quickly and biomagnifies in higher trophic levels; kidney impairment	http://responserestorationnooa.gov /type-subtopic-enty.phy2: Record-ky%28enty-su
Lead (Pb)	Very toxic to aquatic organisms especially fish	http://responserestorationnooa.gov /type-subtopic-enty.phy2: Record-ky%28enty-su
Nickel (Ni)	Rainbow trout toxicity of mixtures of nickel, copper and zinc	http://responserestorationnooa.gov /type-subtopic-enty.phy2: Record-ky%28enty-su
Zinc (Zn)	Elevated concentration toxic to many species of algae, crustaceans and salmonids	http://responserestorationnooa.gov /type-subtopic-enty.phy2: Record-ky%28enty-su
Arsenic (As)	Very toxic to fish	http://www.ciwmb.ca.gov/ LEAAAdvisory/56/attact2.htm
Beryllium (Be)	Toxic to warm water fish	http://www.ciwmb.ca.gov/ LEAAAdvisory/56/attact2.htm
Cadmium (Cd)	Very toxic to a variety of species of fish; causes behavioral, growth and phyprological problems in aquatic life	http://www.ciwmb.ca.gov/ LEAAAdvisory/56/attact2.htm
Chromium (Cr)	Uncertain	http://www.ciwmb.ca.gov/ LEAAAdvisory/56/attact2.htm
Copper (Cu)	Elevated concentrations toxic to many algal species and younger fish	http://www.ciwmb.ca.gov/ LEAAAdvisory/56/attact2.htm

Turner (2001) has reviewed the physiological responses to heavy metals. Some species can evolve tolerance and survive in soils containing metals at normally toxic concentrations. Other important reviews are those of Fodor (2002) and Prasad and Strzalka (2002).

disposal of the metals. In contrast, incineration (Fig. 6.3) centers about burning certain hazardous wastes to destroy them. This technique is not very effective in treating metals.

Acid mine drainage results from the exposure of pyrite (iron plus sulfur, Table 6.7) to air and water. The products are sulfuric acid and iron hydroxide. Pyrite occurs in cool streams and overlying rocks. When the layers are broken, acid mine drainage can result during surface mining (Ripley et al., 1996). This drainage can reduce the pH in water resources and coat stream bottoms with iron hydroxide (http://www.epa.gov/ maia/html/AMD_issue.html). The US EPA has conducted research leading to mechanisms of decontaminating mine wastes (Table 6.8). Methods for analyzing toxic metals in environmental samples are depicted in Table 6.9.

Phytoremediation (Fig. 6.4) is the use of metal-tolerant plants to sequester pollutant levels of toxic metals (Shaw, 1989; Raskin and Ensley, 1999; Glass, 2000; Lasat, 2000; Prasad and Strzalka, 2002; Jankaite and Vesarevicius, 2005; Sripang and Muraoka, 2007) from soils and/or waterways. The targeted metals are arsenic, cadmium, chromium and lead. Harvested plant tissue

methylmercury + cysteine

CH_3Hg^+ + -S-CH$_2$-CH-COO-
                         |
                         NH$_3$+

CH$_3$Hg -S-CH$_2$-CH-COO-
                    |
                    NH$_3$+
methylmercury-cysteine complex

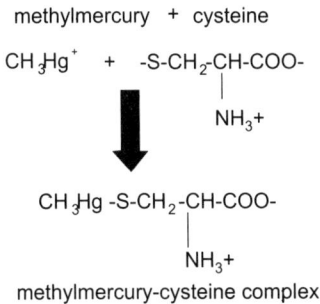

Fig. 6.1 Formation of methylmercury – cysteine complex

containing abundant accumulated metal can be dried, ashed or composted. Then, some of the metals can be reclaimed (Raskin and Ensley, 1999). Subsets of phytoremediation include: phytoextraction (the transport and concentration of metals form soils to above ground shoots are accomplished by high biomass-metal accumulation plant soils amendments), phytostabilization (stabilization of pollutants in soils by plants to render the pollutant harmless), volatilization (extraction of volatile metals from soil and subsequent volatilization from foliage) and rhizofiltration (plants grown in aerated water, precipitate and concentrate toxic metals from effluents that are polluted).

What Are Volatile Organic Compounds (VOCs) and What Are Their Origins?

Volatile organics are those that possess a high vapor pressure and low water solubility (Bloemen, 1993; Harrison, 1995; Wang et al., 1996). They are emitted as gases from some liquids or solids (Fig. 6.7) which can promote both short and long-term adverse

Succiny CoA + Glycine

⇩ Inhibition δ-aminolevulinate synthase

δ-Aminolevulinate (δ-ALT)

⇩ Inhibition δ-aminolevulinate dehydratase

Porphobilinogen

⇩ Prophobilinogen deaminase
   Uroporphyrinogen III cosynthase
Uroporphyrinogen III

⇩ Uroporphyrinogen

Coproporphyrinogen III

⇩ Possible Inhibition Coproporphyrinogen oxidase

Protoporphyrin IX

⇩ Possible Inhibition terrochelatase + Fe^{2+}

Heme

Fig. 6.2 Biochemical site of lead action. Modified from Leslie, E. http://www.unc.edu/courses/2005spring/envr/132/001/toxic%200effec

health effects. Many VOCs are utilized and generated in the production of paints, pharmaceuticals and refrigerants. In addition, VOCs can be found in dry cleaning agents, hydraulic fluids, thinners and petroleum fuels (Table 6.10), common landfills (Fig. 6.5), ground-water (Fig. 6.6) and sometimes soil contaminants. The VOCs' vapor pressure can cause them to vaporize and enter the atmosphere (Koppmann,2007). There, they can react with nitrogen oxides to yield ozone in sunlight.

Volatile organics can produce central nervous system depression at elevated concentrations and may be irritating to mucous membranes when inhaled (Table 6.11). The total volatile emissions of VOCs for the US from 1970–2002 are depicted in Fig. 6.7.

TABLE 6.7 Some Sulfide Minerals[a]

Element	Mineral	Chemical Formulae
Ag	Acanthite	Ag_2S
Cu	Chalcocite	Cu_2S
Fe	Arsenopyrite	FeAsS
	Feuoselite	$FeSe_2$
	Pyrite	FeS_2
Hg	Cinnabar	HgS
	Degassed from Earth's crust and oceans	
Ni	Millerite	Ni S
Pb	Galena	PbS
Sb	Stibnite, geothermal springs	Sb_2S_3
Zn	Sphalerite	ZnS

[a]Adapted, in part from: US-EPA 1994 Technical Document Acid Mine Driainage Predition. US EPA Official Solid Waste, Washington, DC. which cited Ferguson and Erickson (1988) as its source.

TABLE 6.8 Some USA-EPA Research Technologies for Controlling Mine Wastes

Sulfate Reducing Bacteria	Passive Remedial Technology Increases pH and alkalinity of the water to immobilize dissolved metals by precipitating them as metal sulfides or hydroxides	US-EPA
Grout Injection Into a Rock Fracture	Reduce or eliminate the flow of acid mine drainage into underground workings of an abandoned mine	US-EPA
Ecobad TMARD; Magnesium Oxide Passive Technology, Potassium Permanganate Technology, Furfuryl Alcohol Resin Sealant	Prevention of the generation of acid mine drainage from an open-pit highwall	US-EPA
Use cement and added alkali based upon natural alkaline rock	Offers time released higher doses of alkalinity and produce higher pH values for use in passive physical reactors to treat acid mine drainage, precipitate toxic metals	US-EPA

[a]Adapted from US-EPA Mine Waste Technology, Industrial Media, Sustainable Technology, Risk Manor. Kumar (2007) addressed heavy metal pollution research.

TABLE 6.9 Methods for Analysis of Toxic Metals

Method	Reference
Analytical Methods (Szpunas et al., 2004)	
Acid-volatile sulfide and simultaneously extracted metals	Allen et al. (1993)
Atomic Absorption Spectroscopy	US-EPA (1992)
Atomic Emission Spectroscopy	US-EPA (1992)
Cerenkov Counting	Sabbiani and Girardi (1997)
Cold Vapor Atomic Absorption	US EPA (1994)
Coupled Plasma-Mass Spectrometry	Carbon-hamphsive.edu/~dula/teach.html
Graphite Furnace Atomic Absorption	US EPA (1994)

(Contd.)

(Contd.)

High Resolution Gamma-Ray Spectrometry	Sabbiani and Girardi (1977)
Neutron Activation Analysis	Sabbiani and Girardi (1977)
On-line Chelation Preconcentration and	US-EPA
Stabilized Temperature Graphite Furnace Atomic Absorption	US-EPA
Gas Segmented Continuous Flow Colormetric Analysis	US-EPA
X-Ray and Optical Emission Spectroscopy Stripping Voltammetry- Anodic and Cathodic	Cowgill (1973) http://en.wikipedia.org/wiki/anodicstripping-voltammetry, same website for cathodic
Non-Analytical Methods	
Phytoremediation	Powell (1997) Westlake et al. (1998) Glass (2000) Lasat (2000) Pilon-Smits (2005)
Cell Culture	DiToro et al. (2001), Evans et al. (2003), Freshney (2005), Golan – Goldhirsh et al. (2004), Loyola-Vargas and Vasquez-Flota (2005), Mather (2006)

Quilbe et al. (2004) review chemical and biological approaches to investigate metals in agricultural runoff.

TABLE 6.10 Summary of Anthropogenic VOC emissions in the United States Total National Emissions of Volatile Organic Compounds, 1970 – 2002[a] (million short tons)

Source category	1970	1980	1990	1995	2000	2002	Percent of Total, 2002
Highway vehicles	16.91	13.87	9.39	6.75	5.33	4.54	27.5%
Off-highway	1.62	2.19	2.66	2.89	2.64	2.69	16.2%
Transportation total	18.53	16.06	12.05	9.64	7.97	7.23	43.7%
Stationary fuel combustion total	0.72	1.05	1.01	1.07	1.18	1.01	6.1%
Industrial processes total	12.33	12.10	9.01	9.71	10.21	6.96	42.1%
Waste disposal and recycling total	1.98	0.76	0.99	1.07	0.42	0.46	2.8%
Miscellaneous total	1.10	1.13	1.06	0.55	0.73	0.88	5.3%
Total of all sources	**34.66**	**31.11**	**24.11**	**22.04**	**20.51**	**16.54**	**100.0%**

Source: U.S. Environmental Protection Agency, National Emission Inventory Air Pollutant Emission Trends website *www.epa.gov/ttn/chief/trends*.
(Additional resources: *www.epa.gov/oar/oaqps*)
[a]The sum of subcategories may not equal total due to rounding. The EPA's definition of volatile organic compounds excludes methane, ethane and certain other nonphotochemically reactive organic compounds.

TABLE 6.11 General Biological Effects of Volatile Organics

Effect	Reference
At high exposure levels, many VOCs can produce central nervous depression	http://www.bal.NCSU.edu/programs/extension/publicat/wgrom/g473_5.html
All VOCs can be irritating upon contact with skin or mucous membranes when inhaled	http://www.bal.NCSU.edu/programs/extension/publicat/wgrom/g473_5.html

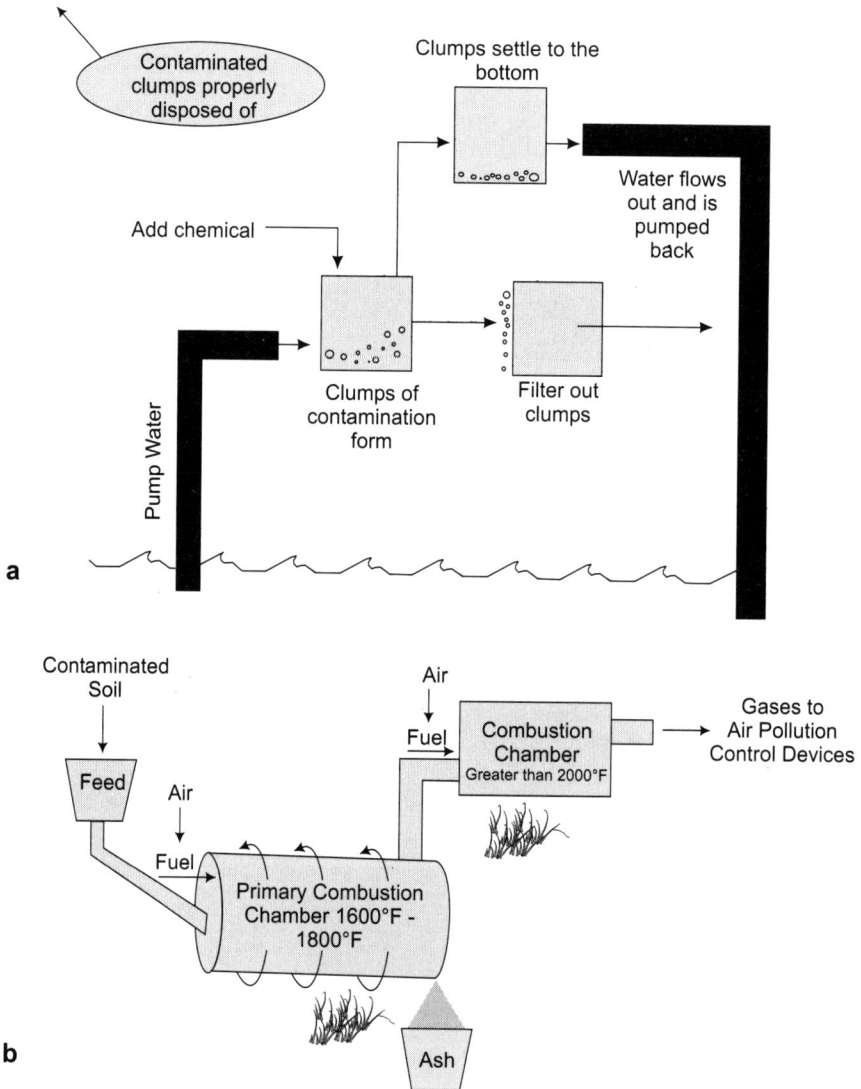

Fig. 6.3 Precipitation Process (a) From: *http://www.epa.gov/superfund/students/clas_act/hoz_act/hoz_ed/ff08.pdf*
Incineration Process (b) From: Source as above

Both analytical methods and research activities for VOCs have been discussed by Bloemen (1993). Table 6.12 presents some analytical methods for VOCs.

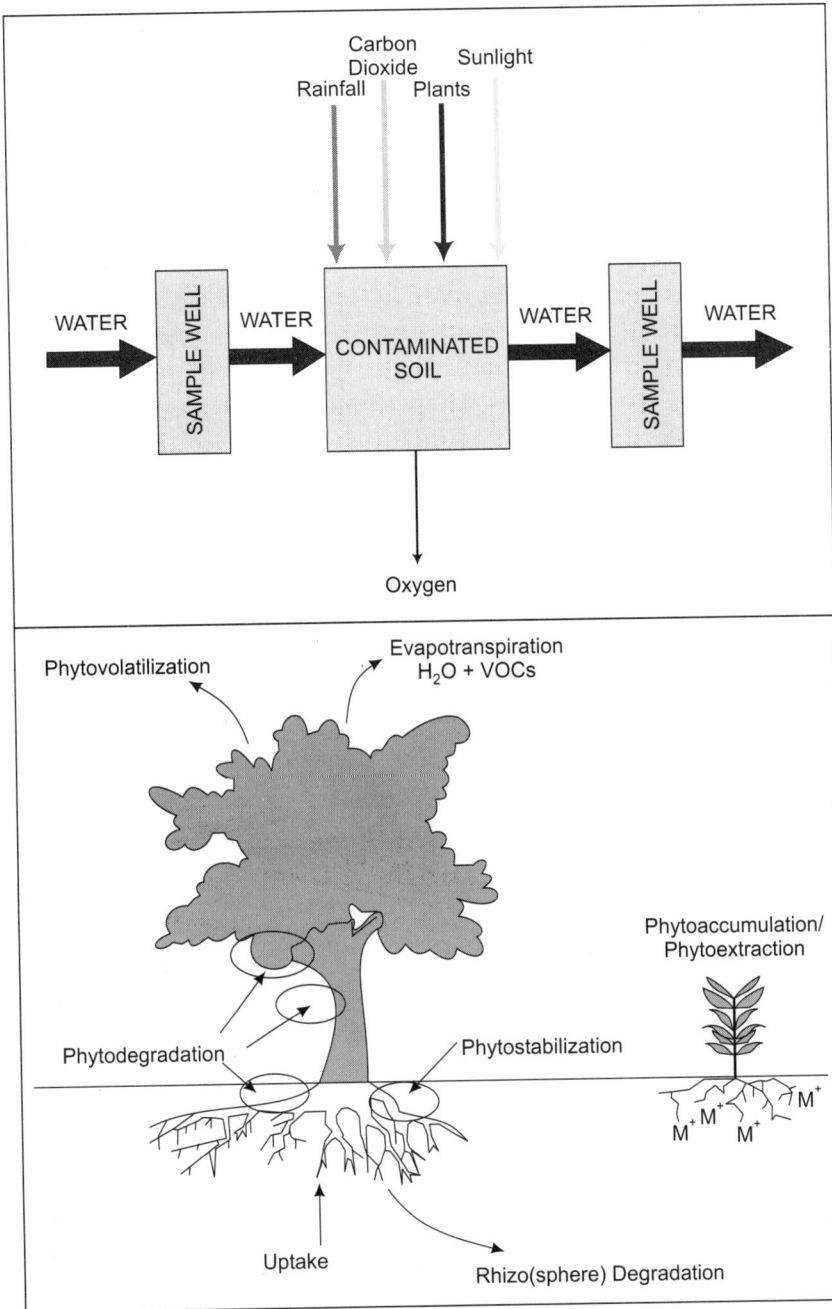

Fig. 6.4 Diagrams of the Phytoremediation Process for Soils. (Top) From: Thiyagarajan, G. and Umadevi, R. *www.techno-preneur.net/sciencetechmag/sept06/phytoremediation.pdf* (Bottom) From: *http:// enviro.nfexc.navy.mil/scripst/webobjects.eye/evbwebwoa/3/wa/determinePageTypeCallDA?pagesho.* A diagram of phytoremediation from water can be found at *http://www.rpi.edu/dept/chem.-eng/Biotech-Environ/MISC/ phytorem.html.*

When landfill is full, layers of soil and clay seal in trash.

topsoil
sand
clay
garbage

Wells and probes to detect leachate or methane leaks outside landfill.

Pipes collect explosive methane gas, used as fuel to generate electricity

Leachate pumped up to storage tank for safe disposal.

Cutaway view of a modern landfill designed to prevent the two main hazard of the dump: explosions or fires caused by methane gas, and leakage of rainwater mixed with dangerous chemicals (or leachate).

Garbage
sand
synthetic liner
sand
clay
subsoil

Clay and plastic lining to prevent leaks; pipes collect leachate from bottom of landfill.

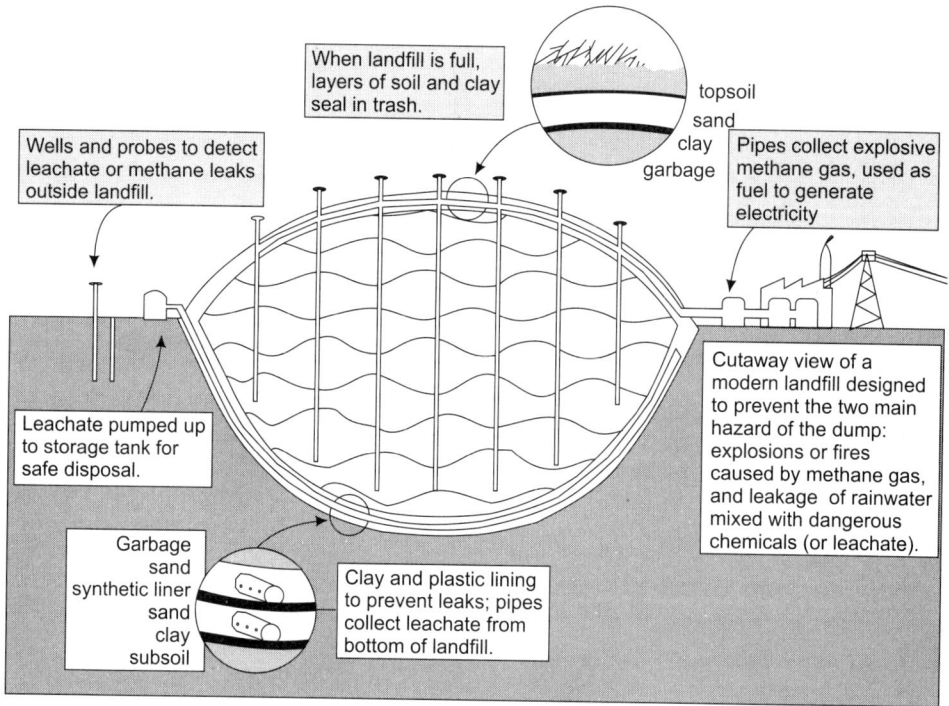

Fig. 6.5 Diagram of Modern Landfill
From: *http://www.epa.gov/superfund/students/clas-act/haz-ed/ff06.pdf*
Another Diagram can be found at *http://www.ccc.gov.N2/publications/Effective Waste Management/Uni.*

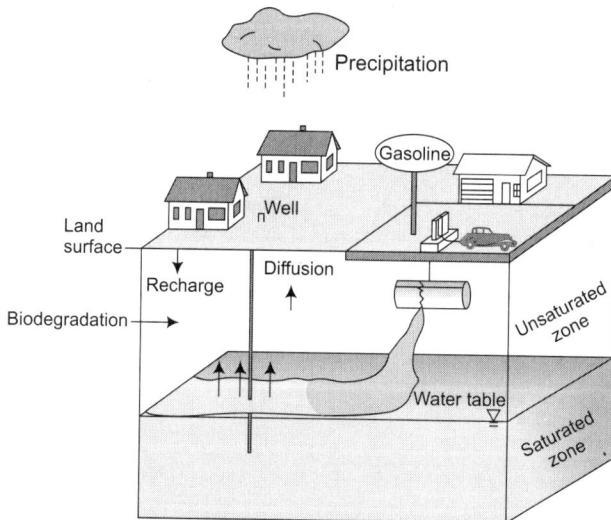

Precipitation

Gasoline

Land surface

Well

Diffusion

Recharge

Biodegradation

Unsaturated zone

Water table

Saturated zone

Fig. 6.6 Leakage of gasoline from an underground storage tank. The water table is the top of the saturated zone. A system for monitoring gas leakage from underground storage tanks can be seen at *http:// www.dec.state.ak.us/SPAR/ipp/images/tanks/musts01.gif.*

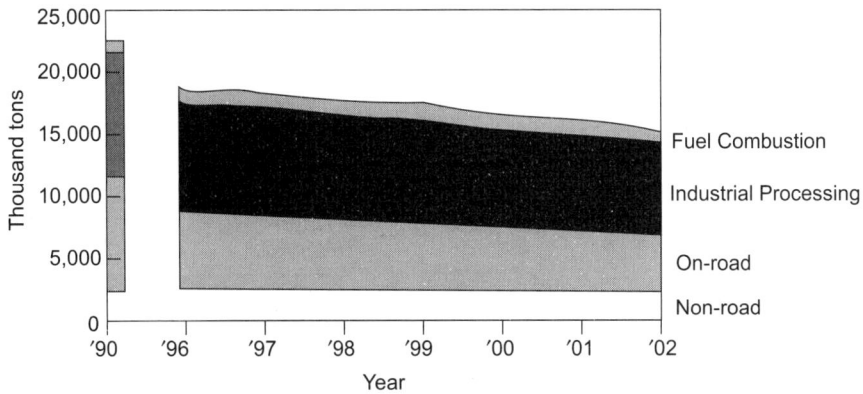

Fig. 6.7 US National VOC Emissions From: US-EPA Airtrends (2006).

| TABLE 6.12 | Analytical Methods for VOCS |

Method/Application	Description	Reference
GC/MS Useful for detecting and monitoring alkanes, alkenes, terpenes	GC separates individual components via vapour pressures and solubility; MS identifies eluted components by ionized molecular fragments	Wagner (1998)
Photoionization Detectors Detect and monitor VOCs such as benzene and toluene	Absorption of energy from a photo-ionization detector; sample gas becomes excited and its molecular content is charged; gas loses an electron and is a positive ion	*http://www.approvedgas masks.com/.Sirius-PID Whitepaper.pdf*
Quadrupole Ion Trap MS Monitor trace levels of VOCs in air (Fig. 6.8)	Confinement of ions and the measurement of their mass/charge ratios	(Palmer et al., 2000)

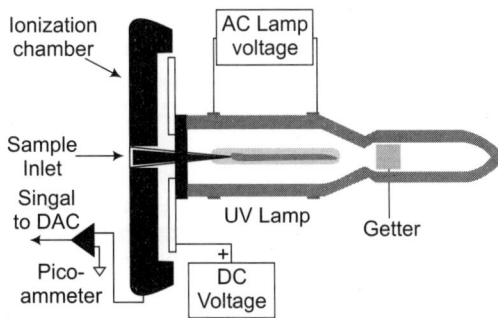

a

(Contd.)

(Contd.)

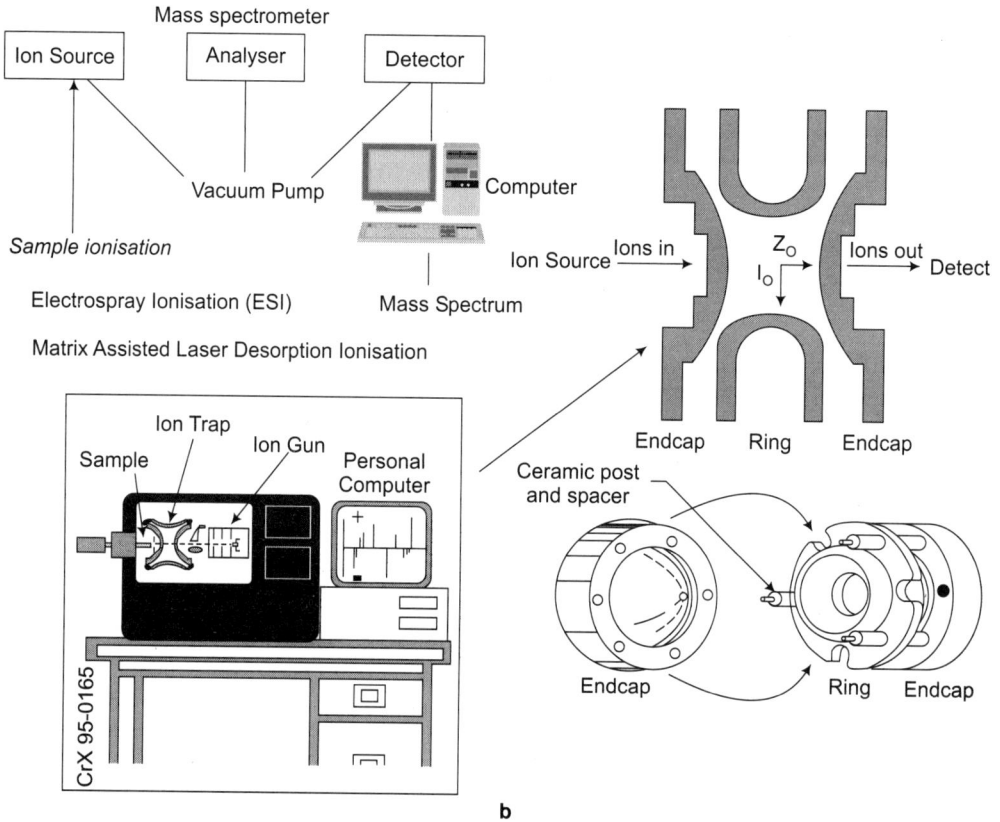

Fig. 6.8 **(a)** Diagram of a mass spectrometer and **(b)** Quadrupole ion trap mass spectrometer. From: *http://www.currentseparations.com/issues/16-3.pdf.* and *http://www.external.ameslab.gov/cmst/CMSTSite/publications/tech_summ_96/cmst5-6pdf.*

References

Adriano, D.C. 2001. *Trace Elements in the Terrestrial Environment. Biogeochemistry. Bioavailability and Risks of Metals.* Springer Verlag, New York, NY, USA.

Allen, H.E., Fu, G. and Deng, B. 1993. Analysis of acid – volatile sulfide (AVS) and simultaneously extracted materials (SEM) for the estimation of potential toxicity in aquatic sediments. Enviro. Toxicol. Chem. 12: 1441-1453.

Allen, H.E. 2002. *Bioavailibity of Metals in Terrestrial Ecosystems. Importance of Patitioning for Bioavalibility to Invertebrates, Micorbes and Plants.* SETAC Foundation, Pensacola, FL, USA.

Antonovics, J. Bradshaw, A.D. and Turner, R.G. 1971. Heavy metal tolerance in plants. . *Ecol. Res.* 7: 1-85.

Basta, N.T., Ryan, J.A. and Chaney, R.L. 2005. Trace element chemistry in residually – treated soil. *J. Environ. Qual.* 34:49-63.

Bloemen, H.J. 1993. *Chemistry and Analysis of Volatile Organic Compounds in the Environment.* Springer, The Netherlands.

Brewer, K. 2006. Basic chemical principles. In: *Plant Cell Biology* (Dashek, W.V and Harrison, M., eds.). Science Publishers, Enfield, NH, USA. pp. 17-32.

Bryan, G.W. and Langston, W.J. 1992. Bioavailability, accumulation and effects of heavy metals in sediments with special reference to United Kingdom estuaries. *Environ. Pollut.* 76:89-131.

Butler, A. 1998. Acquisition and utilization of transition metal ions by marine organisms. *Science* 281: 207-209.

Castillo-Duran, C. and Cassorla, F. 1999. Trace minerals in human growth and development. *J. Pediat. Endocrinol.* 12:589-601.

Clarkson, T. 1995. Health effects of metals. A role for evolution? *Environ. Health Perspect.* 103: 9-12.

Compton, R.G. and Banks, C.E. 2007. *Understanding Voltammetry*. World Scientific, Singapore, Mallaysia.

Cowgill, U.M. 1973. Determination of all detectable elements in the aquatic plants of Linsley Pond and Cedar Lake (North Bearford Connecticut) by x-ray emission and optical emission spectroscopy. *Appl. Spectroscopy* 27:5-9.

Crompton, T.R. 1998. *Occurrence and Analysis of Organometallic Compounds in the Environment.* Wiley, Chichester, England.

DiToro, D.M., Allen, H.E., Bergman, H.L., Meyer, J.S., Poguin, P.R. amd Santore, R.L. 2001. Biotic ligand model of the acute toxicity of metals. Technical Basis. *Enviro. Toxicol. Chem.* 20: 2383-2396.

Evans, D.E., Coleman, J. and Kearns,. 2003. *Plant Cell Culture: The Basics.* BIOS Scientific Publishing, London, England.

Environmental Monitoring Systems 1991 and 1994. Supplement Laboratory. *Methods For the Determination of Metals in Environmental Samples.* USEPA, Cincinnati, OH, USA.

Fodor, F. 2002. Physiological responses of vascular plants to heavy metals. In: *Physiology and Biochemistry of Metal Toxicity and Tolerance in Plants.* (Prasad, M.N. and Strzalka, K., eds.) CPL Press, Newbury, Berks, UK. pp. 149-177.

Freshney, R.I. 2005. *Cultures of Animal Cells. A Manual of Basic Technique.* Wiley, New York, NY, USA.

Glass, D.J. 2000. Economic potential of phytoremediation. In: *Phytoremediation of Toxic Metals: Using Plants to Clean-up the Environment.* (Ensley, B. and Raskin, I., eds.). Wiley, New York, NY, USA. pp. 15-31.

Golan-Goldhirsh, A., Barazani, D., Nepouim, A., Soudek, P., Smreck, S., Dufkova, L., Krehova, S., Yrjala, K., Schroder, P. and Vanek, T. 2004. Plant response to heavy metals and organic pollutants in cell culture and at whole plant level. *J. Soils Sediments.* 4: 133-140.

Hall, J.L. 2002. Cellular mechanisms for heavy metal detoxification and tolerance. *J. Exptl. Bot.* 366:1-11.

Harrison, H. 1995. *Volatile Organic Compounds.* Royal Society of Chemistry, Cambridge, UK.

Hill, M.K. 2004. *Understanding Environmental Pollution.* Cambridge Univ. Press, Cambridge, England.

Hu, H. 2002. Human health metals and heavy metals exposure. www.med.harvard.edu/clge/course/toxic/heavy/mccally.pdf

Jankaite, A and Vesarevicius, S. 2005. Remediation technologies for soils contaminated with heavy metals. *J. Environmental Engineering and Landscape Management* 13: 109a – 113a.

Kabata-Pendias, A. 2000. *Trace Elements in Soils and Plants.* CRC Press, Boca Raton, FL, USA.

Koppmann, R. 2007. *Volatile Organic Compounds in the Atmosphere.* Blackwell, Oxford, England.

Kumar, A. 2007. *Heavy Metal Pollution Research.* Daya Publishing House, New Delhi, India.

Langston, W.J. 1986. Metals in sediments and benthic organisms in the Mersey estuary. *Est. Costal Shelf Sci.* 15: 255-266.

Langston, W.J. and Bebianno, M.J. 1998. *Metal Metabolism in Aquatic Environments.* Springer, New York, NY, USA.

Langston, A. and Turner, D.R. 1995. *Metal Speciation and Bioavailability in Aquatic Systems.* Wiley, New York, NY, USA.

Lasat, M.M. 2000. The use of plants for the removal of toxic metals from contaminated soils. USEPA, Washington, DC, USA.

Logan, T.I. and Chaney, R.L. 1983. Utilization of municipal wastewater and sludge on land-metals In: *Proc. Workshop on Utilization and Municipal Wastewater and Sludge on Land.* Univ. of California at Riverside, Riverside, CA, USA.

Loyola-Vargas,V.M. and Vasquez-Flota,F. 2005. *Plant Cell Culture.* Humana Press, Totowa, NJ, USA.

Malmstroem, S. 2006. Movement of molecules across membranes. In: *Plant Cell Biology.* (Dashek, W.V. and Harrison, M., eds.). Science Publishers, Enfield, NH, USA. pp. 77-106.

Market, B. and Friese, K. 2000. *Trace Elements: Their Distribution and Effects in the Environment.* Elsevier, New York, NY, USA.

Mason, C.F. 1996. *Biology of Freshwater Pollution.* 3rd ed. Longman, Upper Saddle River, NJ, USA.

Mather, J.P. 2006. *Animal Cell Culture Methods.* Elsevier, New York, NY, USA.

Nriagu, J.O. 1979. Global inventory of natural and anthropogenic emissions of trace metals to the atmosphere. *Nature* 279: 409-411.

Pacyna, J.M. and Pacyna, E.G. 2001. An assessment of global and regional emissions of trace metals to the atmosphere from anthropogenic sources worldwide. *Env. Rev.* 9: 269-298.

Palmer, P.T., Remigi, C. and Karr, D. 2000. Evaluation of two different direct-sampling ion-trap mass spectrometry methods for monitoring halocarbon compounds in air. *Field Anal. Chem. Tech.* 4: 14-30

Pilon-Smits, E. 2005. Phytoremediation. *Annu. Rev. Plant Biol.* 56: 15-39.

Powell, R. 1997. The use of vascular plants as field monitors. In: *Plants for Environmental Studies.* (Wang, W., Lower, W.R., Gorsuch, J.W. and Hughes, J.G., eds.). Lewis Publishers, Boca Raton, FL, USA. pp. 335-365.

Prasad, M.N.V. and Strzalka, K. 2002. *Physiology and Biochemistry of Metal Toxicity and Tolerance in Plants.* CPL Press, Newbury, Berks, UK.

Quilbe, R., Pieri, E., Wicherek, S., Dugas, N., Tasteyre, A., Thomas, Y. and Oudent, J.P.

2004. Combinatory chemical and biological approaches to investigate metal elements in agricultural runoff water. *J. Environ. Qual.* 33:149-153.

Raskin, I and Ensley, B.D. 1999. *Phytoremediation of Toxic Metals Using Plants to Clean Up the Environment.* Wiley-Interscience, New York, NY, USA.

Ripley, E.A. Redmann, R.E. and Crowder, A.A. 1996. *Environmental Effects of Mining.* St. Lucie Press, Boca Raton, FL, USA.

Ross, S.M. 2001. *Toxic Metals in Soil – Plant Systems.* Wiley, New York, NY, USA.

Sabbiani, E. and Girardi, F. 1997. Metallobiochemistry of heavy metal pollution: nuclear and radiochemical techniques for long term – low level exposure (LLE) experiments. *Sci. Total Environ.* 7:145-179.

Sadiq, M. 1992. *Toxic Metal Chemistry in Marine Environments.* Marcel Dekker, New York, NY, USA.

Shaffer, M. 2001. *Waste Lands. The Threat of Toxic Fertilizer. http://www.pirg.org/toxics/reports/wastelands.*

Shaw, A.J. 1989. *Heavy Metal Tolerance in Plants.* CRC Press, Boca Raton, FL, USA.

Shukla, G.S. and Singhal, R.L. 1984. The present status of biological effects of toxic metals in the environment: lead, cadmium and manganese. *J. Physiol. Pharmacol.* 62:1015-103

Shkolink, M. Ya. 1984. *Trace Elements in Plants.* Elsevier, Amsterdam, Holland.

Siegel, F.R. 2002. *Environmental Geochemistry of Potentially Toxic Metals.* Springer, New York, NY, USA.

Singh, S.N. and Tripathi, R. 2007. 'Environmental Bioremediation Technologies.' Springer, Berlin, Germany.

Smith, S.R. 1996. *Agricultural Recycling of Sewage Sludge and the Environment.* CABI Publishing, Wallingford, England.

Sripang, R. and Muraoka, Y. 2007. Accumulation and detoxification of metals by plants and microbes. In: *Environmental Bioremediation Technologies.* (Singh, S.N. and Tripathi, R.D., eds.) Springer, Berlin, Germany. pp. 77-100.

Szpunas, J.; Bouyssiere, B. and Lobinski, R. 2004. Advances in analytical methods for speciation of trace elements in the environment. In: *Organic Metal and Metalloid Species in the Environment*. (Hirner, A.V. and Emons, H., eds.). Springer, Berlin Germany. pp. 17-40.

Taylor, E.W. 1996. *Toxicology of Aquatic Pollution: Physiological, Molecular and Cellular Approaches*. Cambridge Univ. Press, Cambridge, England.

Tessier, A. and Turner, D.R. 1995. *Metal Speciation and Bioavailability in Aquatic Systems*. Wiley, Chichester, UK.

Turner, A.P. 2001. The responses of plants to heavy metals. In: *Toxic Metals in Soil-Plant Systems*. (Ross, S.M., ed.). Wiley, New York, NY, USA. pp. 153-187.

US EPA. 1992. *Methods for the Determination of Metals in Environmental Samples*. CRC Press, Boca Raton, FL, USA.

Wagner, R.E. 1998. *Guide to Environmental Analytical Methods*, 4th Ed. Genian Publishing Corporation, Schenectady, NY, USA.

Wang, W., Schnoor, J.L. and Doi, J. 1996. *Volatile Organic Compounds in the Environment*. ASTM International, Philadelphia, PA, USA.

Westlake, D.F., Kvet, J, and Szcleposki, A. 1998. *The Production Ecology of Wetlands*. Cambridge Univ. Press, Cambridge, UK.

CHAPTER 7

Pesticide Pollution

What are Pesticides?

Pesticides are synthetic or natural chemical compounds which kill unwanted plants, animals, microbes or fungi (Milne, 1994; Ritter et al., 1995; Ware, 1999; Clark and Ohkawa, 2005; Greene and Pohanish, 2005 and Manahan, 2005). The pesticides include: Algaecides (kill algae), insecticides (kill insects), herbicides (kill weeds), fungicides (kill fungi and molds), mitocides (kill mites) and rodenticides (kill rodents).

The World Health Organization (WHO) has classified pesticides based upon their toxicities and hazards (Table 7.1). Pesticides can also be classified by target (Table 7.2) or chemistry (Buchel, 1983; Pesticides Biochemistry online) and mode of action (Stenersen, 2004) (Table 7.3). Algaecides are, in the main, copper compounds which are commonly employed in swimming pools (Table 7.4). At certain levels, copper can be toxic (see chapter on 'Metals'). Pesticides can significantly affect the environment (Stephenson and Solomon, 1993).

Two of the common classes of fungicides (US Government office, 1990; Hewitt, 1998) are the phthalimides and dithiocarbamates (Table 7.3). Chloranil is an example of a fungicide (Fig. 7.1).

Herbicides are employed to control weeds (Masuinas, 2005) in agriculture and rangeland, lakes, ponds and forests (Hatzios, 1982). Two of the most widely used herbicides are 2, 4-D and 2, 4, 5 – T (Fig. 7.1). The top ten herbicides employed in the USA in 1992 were butylate, EPTc, glyphosphate, pendimethalien, trifluralin, cyanazine, 2, 4-dichloropheoxyacetic acid, alachlor, metolachlor and atrazine (Gianessi and Anderson, 1995).

Insecticides currently in use, are synthetic organic compounds and may be nerve poisons (Dev and Koul, 1997; Roberts, 1999). Although there are many insecticides, three main groups are chlorinated hydrocarbons (organochlorines) organophosphorous and carbamates (Fig. 7.1). The organophosphorous compounds are the most toxic insecticides.

There are both fumigant and non-fumigant pesticides. The former are gases which volatilize quickly. Their movement through the soil as gases is affected by temperature, moisture and soil texture. Telone and Vapan, examples of pesticides, act by interfering with life processes after penetrating the cuticle. In contrast, non-fumigant nematicides are organophosphates and carbamates which are cholinesterase

TABLE 7.1 WHO Classification of Pesticides Based on Toxicity and Hazard

Class		Oral Solids[a]	Oral Liquids[a]	Dermal Solids[a]	Dermal Liquids[a]
Ia	Extremely hazardous	5 or less	20 or less	10 or less	40 or less
Ib	Highly hazardous	5 – 50	20 – 200	10 – 100	40 – 400
II	Moderately hazardous	50 – 500	200 – 2000	100 – 1000	400 – 4000
III	Slightly hazardous	Over 500	Over 2000	Over 1000	Over 4000

LD_{50} for the rat (mg/kg body weight)

[a] The terms "solids" and "liquids" refer to the physical state of the active ingredient being classified. From: 1975 WHO Chronicle 29: 397-401. Updated in 2004.

TABLE 7.2 Types of Pesticides: There are many classes of pesticides, each designed to destroy or repel certain species.

Type	Targets
Insecticides	Flying and crawling insects
Herbicides	Undesirable plants/weeds
Rodenticides	Mice, rats and other rodents
Fungicides	Fungi that cause plant disease/wood rot, etc.
Nematicides	Invertebrates (worms)
Fumigants	Insects/fungi
Antimicrobials	Microorganisms such as bacteria, molds fungi
Biopesticides	Natural materials such as animals, plants, bacteria, and certain minerals that target a variety of pests
Plant or insect growth regulators	Plant (accelerate or retard the rate of growth of a plant), insect (affect the growth of insects)

[a] http://www.epa.gov/pesticides/factsheets/registration.htm

TABLE 7.3 Herbicide Groups Based on Sites of Action

This list is based on the Herbicide Classification of the Weed Science Society of America (Weed Technology, 1997, 11:384-393). Microbial herbicides are not included.a.)

Group	Site of Action	Chemical Family
1	Inhibitors of acetyl CoA carboxylase ACCase	Aryloxyphenoxy propionates / Cyclohexanediones
2	Inhibition of acetolactate synthase (ALS) and also called acetohydroxyacid synthase (AHAS)	Sulfonylureas
3	Microtubule assembly inhibitors	Dinitroanilines / Pyridazine
4	Synthetic auxins (action like indoleacetic acid)	Phenoxys / Benzoic acids / Carboxylic acids / Quinoline carboxylic acid / Semicarbazone

(Contd.)

(Contd.)

5	Inhibitors of photosynthesis at photosystem II Site A	Triazines Triazinones Uracils Pyridazinone Phenyl-carbamates
6	Similar to group 5, but different binding behavior	Nitriles Benzothiadiazoles Phenyl-pyridazine
7	Inhibitors of photosynthesis at photosystem II Site B	Ureas Amide
8	Inhibition of lipid synthesis, not ACCase inhibition	Thiocarbamates None
9	Inhibitors of 5-enolpyruvylshikimate-3-ogisogate (EPSP) synthase	None
10	Inhibitors of glutamine synthetase	None
11	Bleaching: Inhibitors of carotenoid biosynthesis (unknown target)	Triazole
12	Bleaching: Inhibitors of carotenoid biosynthesis at the phytoene desaturase step (PDS)	Pyridazinone Nicotinanilide Others
13	Bleaching: Inhibition of all diterpenes	Isoxazolidinone
14	Inhibitors of protoporphyrinogen oxidase (PPO)	Diphenylethers N-phenylphthalimides * Oxadiazole Triazolinone *
15	Unknown	Chloroacetamides Acetamides Oxyacetamides *
16	Unknown	Benzofuran
17	Unknown	Organoarsenicals
18	Inhibits dihydropteroate (DHP) synthase step	Carbamate
19	Inhibits indoleacetic acid action	Phthalamate
20	Inhibits cell wall synthesis Site A	Nitrile
21	Inhibits cell wall synthesis Site B	Benzamide
22	Photo system I-electron diverters	Bipyridyliums
23	Inhibitors of mitosis	Carbamates
24	Uncoupling membrane disruptors	Dinitrophenol
25	Unknown	Arylaminopropionic acid
26	Unknown	Various
27	Inhibition of 4-hydroxyphenyl-pyruvateddioxygenase (4-HPPD)	Benzoylisoxazole * Isoxazole * Pyrazole * Triketone *

*Not registered in Canada at the time of publication.

Fungicide/Bactericide Groups Based on Sites of Action

Group	Site of Action	Chemical Family
1	Inhibition of tubulin formation	Benzimidazole
2	Affect cell division, deoxyribonucleic acid (DNA) and ribonucleic acid (RNA) synthesis, and metabolism	Dicarboximide

(Contd.)

(Contd.)

3	Demethylation Inhibitor (DMI): Inhibition of demethylation in sterol biosynthesis	Imidazoles * Piperazine Pyridine * Pyrimidines * Triazoles (includes conazoles)
4	Phenylamides affect RNA synthesis	Acylamines Oxazolidinones * Butyrolactones *
5	Morpholines inhibition of an isomerase in sterol biosynthesis	Morpholines Piperidine * Spiroketalamine *
6	Phosphorothiolate inhibition of chitin and phospholipids synthesis	Organophosphorous *
7	Mitochondrial transport chain	Anilide (Oxathiin)
8	Hydroxyprimidine	Pyrimidinol *
9	Anilinopyrimidine Inhibition of amino acid synthesis	Anilinopyrimidine
10	N-Phenyl carbamated interfere with cell division	Diethofencarb
11	Strobilurin Type Action and Resistance (STAR). Inhibit mitochondrial respiration	Strobilurin * Oxazolidinedione *
12	Phenylpyrroles	Phenylpyrroles
13	Quinolines	Quinoline
14	Aromatic hydrocarbons	Chlorophenyl Quintozene (PCNB) Thiadiazole
15	Cinnamic acids	Cinnamic acid *
16	Melanin Biosynthesis Inhibitors (MBI)	Reductase inhibitors * Dehydratase inhibitors *
17	Hydroxyanilide	Hydroxyanilide *
18	Antibiotics	Antibiotics
19	Polyoxins	Polyoxin *
20	Phenylurea	Phenylurea
21	Plant host defence inducers	Benzothiadiazole (BTH) *
U[1]	Unknown Miscellaneous	Amino acid amide * Carbamate Cyano-acetamide oxime *
M[2]	Multi-site activity	Inorganics Dithiocarbamates and relatives Phthalimide Chloronitrile Sulphamide * Guanidine Triazine Phenyl-pyridinamine * Quinoxaline

(Contd.)

(Contd.)

Fungicide Group	Site of Action
Strobiturins	Interfere with ATP synthesis (ATP is adenosine triphosphate-energy currency of the cell, Bowlby 2006)
Triazoles	Prevents Sterol Production (Sterols are significant components of cellular membranes – Dashek, 2006)
Chloronitriles	Interferes with ATP Synthesis
Carboxamides	Interferes with ATP Synthesis

[1] The unknown group, designated by symbol "U", comprises a set of miscellaneous compounds for which that biochemical mode of action may or may not be known, but are not able to be placed with certainty in any other groupings.

[2] The multi-site activity grouping, designated by symbol "M", comprises a collection of various chemicals that act as general toxophores with several sites of action. These sites may differ between group members.

*Not registered in Canada at the time of publication

Insecticide and Acaricide Groups Based on Sites of Action

1A[1]	Acetylcholinesterase inhibitors. Inhibition of the enzyme acetylcholinesterase, interrupting the transmission of nerve impulses	Carbamates
1B[1]		Organophosphates
2A[1]	Gamma-aminobutyric acid (GABA)-gated chloride channel antagonist. Interferes with GABA receptors of insect neurons leading to repetitive nervous discharges repetitive nervous discharges	Chlorinated cyclodienes Polychlorocycloalkanes
2B[1]	GABA-gated chloride channel antagonists. Interferes with GABA receptors of insects neurons, leading to repetitive nervous discharges – fiprole site	Phenylpyrazoles *
3	Sodium channel modulators. Acts as axonic poisons by interfering with the sodium channels of both the peripheral and central nervous system stimulating repetitive nervous discharges, leading to paralysis	Diphenylethanes Synethetic pyrethroids Pyrethrins
4	4 Acetylcholine receptor agonists/antagonists. Binds to nicotinic acetylcholine receptor, disrupting nerve transmission	Chloronicotines (nitroguanidines) Nicotine Cartap * Bensultap *
5	Acetylcholine receptor modulators. Alters acetylocholine receptor site and disrupts binding	Spinosyns
6	Chloride channel activators. Interferes with the GABA nerve receptor of insects.	Avermectin Milbemycin *
7	Juvenile hormone mimics (insect growth regulator). Mimic juvenile hormones, which prevent moulting from the larval to the adult stage	Juvenile hormone analogues
8A[1]	Unknown or non-specific site of action	Fumigant
8B[1]	(fumigants)	
9A[1]	Compounds of unknown or non-specific site of action (feeding disruptors)	Feeding disruptors * (pymetrozine, cryolite)
9B[1]		
10	Compounds of unknown or non-specific site of action (mite growth inhibitors)	Mite growth inhibitors (ovicide)
11	Microbial disruptors of insect mid-gut membranes (includes Cry proteins expressed in transgenic plants). Organism has protein inclusions that are released in the gut of the target pest resuting in gut paralysis and a cessation of feeding	Bt microbials (biological insecticide/larvicide)

(Contd.)

(Contd.)

12	Inhibition of oxidative phosphorylation at the site of dinitrophenol uncoupling (disrupt adenosine triphosphate (ATP) formation)	Organotin matricides
13	Uncoupler of oxidative phosphorylation (disrupt H proton gradient formation)	Pyrrole compound * (broad spectrum contact and stomach poison)
14	Inhibit magnesium-stimulated ATPase	Sulfite ester matricides
15	Inhibit chitin biosynthesis	Substituted benzoylurea
16	Inhibit chitin biosynthesis type 1 – Homopteran	Thiadizine *
17	Inhibit chitin biosynthesis type 2 - Dipteran	Triazine
18	Ecdysone agonist/disruptor. Disrupts insect molting by antagonizing the insect hormone ecdysone	Benzoic acid hydrazide Botanical * (Neem oil or azadirachtin)
19	Octopaminergic agonist	Trizapentadiene
20	Site II electron transport inhibitors	None
21	Site I electron transport inhibitors	Botanical pyridazinone

[1] Other resistance mechanisms that are not linked to site of action, such as enhanced metabolism, are common for this group of chemicals. All members of this class may not have developed significant crossresistance. When only this group of products is available, alternation of compounds form subgroup A and subgroup B are recommended.

*Not registered in Canada at the time of publication a.) Mallory-Smith, C.A. and Retzinger, E.J. 2003. Revised classification of herbicides by site of action for weed resistance management strategies. Weed Technology 17:605-619 have updated the classification.

TABLE 7.4 Some Copper Compounds Used as Algaecides

Compound
Copper Sulfate
Copper II Alkanolamine Complex
Copper Ethylenediamine Complex
Copper Triethalanolamine Complex
Copper Citrate

inhibitors. The soil distribution of non-fumigant pesticides is dependent upon an aqueous environment as they are water-soluble. Thus, the 'killing attributes' of these nematicides is affected by the availability of soil water. Figure 7.1 depicts two commonly used nematicides, Avermectin and Benomyl.

Microbial pesticides possess the ability to control a variety of pests (Clark and Ohkawa, 2005). However the ingredients are relatively specific for target pests. *Bacillus thuringiensis* is the most widely utilized microbial pesticide. There is a number of

B.thuringiensis strains which synthesize crystalline proteins that kill ceratain insect larvae by binding to larval receptors thereby causing starvation by disruption of intestinal cells.

An example of plant-incorporated protectants is the incorporation of a gene responsible for the synthesis of a *B. thuringiensis* pesticidal protein into a plant's genome (Lopez et al., 2005). The result is that the plant itself provides a pest-destroying substance. Because the *B. thuringiensis* proteins act upon a narrow range of insects (catterpillars, mosquito larvae and beetle larvae) they are less likely to destroy non-target species. This offers an environmental advantage over many conventional synthetic insecticides. Other advantages of biopesticides include effectiveness in small quantities, rapid decomposition and less toxicity than conventional pesticides and often decompose quickly. However, thorough documentation of the possible *B. thuringiensis* human health effects needs to

INSECTICIDES

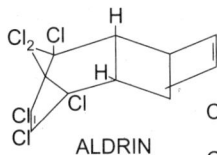

ALDRIN

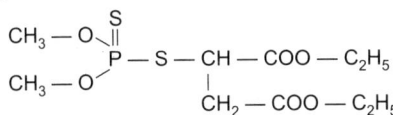

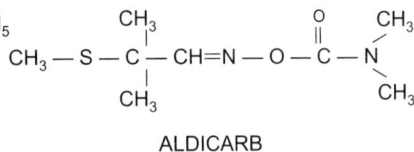

ALDICARB

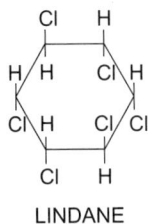

LINDANE

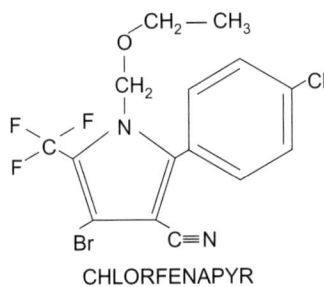

PARATHION

MITICIDE

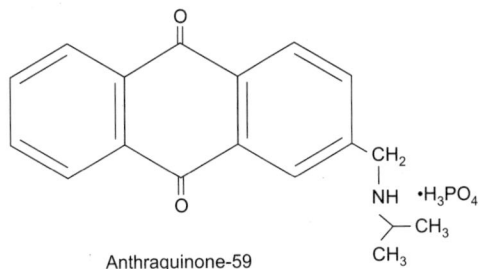

CHLORFENAPYR

ALGAECIDE

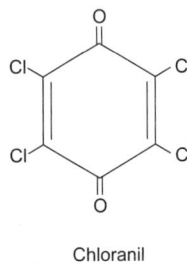

Anthraquinone-59

FUNGICIDE

Chloranil

(Contd.)

(Contd.)

HERBICIDES

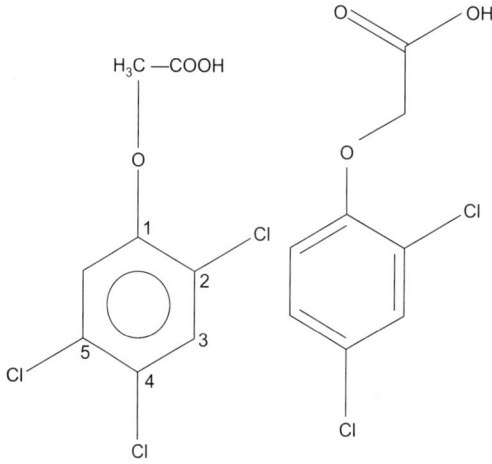

2,4,5 T 2,4 Dichlorophenoxyacetic Acid

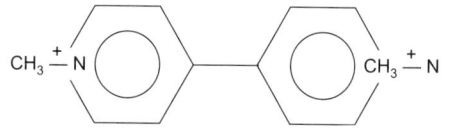

Paraquat

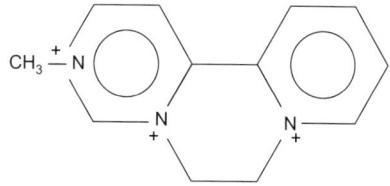

Diquat

NEMATICIDES

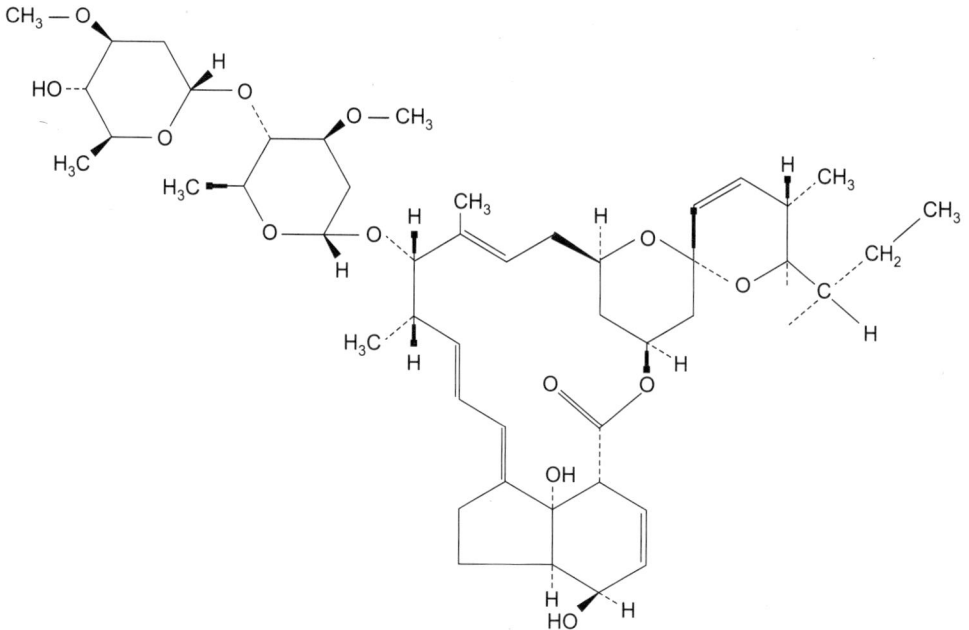

AVERMECTIN

(Contd.)

(Contd.)

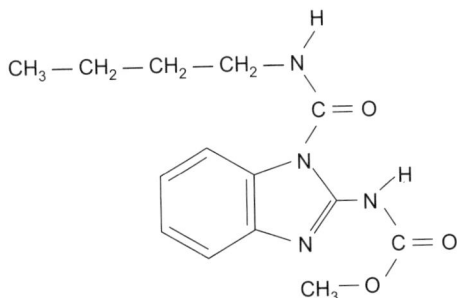

BENOMYL

RODENTICIDES

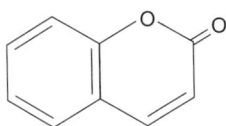

COUMARIN

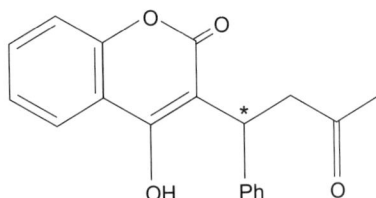

WARFARIN

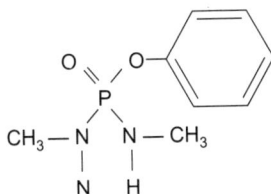

Phenyl N,N'- Dimethylphoshorodiamidate

Natural Pesticides From Plants

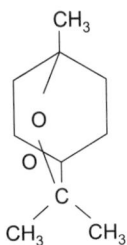

Cineole

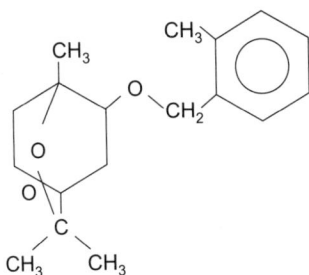

Cinmethylin

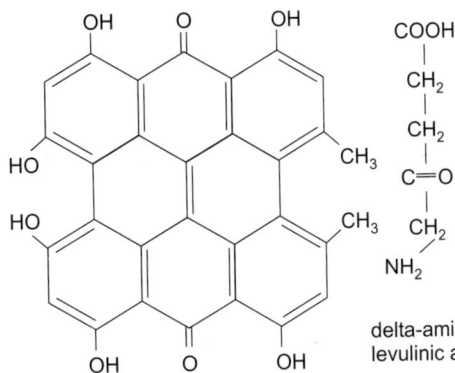

delta-amino-levulinic acid

Hypericn

(Contd.)

(Contd.)

Camphene

Nicotine

Anabasine

Rotenore

Pisatin

Juglore

Strychine

α-terhienyl

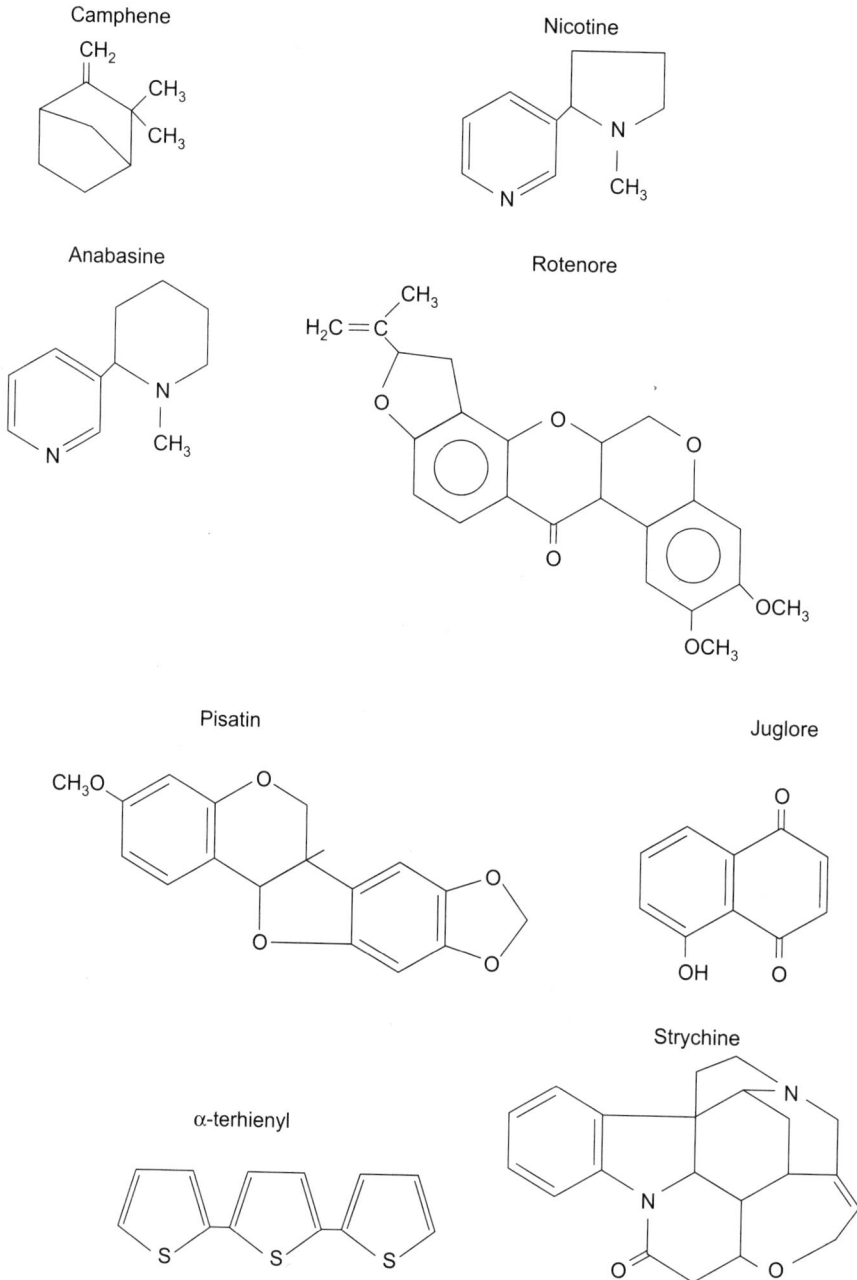

Fig. 7.1 Chemistry of Pesticides. Natural pesticides from: *http://www.hort.purdue.edu/Newcrop/proceeding 1990/vl-5//html.*

be accomplished since some individuals have experienced respiratory, eye and skin irritation. In addition, there are ecological concerns as *B. thuringiensis* can decrease the number of moth and butterfly species. Biochemical pesticides regulate pests by non-toxic mechanisms. Insect sex pheromones which interfere with mating are examples of this type of pesticide. A thorough description of biopesticides can be found at *http://www.epa.gov/pesticides/ biopesticides/whatarebiopesticides.htm.*

Most rodenticides (Allen, 1997; Moorman, 1983) are highly toxic compounds (Fig. 7.1) and some inhibit blood coagulation (Tasleva, 1995). They can be harmful to children, birds and other non-target wildlife. Scavengers and predators are at risk of secondary poisoning from dead rodents (*http://www.pa_uk.org/pestnews /pm52/pm52p18.htm*). There are many rodenticides (Fig. 7.1), e.g., brodifacoum (bromodiolone) and flocomafen. These hydroxycoumarin pesticides are chemically related and are examples of the anticoagulant rodenticides. Unfortunately, the two compounds exhibit secondary poisoning and have been responsible for the deaths of birds of prey (World Health Organization, 1995). Other rodenticides are alphachloralase, aluminum phosphide, caliciferol/difenocoum, chlorophocinone and coumateetratyl.

Finally, there are plant-derived compounds which appear to possess pesticidal activity (Metcalf and Metcalf, 1992; Dev and Koul, 1997). Duke (1990) has reviewed these compounds (Fig. 7.1) which possess herbicidal, insecticidal, fungicidial, nematicidal and rodenticidal activities.

What Are the Expenditures For Pesticides?

The 1998 and 1999 estimates of USA expenditures (Pimentel, 2005) for pesticides are depicted in Table 7.5. It is apparent that the greatest expenditures are in the agricultural sector.

DISTRIBUTION OF PESTICIDES

Pesticides in the Atmosphere

The United States Geological Survey has gathered data regarding pesticides in the atmosphere for the USA. The data revealed that most of the pesticides which have been investigated were detected in either rain or snow. Furthermore, pesticides were found in the atmosphere in all regions of the USA's sampled atmosphere. The highest atmospheric concentrations occurred seasonally in high-use regions. In contrast, low levels of long-lived pesticides occurred throughout the year. (*http://ca.water.usgs. gov/pnsp/atmos*).

Pesticides can enter the atmosphere via volatilization from leaves and stems. While in the atmosphere pesticides can be transported long distances by wind (Kurtz, 1990). They may be transformed or degraded by sunlight.

Pesticides in the Terrestrial Ecosystems

The fate of pesticides in tropical ecosystems is not as well understood as that in temperate ones. Dissipation appears to occur more rapidly in the former (Racke, 2003). There are various dissipation pathways (Fig. 7.2) for applied pesticides including spray drift, surface runoff, lateral flow, sorption, retention, leaching, microbial and chemical transformation as well as plant uptake and wash-off (Hemond and Fechner-Levy, 2000; *http://www.epa.gov/appelfed/ecorisk-ders/ tolra_analysis_exp.htm*).

Phelps et al. (2002) compared laboratory and field investigations for dissipation of pesticides. While lab studies concentrate

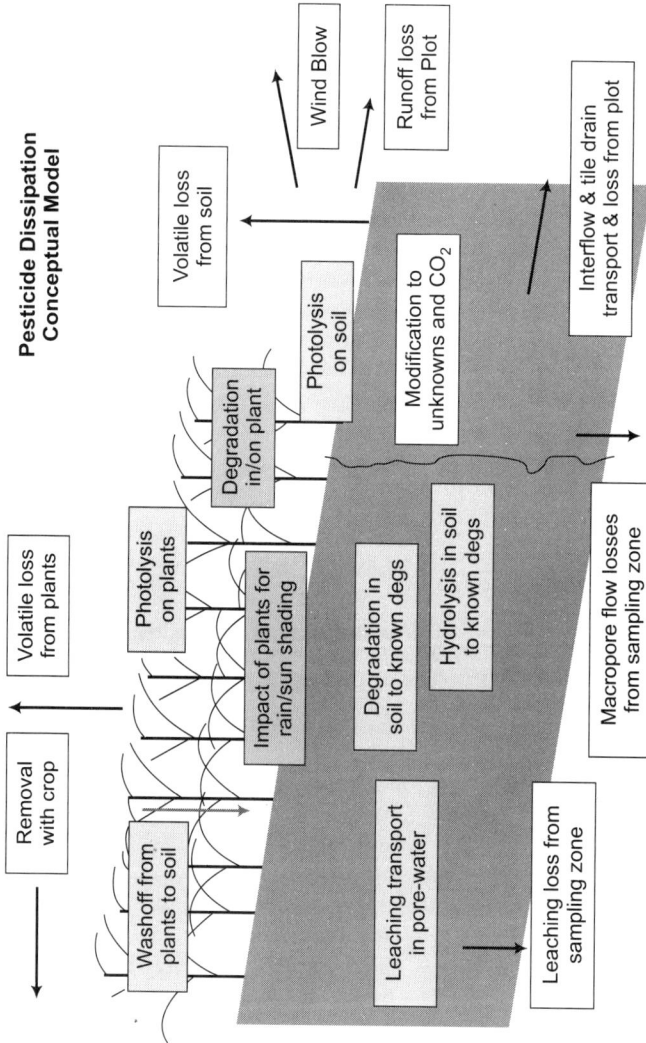

Fig. 7.2   Conceptual model for pesticide terrestrial field soil dissipation. Reprinted with permission from Paul Hendley, Syngenta Crop Protection, Inc. Terrestrial field soil dissipation in pesticide environmental risk assessment. P. Hendley pp. 45-53 in *Terrestrial Field Dissipation Studies: Purpose, Design and Interpretation.* (Eds. Arthur, E.L., Barefoot, A.C. and Clay, V.E.) ACS 2003, ACS Symposium Series 842. Washington, DC, USA. Howard (1991) has published a handbook of environmental fate of pesticides. Another website figure depicting pesticide dissipation pathways is *http://ewr.cee.vt.edu/environmental/teach/gwprimer/gov/pest.*

TABLE 7.5 U.S. User Expenditures for Pesticides[a]
By Pesticide Type and Market Sector, 1998 and 1999 Estimates

Year	Herbicides/Plant Growth Regulators		Insecticides/Miticides		Fungicides		Other (1)		Total	
Market Sector	Mil $	%	Mil $	%	Mil $	%	Mil $	%	Mil $	%
1998										
Agriculture	$5,632	11%	$1,427	50%	$695	74%	$514	68%	$8,268	72%
Ind/Comm./Gov.	$728		$428	15%	$215	23%	$77	10%	$1,445	13%
Home & Garden	$493	7%	$1,020	36%	$26	3%	$164	22%	$1,703	15%
Total	$6,853	100%	$2,872	100%	$936	100%	$755	100%	$11,416	100%
1999										
Agriculture	$5,012	79%	$1,370	45%	$660	73%	$583	70%	$7,625	68%
Ind/Comm./Gov.	$794	12%	$463	15%	$215	24%	$74	9%	$1,546	14%
Home & Garden	$562	9%	$1,213	40%	$35	4%	$174	21%	$1,984	18%
Total	$6,368	100%	$3,046	100%	$910	100%	$831	100%	$11,155	100%

*Note: Totals may not add due to rounding. Table does not cover industrial wood preservatives, specialty biocides, and chlorine/hypochlorites.

Source: EPA estimates based on Croplife America annual surveys and EPA proprietary data. See Table 5.1 to Table 5.4 for 1980-1999 estimates.

(1) "Other" includes nematicides, fumigants, rodenticides, molluscicides, aquatic and fish/bird pesticides, other miscellaneous conventional pesticides, plus other chemicals used as pesticides, e.g., sulfur and petroleum.

[a] http://www.epa.gov/oppbead1/pestsales/99pestsales/sales1999.html

on a single route (pH, photolysis, volatilization), field investigations focus on dissipation pathways such as groundwater leaching, field volatility and surface water funoff.

Pesticides in Soil

There are several parameters which can affect the persistence of a pesticide in the soil (Table 7.6). These include: 1. solubility, volatility and stability of the pesticide, 2. reaction of the pesticide with inorganic and organic soil colloids and 3. inability of micro-flora to induce enzymes capable of metabolizing a pesticide (Evans, 1986). Some microorganisms can utilize certain pesticides as an energy source and nutrients for growth (Fournier et al., 1997). The main mechanism by which soils can retain pesticides preventing the chemicals' transport

centers about absorption, i.e., accumulation of pesticide on soil particle surfaces. The soil's properties, e.g., permeability, surface charges characteristics, pH as well as organic matter and clay contents (Pepper et al., 2006) (http://soils.usda.gov). Some pesticides can behave as either cations or anions while others are neutral. Finally, some pesticides possess limited soil mobility while others exhibit moderate to high or very high mobility (Fellenberg and Wier, 2000).

Pesticides in Waterways

Gillom et al. (1999) reported that one or more pesticides occurred in nearly every stream sampled. Pesticides were detected in urban, agricultural and mixed land use for streams. However, the frequency in groundwater was less but significant (Meyer and Thurman, 1996). Pesticides can

TABLE 7.6 Summary of Pesticide Distribution Processes[a]

Process	Description
Retention (absorption)	Capacity of soil to hold or retain a pesticide on its surface. Factors which affect retention are pH, moisture content, clay content, oxide content, cation exchange capacity, specific surface area and organic matter content
Transformation	Following entrance of pesticide into the soil and groundwater, it can be degraded, transformed or stored by micro-organisms, plant and animals; Photochemical, chemical and microbial processes are the chief transformation mechanisms
Transport	A pesticide can be transported in the soil via mass flow and diffusion. Mass flow – transport of a pesticide by a flow of water Diffusion – pesticide movement from a high concentration to a lower area by random molecular movement
Plant uptake (Zablotowicz et al., 2005)	Process by which pesticides are transported into and with the plant's structure

[a]Modified from: *http://www.ewr.cee.vt.edu* (*see www.vtpp.ext.vt.edu/*)

Parameters Which Can Affect Persistence of a Pesticide in Soils

Pesticide Class	% Persistence After 15 Years	Solubility	Mobility	Soil Adsorption Behavior
Organochlorines	3 – 20	Very low	Very low	Strongly
Organophosphorous	0.1	Wide range	Low	Adsorbed

infiltrate the water table and the groundwater (Crowe and Mutch, 1994; Crowe and Booty, 1995; Lampman, 1995). The latter can penetrate wells which are used as drinking water (Prevost et al., 2005; Hamilton and Crossley, 2004; *http:ga.water.ugs.gov/edu/ pesticidesgw.html*).

To reduce well water contamination, geosynthetics (synthetic materials) can create barriers, control erosion and provide containment filtration, separation and drainage (Peggs, E.J., *www.geosynthetics. net*).

Pesticides in Plants

Following uptake of pesticides together with water and minerals, soluble pesticides can enter the xylem of the vascular tissue (Fig. 7.3) and be subsequently transported to the leaf for transpiration. Transpiration occurs via water loss as vapor through the stomata (Fig. 7.4) or the cuticle. Whereas 90% of water vapor exists through the stomata, 10% leaves through the cuticle (*http://www.dwevans.com/advancedbiology/ transpiration.html*). The movement of pesticides through the plant cuticle has been reviewed by Schreiber (2005).

There also appears to be some transport within the phloem. In addition, pesticides can migrate to other cells via plasmodesmata, i.e., channels within cell walls (Fig. 7.6). Within cells, pesticides have been reported to occur in mitochondria, chloroplasts and vacuoles (Fig. 7.5). Pesticides appear to bind to DNA preventing appropriate replication of the macromolecules (Boerth, *http://www.unmasd.edu/com-munications/articles/slowarticles.fm?a-key=480*).

How are Pesticides Metabolized?

The metabolism of pesticides by plants consists of three phases, oxidative, reductive and hydrolytic (Hatzios, 1982; Van Eerd et al., 2003; Zablotowicz, et al., 2005). Phases I

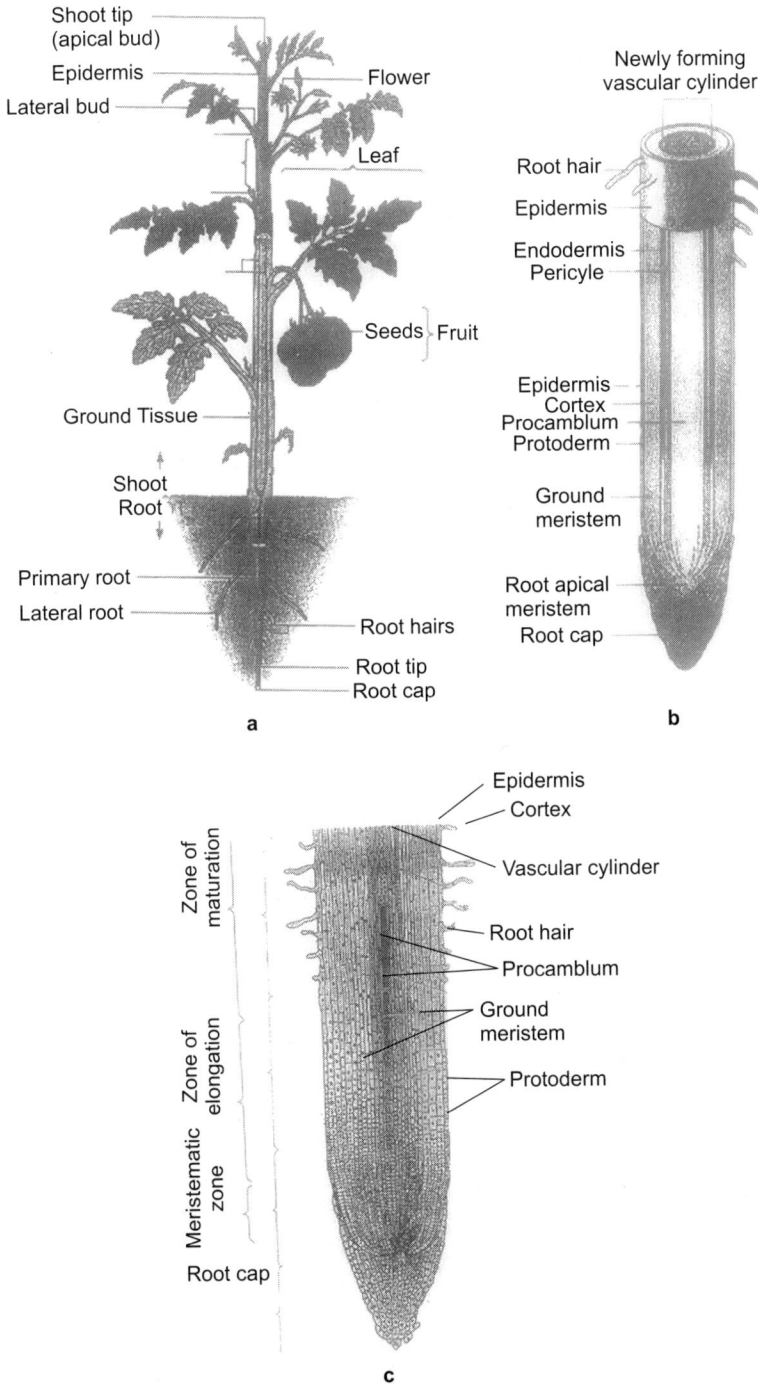

Fig. 7.3 A – C. Root Structure
a. Gross morphology, **b.** Schematic of root organization, **c.** Cellular tissue organization of root. Gregory, P. 2006. *Plant Roots: Growth, Activity and Interaction With Soils.* Blackwell Publishing, Ames, Iowa, USA.
b. From: *http://www.ua/r.edu/botany/root-anatomy.jpg.*

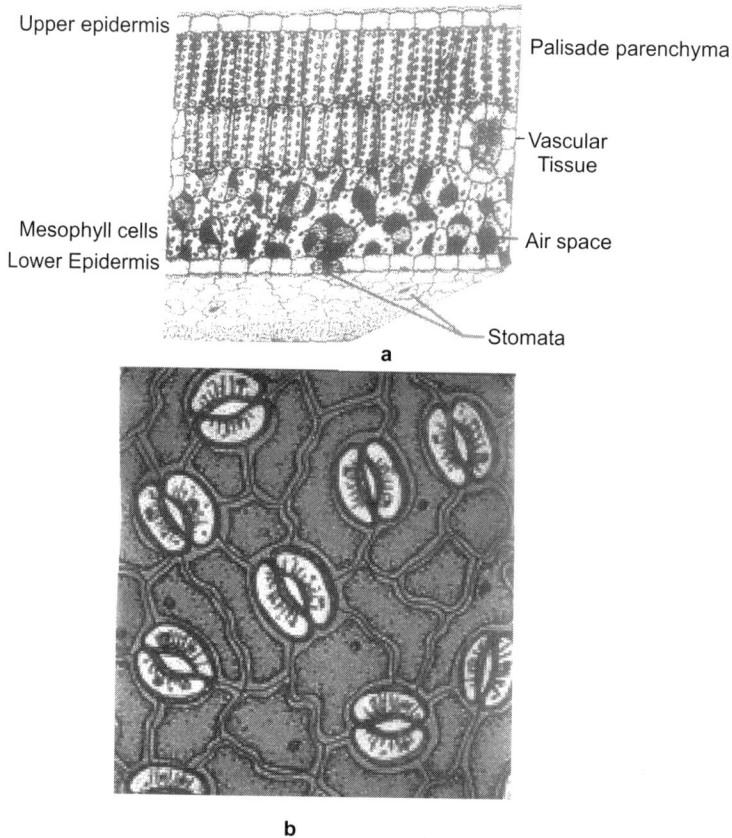

Fig. 7.4 a, b
a. Diagram of leaf structure. Modified from: *http://museum.utep.edu/archive/plants/DDtranspiration.htm*
b. Photomicrograph of leaf stomata. From: *http//users:rcn.com*

or the oxidative steps utilizes multiple forms of the cytochrome P 450s which catalyze oxidative deamination, sulfoxidation and nitrogen oxidation as well as alkyl hydroxylations and O and N dealkylations. Phase II (conjugative process, Paulson, 1986) is catalyzed by glutathione S-trasferased or sugar, amino acid or organic acid trasferased. Phase III is characterized by the formation of secondary conjugates via malonyl – COA transferases and the utilization of peroxidases (Zablotowicz et al., 2005). Do plant parts, e.g., roots and leaves,

metabolize pesticides differently? This question may be answered through the use of plant tissue cultures (Mumma and Hamilton, 1982). Komoba et al. (1995) consider metabolic processes for organic chemicals in plants.

In contrast to plants, certain prokaryotes can completely metabolize some pesticides. Indeed, some prokaryotes utilize pesticides as carbon sources. Van Eerd et al. (2003) compare pesticide metabolism in plant and microbial systems. Because the types of pesticides are quite numerous, the

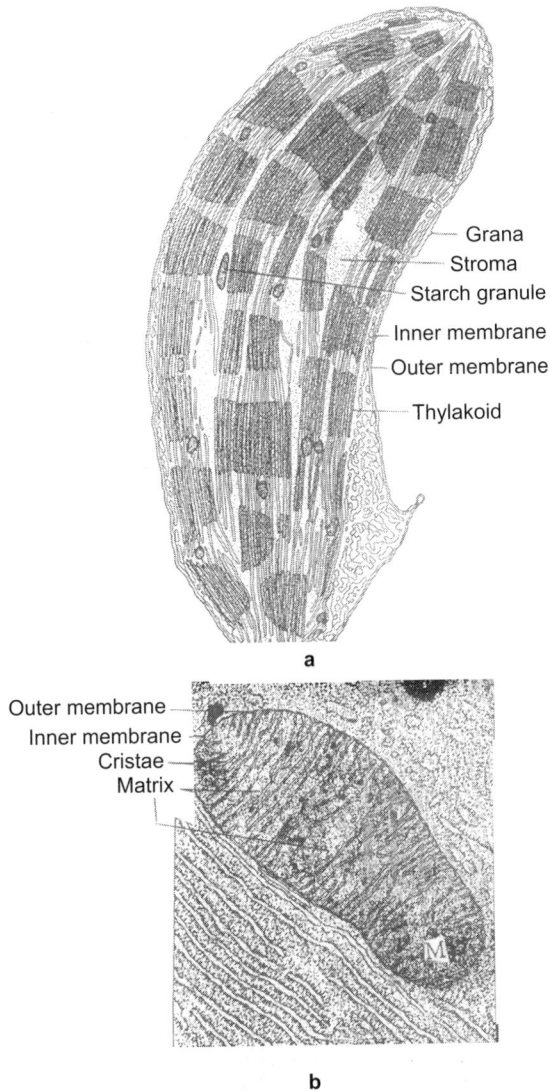

a

b

Fig. 7.5 a,b Election Micrographs of a Mitochondrion (a) and Chloroplasts (b) From: Maseuth (1991).

pesticide metabolic pathways of the phases resulting in detoxification (Gan, 2004) of inhaled, absorbed or ingested are quite varied in humans (Hutson and Paulson, 1995). However, because pesticides are foreign to the body, i.e., xenobiotics (Franklin and Yost, 2003), they are generally metabolized by liver enzymes (Wilkinson and Denison, 1982). Isoforms of the flavin-containing monooxygenases and the microsomal cytochrome P 450s (McKinnon and McManus, 1996) carry-out metabolism of pesticides (Hodgson et al., 1998). A thorough review of the structure and reactions catalyzed by cytochrome. P 450 enzymes has been published by Guengerich (2001). Gene

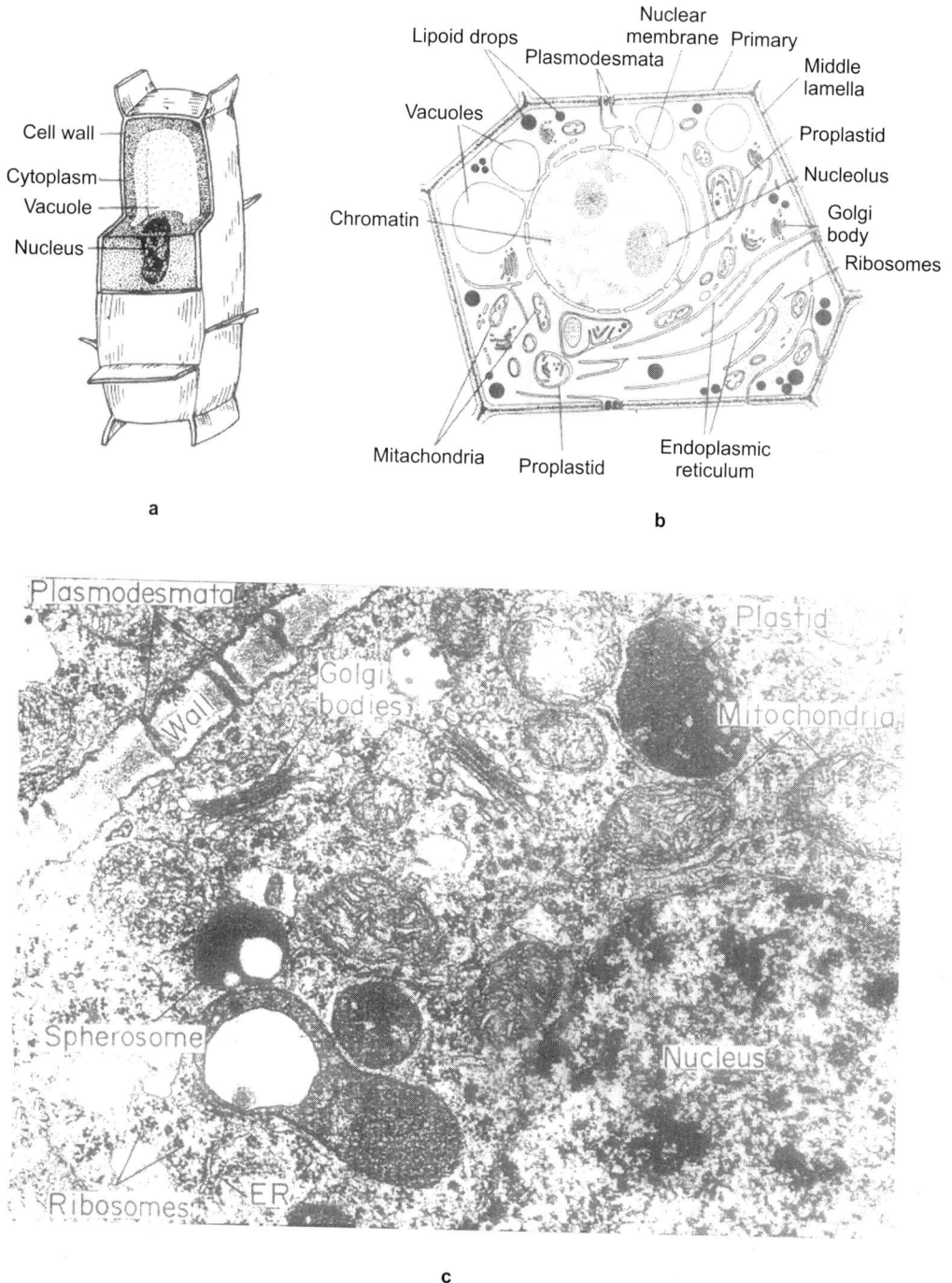

a

Cell wall
Cytoplasm
Vacuole
Nucleus

Lipoid drops
Vacuoles
Plasmodesmata
Nuclear membrane
Primary
Middle lamella
Proplastid
Nucleolus
Golgi body
Ribosomes
Chromatin
Mitachondria
Proplastid
Endoplasmic reticulum

b

Plasmodesmata
Wall
Golgi bodies
Plastid
Mitochondria
Spherosome
Nucleus
Ribosomes
ER

c

Fig. 7.6 a-c Diagrammatic structure of plant cells (**a**, **b**) and an electron micrograph (**c**) of Plasmodesmata **a**. Shows vacuoles, **b**. Depicts plasmodesmata in cell wall. Reprinted from: Fahn (1982). *Plant Anatomy*. 3rd Ed. Elsevier, With permission of Elsevier.

expression in multi-gene families of cytochrome P 450 has been reviewed by Eaton (2000).

The P450 and FMO systems are multigene families involved in the oxidative metabolism of a wide range of xenobiotics. The P450 family is a superfamily of monooxygenases with at least 57 isoforms identified in human (*http://drnelson.utmem.edu/hum.html*). They are heme-containing proteins that exist in a multienzyme system that also includes NADPH-cytochrome P450 reductase and cytochrome b_5. The FMO family is genetically less diverse than the P450s, with only five functional genes (FMO1-5) currently identified in humans (Lawton et al., 1994). Although both systems are dependent on NADPH as a cofactor, the FMO family accepts reducing equivalents directly from NADPH without

the need for NADPH reductase (Ziegler 1990).

The P450 and FMO families have partially overlapping substrate specificities and may or may not form the same metabolites. Differences in stereoselectivity also are observed between the two groups as well as between the isoforms (Levi and Hodgson, 1998). The mammalian P450s are catalytically more diverse than the FMOs and are associated with the metabolism of important endogenous compounds, such as arachidonic acid and testosterone, as well as bioactivation and detoxification of numerous xenobiotics. The FMO-catalyzed pathway is generally thought to be a detoxification process in which lipophilic compounds are converted to more polar compounds by introducing molecular oxygen.

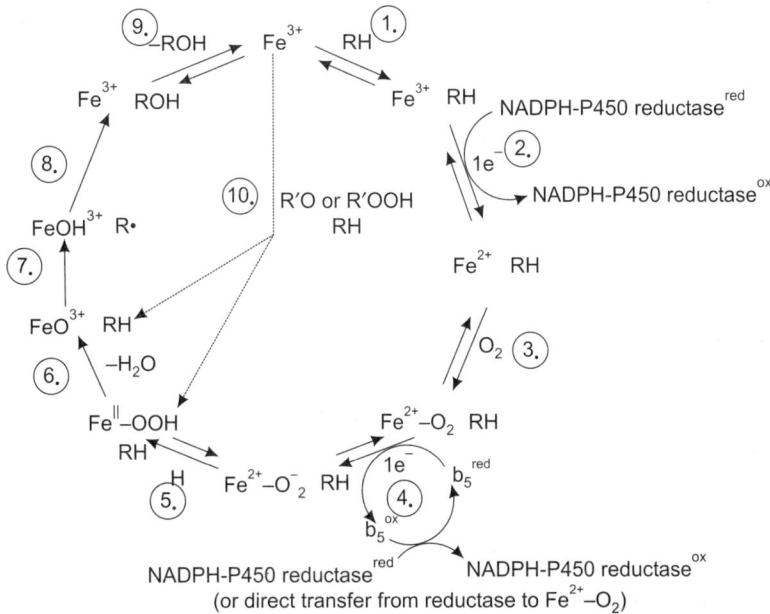

^{a} Fe = iron atom in P450 heme, RH = substrate, ROH = product, ox and red indicate the reduced and (1-electron) oxidized states of the reductase involved in the electron transfers

Fig. 7.7 Generalized Cytochrome P450 Catalytic Cycle Reprinted with permission from Guengerich, F.P. 2001. Chemical Research in Toxicology 14:612. Copyright 2001. American Chemical Society.

Can glutathione-S-transferases (GSTs) conjugate the glutathione tripeptide (g-Glu, Gys, Gyl, SH) to various xenobiotics. The GSTs are ubiquitous and the most abundant non-photosynthetic enzymes in plant cells. These appear to be at least 38 plant GSTs (See literature in Alfenito et al., 1998). "Numerous genetic polymorphisms in the mulit-gene families of cytochromes P450 (CYPs and GSTs have been described in the human population" (Eaton, 2000). Certain isoforms of the human liver CYPs have been cloned and examined for their abilities to metabolize pesticides (Hodgson and Levi, 1997). In plants CYPs are soluble proteins which serve for chemical defense (Werck-Reichart and Feyereisen, 2000).

Pesticide Resistance

The development of pesticide resistance (Scott, 1999; Soberon et al., 2007) is an example of artificial selection by which populations gradually achieve resistance (Fig. 7.8). In certain instances, resistance can be obtained via either conventional breeding or genetic engineering, i.e., incorporation of a gene(s) for a beneficial trait into a host's genome.

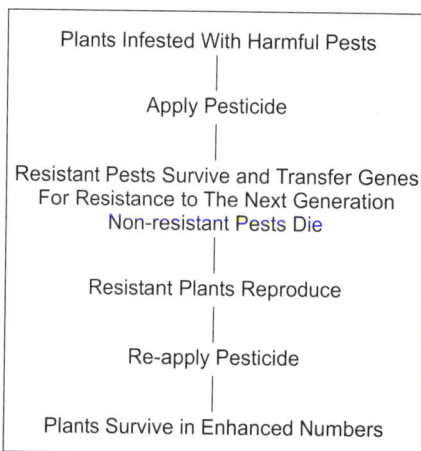

Plants Infested With Harmful Pests

|

Apply Pesticide

|

Resistant Pests Survive and Transfer Genes
For Resistance to The Next Generation
Non-resistant Pests Die

|

Resistant Plants Reproduce

|

Re-apply Pesticide

|

Plants Survive in Enhanced Numbers

Fig. 7.8 Selection For Pesticide Resistance

There have been attempts to slow the spread of resistant genes (*http://www.understandingevolution.com/evolibrary/article/0_0_0_agu.*)

A program called refuge, i.e. "fields without pesticides or plant-products/located near fields planted with pesticide producing crops." The above website presents a summary diagram which depicts how "refuge slows down the evolution of pesticide-resistant pests by allowing non-resistant pest strains to survive."

BIOLOGICAL EFFECTS

Human Health and Other Organisms

Chronic effects include carcinogenesis, immunotoxicity, neurotoxicity and reproductive as well as developmental effects (Baker and Wilinkson, 1990; Hodgson and Levi, 1997). Tables 7.7, 7.8 offer details for certain of these effects (Hayes and Laws, 1991; Mathews, 2006). However, there are not enough data to precisely establish the true impact on human health.

The lethality effects of certain pesticides on aquatic (Metclaf, 1977), mammalian and avian species occur in Table 7.9. The specific effects on fish, amphibians, reptiles and avian species are presented in Tables 7.10, 7.11 and 7.12, respectively. Karthikeyan et al. (2003) and Veleminsky and Gichner (2006) reported the effects of pesticides on non-target plants.

Are There Pesticide Regulations?

Finally, there are both national and international pesticide regulations (Marrs and Ballantyne, 2004; Carlile, 2006; Table 7.13) based upon research (Table 7.14) and analyses (Table 7.15). These exist because of the known adverse health effects, disasters

TABLE 7.7 Reported Health Effects of Pesticides[a]

Carcinogenesis

Epidemiological studies suggest that organochlorine insecticides can cause lung, bladder, soft-tissue sarcomas, lymphatic and hematopoetic cancers.

Developmental effects

Certain pesticides can affect thyroid and spleen development.

Immunotoxicity

Occupational exposures can induce hypersensitivity responses. Reductions of white blood cells and lymphocytes have been reported following exposure to certain pesticides.

Neurotoxicity

Insecticides designed to attack the insect nervous system are capable of causing acute and chronic neurotoxic effects. There are alterations in sensory, motor autonomic, cognitive and behavioral functions for individuals who are accidentally exposed to high concentrations of certain pesticides.

Reproductive Toxicity

Several pesticides have been shown to affect male reproduction after occupational exposures.

[a] Further details are available in Hallenbeck and Cunningham-Burns (1985); Costa et al., (1986); WHO (1990); Ecobichon (1999); Lappe (2000); Solomon et al. (2000) and Yu Ming, Ho (2005).

TABLE 7.8 The US-EPA has categorized pesticides into four groups based upon LD50s, LC50s as well as eye and skin effects.

	I	II	III	IV
Oral LD50	Up to and including 50 mg/kg	From 50 thru 500 mg/kg	From 500 thru 5000 mg/kg	Greater than 5000 mg/kg
Inhalation LC 50	Up to and including 0.2 mg/liter	From 0.2 thru 2 mg/liter	From 2.0 thru 20 mg/liter	Greater than 20 mg/liter
Dermal LD 50	Up to and including 200 mg/kg	From 200 thru 2000 mg/kg	From 2000 thru 20,000 mg/kg	Greater than 20,000 mg/kg
Effect on eyes	Corrosive; corneal opacity not reversible within 7 days	Corneal opacity reversible within 7 days; irritation persisting for 7 days	No corneal opacity; irritation reversible within 7 days	No irritation
Effect on skin	Corrosive	Severe irritation at 72 hours	Moderate irritation at 72 hours	Mild or slight irritation at 72 hours

[a] From: http://www.epa.gov/pesticides/health/tox_categories.htm.

TABLE 7.9 Lethality Effects and Other Parameters of Certain Pesticides

Chemical	Aquatic LC$_{50}$ (μg/L)[a]	Mammalian Oral LD$_{50}$ (mg/kg body weight)	Avian Oral LD$_{50}$ (mg/kg body weight)	Bioconcentration Factor	Half-Life in Soil	Half-Life in Atmosphere
Aldrin	1.3 – 8.9	33 – 320	6.6 – 520	1550 – 2000	Several months	35 minutes
Chlordane	0.4 – 52	335 – 1720	1200	200 - 18500	Up to 20 years	1.3 days
Dieldrin	0.5 – 330	37 – 330	26.6 – 381	4860 – 14500	1 month to 5 years	Unknown
DDT	0.4 – 380	113 – 1770	386 – 2240	12000	2 – 5 years	2 days
Mirex	NA	125 – 1000	1400 – 10000	2600 – 51400	Up to 10 years	Unknown
Toxaphene	2.2 – 21	46 – 365	70.7 – 250	4200 – 90000	2 months to 11 years	4 – 5 days

From: http://www.epa.gov/pesticides

TABLE 7.10 Effects of Certain Pesticides on Fish

Pesticide	Effect
Chlordane	Reproductive Developmental, Neurological Behavior, Mortality
DDT	Neurological Behavior
Toxaphene	Reproductive Development, Neurological Behavior, Decreased Growth, Mortality

TABLE 7.11 Effects of Some Pesticides on Amphibians/Reptiles

Pesticide	Effect
Chlordane	Reproductive Development, Mortality
Dieldrin	Reproductive Development
DDT	Reproductive Development
Toxaphene	Neurological Behavior, Mortality

TABLE 7.12 Effects of Certain Pesticides on Avian Species

Pesticide	Effect
Chlordane	Reproductive Development, Neurological Behavior, Mortality
Dieldrin	Reproductive Development, Mortality
DDT	Reproductive Development, Neurological Behavior
Toxaphene	Reproductive Development, Neurological Behavior, Decreased Growth, Mortality

Tables adopted from Great Lakes Pesticide Report December 30, 1998 – Draft Final and EPA (1997).

TABLE 7.13 Summary of Some Pesticide Regulations[a]

Regulation	Purpose
Federal Insecticide, Fungicide and Rodenticide Act of 1947 (FIFRA)	Requires that pesticides be registered by the severity of agriculture before being marketed.
Federal Pesticide Control Act 1972 with 1975, 1978 and 1980 amendments	Refocused Emphasis on the safety of pesticides and making the registration process for chemical pesticides more vigorous
Pesticide Monitoring Improvements Act of 1988	Furthers the control of pesticides in order to assure public health. Establishes a data management system.
Pesticide Laws	
National USA – Federal Insecticide, Fungicide and Rodenticide Act (FIFRA)	Provides the basis for regulation, sales, distribution and use of pesticides in the US.
Federal Food, Drug and Cosmetic Act (FFDCA)	Authorizes US-EPA to set maximum residue levels, or tolerances, for pesticides used in or on foods or animal feed
Food Quality Protection Act of 1996 (FQPA)	Amended FIFRA and FFDCA Setting tougher safety standards for new and old pesticides

(Contd.)

(Contd.)

Agreements

| International Program on Chemical Safety | Provides toxicity evaluations in support of WHOPES (WHO Pesticide Evaluation Scheme) and WHO guidelines for drinking water. (*http://www.who.int/whopes/recommendations/wgn/en* and Maroni et al., 1999) |

[a]Marrs and Ballantyne (2004) discuss international pesticide regulations.

TABLE 7.14 Summary of Pesticide Research Activities

Agency	Research Activity
National	
US – EPA Office of Pesticide programmes Analytical chemistry Laboratory	Develop new mulit-residue analytical methods
Environmental chemistry lab	Test methods for the evaluation of pesticides in soil and water
USDA Soil Physics and Pesticide Research	Minimize air and water contamination for agricultural pesticides
US Food and Drug Administration Total Diet and Pesticide Research Center	Develop analytical procedures to detect various pesticides and chemicals in foods
In situ assessment of small, headwater streams impounded by suburban pesticide runoff	University of Georgia, Atlanta, GA, USA. *http://www.uga.edu/publichealth/ehs/researchareas.html.*
International	
Epidemiological, toxicological and clinical research on the effects of pesticides on man	International Center for Pesticide Application Research Centre Imperial College, London http://www.dropdata.net/iparc
Development of improved forms of entomopathologic fungi for controlling coffee pests	National Coffee Research Centre Columbia
Development of improved systems for managing insect pests of tomatoes	Egyptian Ministry of Agriculture and Land Reformation
Development of improved pest management systems on grapes, cotton and faba beans in middle Egypt	USAID University Linkage Project

TABLE 7.15 Some Analytical Methods for Detecting and Quantifying Pesticides in Food[a]

Method	Reference
Capillary electrophoresis	Katrice (2002)
ELISA	Morozova et al. (2005)
FT-Raman spectroscopy	*http:www.thescitntist.com/article/display10726/-35k*
Gas chromatography (GC) with photometric detection or mass detection (GC-MS)	Aprea et al. (2002) Rosh (1987)
HPLC/DAD	Parilla et al. (1994)
Immunoassay	Kolosova et al. (2003)

(Contd.)

(Contd.)

Liquid chromatography/atmospheric-pressure chemical ionization mass spectrometry	Doergo et al. (2005)
Atomic absorption	*http://www.epa.gov/oppbead/methods/glossary.*
High performance thin layer chromatography	*http://www.epa.gov/oppbead/methods/glossary.*
Ion chromatography	*http://www.epa.gov/oppbead/methods/glossary.*
Liquid scintillation counting	*http://www.epa.gov/oppbead/methods/glossary.*
Thin layer chromatography	*http://www.epa.gov/oppbead/methods/glossary.*
Ultraviolet spectrophotometry	*http://www.epa.gov/oppbead/methods/glossary.*
Optimal biotest battery (crustaceans, fish, algae and protozoans)	Fochtman et al. (2000)
Plant tissue cultures	Mumma and Hamilton (1982)
Bioassay	
Cell culture test	
Inhibitory effects of pesticides on mouse hybridoma cell line IE6	Bertheussen et al. (1997)
In vitro skin absorption models for dermal penetration	Ecobichon (1999)
Benthic Invertebrates	Reynoldson and Metclafe-Smith (1992)

[a]Thiev and Zeumer (1987) have published a manual of pesticide residue analysis as has Tadeo, J.L. (2008) *Analysis of Pesticides in Food and Environmental Samples*. CRC Press, Boca Raton, FL, USA.

TABLE 7.16 Examples of Pesticide Disasters

Event	Result
1958 India – ingestion of wheat flour contaminated with parathion	100 deaths of humans
Bhopal, India – release of methyl isocyanate from a pesticide plant	4,000 people killed, 150,000 injured effects – emphysema alveolar and interstital edema, (*http://www.nih.gov/entrez/query.fcgi?*)

TABLE 7.17 Costs of Pesticide Damage[a]

Costs	Million of $/yr. USA	
	1992	2005
Public health impacts	787	1140
Domestic animal and fish deaths	54	30
Loss of natural enemies and beneficial organisms/biodiversity	520	520
Pesticide resistance	1400	1500
Crop and fishery losses	1062	1500
Bird losses	2100	1491
Groundwater contamination	1800	2160
Government regulation	200	2000
		470

[a] Adopted from Pimentel (1992) and (2005).

(Contd.)

(Contd.)

International Damage Costs of Chemical Pesticides[a]

Type of Costs	Thailand	West Germany	West Germany
Public health	0.6	13.6	23
Domestic animals and fish	n.a.	n.a.	n.a.
Loss of beneficial organisms/biodiversity	2.4	5.9	10
Residues in food	209.0	n.a	23
Pesticide resistance	n.a	n.a.	n.a.
Production loss	n.a.	1.2	n.a.
Bird losses	n.a.	n.a.	n.a.
Groundwater contamination	n.a.	75.3	128 – 186
Government regulation & research	16.8	52.2	66

	Sri Lanka	Thailand	Philippines	Ecuador
Health costs related to farmers	10 wks	127.7 health, monitoring, research, regulation and extension	61% higher health costs for farmers expensed to pesticides	6 worker days for poisoning of a potato farmer

[a] For Egypt the % of GDP in damage is 21% (air), 3% (costal zones and cultural heritage), 12% (soil), 2% (wastes) and 10% (water). http://www.iup

[b] Adopted from Pimentel (2005)

[c] Data for Germany, Sri Lanka, Thailand, Philippines and Ecuador adopted from http://www.pan-uk/org/postnews/pn61p3.htm. 1,689 DM = 45 $ 3,000 million.

(Table 7.16) and costs (Table 7.17) involving certain pesticides. The question is will developing countries use conventional pesticides at the same rate as in advanced countries? What impact will biotechnology have on the employment of conventional pesticides?

The control of pests is the concern of IPM or Integrated Pest Management (http:www.com). This program responds "to pest problems with the most effective least-risk option."

References

Alfenito, M.R., Souer, E., Goodman, C.D., Buell, R., Mol, J., Kues, R. and Walbot, V. 1998. Functional complementation of anthocyanin sequestration in the vacuole by widely divergent glutathione s-transferases. *Plant Cell* 10: 1135-1150

Allen, S.L. 1979. *Chemical Rodenticides*. NC Agricultural Extension Service. North Carolina State Univ., Raleigh, NC, USA.

Andreoli, C.V. 1993. The influence of agriculture on water quality. In: *Prevention of Water Pollution by Agriculture and Related Activities*. Proceedings of the FAO Expert Consultation, Santiago, Chile, 20-23 Oct. 1992. Water Report 1. FAO Rome, pp. 53-63.

Aprea, C., Colosio, C., Mammone, T., Minioa, C. and Maroni, M. 2002. Biological monitoring of pesticide exposure: A review of all articles. *J. Chromatogr. B. Analyt. Technol. Biomed. Life Sci.* 769: 191-219.

Baker, S. and Wilinkson, C.F. 1990. *The Effect of Pesticides on Human Health*. Princeton Publishing Co., Princeton, NJ, USA.

Bertheussen, K., Yousef, M.I. and Figenschau, Y. 1997. A new sensitive cell culture test for the assessment of pesticide toxicity. *J. Environ. Sci. Health B.* 32: 195-211.

Boerth, D. 2005. *http://www.umassd.edu/communications/articles/showarticles. cfm?a_key=480.*

Bowlby, N. 2006. plant metabolism-respiration. In: *Plant Cell Biology*. (Dashek, W.V. and Harrison, M., eds.), Science Publishers, Enfield, NH, USA, pp. 359-398.

Buchel, K.H. 1983. *Chemistry of Pesticides*. Wiley – Interscience, New York, NY, USA.

Carlile, B. 2006. *Pesticides Selectivity, Health and The Environment.* Cambridge Univ. Press, Cambridge, England.

Clark, J.M. and Ohkawa, H. 2005. *New Discoveries in Agrochemicals.* Oxford Univ. Press, Oxford, England.

Costa, L.G., Galli, C. and Murphy, S.D. 1986. *Toxicology of Pesticides: Experimental, Clinical and Regulatory Perspectives.* Springer-Verlag, Berlin, Germany.

Crowe, A.S. and Mutch, J.P. 1994. An expert systems approach for assessing the potential for pesticide contamination of ground water. Ground Water 32:487-498.

Crowe, A.S. and Booty, W.G. 1995. A multi-level assessment methodology for determining the potential for groundwater contamination by pesticides. *Environ. Maint. Assess.* 35:239-261.

Dev, S. and Koul, D. 1997. 'Insecticides of Natural Origin.' CRC Press, Boca Raton, FL, USA.

Doergo, D.R., Bajic, S. and Millington, D.S. 2005. Analysis of pesticides using liquid chromatography/atmospheric-pressure chemical ionization mass spectrometry. *Rapid Commun. Mass Spec.* 6: 663-666.

Duke, S.O. 1990. National pesticides from plants In: *Advances in New Crops.* (Janick, J. and Simon, J.E.; eds.). Timber Press, Portland, OR, USA.

Eaton, D.L. 2000. Biotransformation enzyme polymorphism and pesticide susceptibility. Neurotoxicology 21: 101-111.

Ecobichon, D.J. 1999. *Occupational Hazards of Pesticide Exposure. Sampling, Monitoring and Measuring.* Taylor and Francis, Philadelphia, PA, USA.

Esteban, A.M., Fernandez, P. and Carmarci, C. 1997. *Immunosorbents: A new tool for pesticide sample handling in environmental analysis. Fresenuis J. or Anayl. Chem.* Springer, Berlin, Germany.

Evans, R. 1986. *Population biology of pesticide resistance bridging the gap between theory and practical applications.* National Academies Press, Washington, DC, USA.

Franklin, M.R. and Yost, G.S. 2003. Biotransformation: A balance between bioactivation and detoxification. In: *Principles of Toxicology* (Williams, P.L., James, R.C. and Roberts, S.M., eds.). Wiley, New York, NY, USA.

Fellenberg, G. and Wier, A. 2000. *The Chemistry of Pollution.* Wiley, New York, NY, USA.

Fochtman, P., Raska, A. and Nierzedska, E. 2000. The use of conventional bioassays, microbiotests and some 'rapid' methods in the selection of an optimal test battery for the assessment of pesticide toxicity. *Environ. Toxicol.* 15: 376-384.

Fournier, J.C., Soulas, G. and Parekh, N.R. 1997. Main microbial mechanisms of pesticide degradation in soil. In: *Soil Ecotoxicology.* (Tarrodellos, J., Bitton, G. and Rossel, D., eds.) CRC Press, Boca Raton, FL, USA pp. 85-116.

Gan, J.J. 2004. *Pesticide Decontamination and Detoxification.* Oxford Univ. Press, Oxford, England.

Gianessi, L.P. and Anderson, J.E. 1995. Pesticide use in U.S. crop production. National Data Report. 1992. National Center for Food and Agriculture Policy, Washington, DC, USA.

Gillom, R.J., Barbash, J.E., Kolpin, D.W. and Larson, S.J. 1999. Testing water quality for pesticide pollution. *Environ. Sci. Techno.* 33: 164a-169a.

Greene, S.A. and Pohanish, R.P. 2005. *Sittig's Handbook of Pesticides and Agricultural Chemicals.* William Andrew Pub., Norwich, NY, USA.

Guengerich, F.P. 2001. Common and uncommon cytochrome P450 reactions related to metabolism and chemical toxicity. *Chem. Res. Toxicol.* 14: 612-

Hallenbeck, W.W. and Cunningham-Burns, K.M. 1985. *Pesticides and Human Health.* Springer-Verlag, New York, NY, USA.

Hamilton, D. and Crossley, S. 2004. *Pesticide Residues in Food and Drinking Water – Human Exposure and Risks.* Wiley, New York, NY, USA.

Hatzios, K.K. 1982. *Metabolism of Herbicides in Higher Plants.* Burgess Intl. Group, Minneapolis, MN, USA.

Hayes, W.J. and Laws, E.R. 1991. *Handbook of Pesticide Toxicology*. Academic Press, San Diego, CA, USA.

Hemond, H.F. and Fechner-Levy, E.J. 2000. *Chemical Fate and Transport in the Environment*. Academic Press, San Diego, CA, USA.

Hewitt, H.G. 1998. *Fungicides in Crop Protection*. CAB International Publication, Wallingford Oxfordshire, UK.

Hodgson, E. and Levi, P.E. 1997. *A Textbook of Modern Toxicology*. 2nd ed. Appleton and Lange, Stamford, CT, USA.

Hodgson, E. Cherrington, N.J., Coleman, S.C., Lui, S., Falls, J.G., Cao, Y., Goldstein, J.E. and Rose, R.L. 1998. Flavin – containing monoxygenase and cytochrome P450 mediated metabolism pesticides: From mouse to human. *Rev. Toxicol.* 2: 231-243.

Howard, P.H. 1991. *Handbook of Environmental Fate and Exposure Data for Organic Chemicals. Volume 3 – Pesticides*. Lewis Publishers, Chelsea, MI, USA.

Hutson, D.H. and Paulson, G.D. 1995. *The Mammalian Metabolism of Agrochemicals, Volume 8, Progress in Pesticide Biochemistry and Toxicology*. Wiley, New York, NY, USA.

Karthikeyan, R., Davis, L.C., Erickson, L.E., Al-Khatb, K., Kulakow, P.A., Barnes, P.L., Hutchinson, S.L and Nurzhanovan, A.A. 2003. *Studies on responses of non-target plants to pesticides: A review*. Hazardous Substance Research Center, Kansas State University

Katrice, L.M. 2002. *Online Immunoassay for Pesticides Analysis Using Capillary Electrophoresis Antibody Immobilization Strategies*. M.S. Thesis, VCU, Richmond, Va, USA.

Kolosova, A.Y., Park, J.H., Eremin, S.A., Kang, S.J. and Chung, D.H. 2003. Fluorescence polarization immunoassay based on monoclonal antibody for the detection of the organophosphorus pesticide parathion-methyl. *J. Agric. Food Chem.* 51: 1107-1114.

Komoba, D., Langebartels, C. and Sanderman, H. 1995. Metabolic processes for organic chemicals in plants. In: *Plant Contamination Modeling and Simulation of Organic Chemical Process*. (McFarlane,C. and Trapp, S., J.C., eds.). CRC Press, Boca Raton, Fl., USA

Kurtz, D.A. 1990. *Long Range Transport of Pesticides*. Lewis Publishers, Chelsea, MI, USA.

Lampman, W. 1995. Susceptibility of groundwater to pesticides and nitrate contamination in predisposed areas of southwestern Ontario. *Water Qual. Res. Jour.* Canada 30:443-468.

Lappe, M. 2000. Secondary effects of pesticide exposures in : *Environmental Toxicology Current Developments*. (Rose, J., ed.). Gordon and Breach Amsterdam, The Netherlands pp. 151-158.

Lawton, M.P., Cashman, J.R., Cresteil, T., Dolphin, C.T., Elfarra, A.A., Hines, R.N., Hodgson, E., Kimura, T. Ozols, J. and Phillips, I.R. 1994. A nomenclature for the mammalian flavin-containing monooxygenases gene family based on amino acid sequence identities. *Arch. Biochem. Biophys.* 308: 254-257.

Levi, P.E. and Hodgson, E. 1998. Stereospecificity in the oxidation of phorate and phorate sulphoxide by purified FAD-containing mono-oxygenase and cytochrome P-450 isozymes. *Xenobiotica*. 18: 29-39.

Levine, M.J. 2007. *Pesticides: A Toxic Time Bomb in Our Midst*. Praeger, Westport, CT, USA.

Lopez, O., Olmedor, G., and Guevora, F. 2005. *Introduction to Molecular Biotechnology for Food and Agricultural Sciences*. CRC Press, Boca Raton, FL, USA.

Manahan, S.E. 2005. *Environmental Chemistry*. CRC Press, Boca Raton, Fl, USA.

Marvel, J.T., Brightwell, B.B., Malik, J.M., Sutherland, M.L. and Rueppel, M.L. 1978. A simple apparatus and quantitative method for determining the persistence of pesticides in soil. *J. Agric. Food Chem.* 26: -1116.

Maseuth, J.O. 1991. *Botany. An Introduction to Plant Biology*. Saunders College Publ. Philadelphia PA, USA.

Masuinas, J. 2005. Weed control for commercial vegetable crops. *www.aces.uiuc.*

Maroni, M., Fait, A. and Colosiv, C. 1999. Risk assessment and management of occupational exposure to pesticides. *Toxicol. Lett.* 107: 145-153.

Marrs, T.C. and Ballantyne, N.B. 2004. *Pesticide Toxicology and International Regulations.* (Current Toxicology). Wiley, New York, NY, USA.

Mathews, S.G. 2006. 'Pesticides: Health, Safety and Environment.' Blackwell, Upper Saddle River, NJ, USA.

McKinnon, R.A. and McManus, M.E. 1996. Localization of cytochromes in P450 in human tissues: implications for chemical toxicity. *Pathology* 28: 148-155.

Metcalf, R.L. 1977. Model ecosystem approach to insecticide degradation: A Critique. *Ann. Rev. Entomol.* 22: 241-261.

Metcalf, R.L. and Metcalf, E.R. 1992. *Plant Kairomones in Insect Ecology and Control.* Chapman and Hall, New York, NY, USA.

Meyer, T.T. and Thurman. 1996. *Herbicide Metabolites in Surface Water and Groundwater.* ACS Symposium Series 630. ACS, Washington, DC, USA.

Milne, G.W.A. 1994. *CRC Handbook of Pesticides.* CRC Press, Boca Raton, FL, USA.

Moorman, R.B. 1983. Rodenticides (WL-37). Cooperative Extension Service, Iowa State Univ., Ames, Iowa, USA.

Morozova, V.S., Levashova, A.L. and Eremin, S.A. 2005. Determination of pesticides by enzyme immunoassay. *J. Anal. Chem.* 60: 202-217.

Mumma, R.O and Hamilton, R.H. 1982. Advances in pesticide identification through the use of plant tissue cultures. *J. Toxicol. Clin. Toxicol.* 19: 535-55.

Onlgey, E.D., Krishnappan, B.G., Droppo, I.G., Rao, S.S., and Maguire, R.J. 1992. Cohesive sediment transport: Emerging issues for toxic chemical management. *Hydrobiologia* 235/236: 177-187.

Parrilla, P., Martinez, J.L., Martinez, Galera, M. and Frenich, A.G. 1994. Simple and rapid screening procedure for pesticides in water using SPE and HPLC/DAD detection. Fresenius', *J. Analy. Chem.* 60: 202-217.

Paulson, G.D. 1986. *Xenobiotic Conjugation Chemistry.* American Chemical Society, Washington DC, USA.

Pepper, I.L., Gerba, C.P. and Brusseau, M.L. 2006. *Abiotic Soil in Environmental and Pollution Science.* Academic Press, New York, NY, USA.

Pesticide Biochemistry and Physiology (online) Academic Press, Orlando, FL, USA.

Phelps, W., Winton, K. and Effland, W.R. 2002. *Pesticide Environmental Fate. Bridging the Gap Between Laboratory and Field Studies.* American Chemical Society. Washington, DC, USA.

Philip, R.B. 2001. *Ecosystems and Human Health. Toxicology and Environmental Standards.* Lewis, Boca Raton, FL, USA.

Phipps, R.H. and Park,J.R. 2002. Environmental benefits of genetically modified crops: global and European perspectives on their ability to reduce pesticide use. *J. Anim. Feed Sci.* 11:1-18.

Pimentel, D. 2005. Environmental and economic costs of the application of pesticides primarily in the United States. *Environment, Development and Sustainability* 7: 229-252.

Prevost, M., Laurent, P., Sevvais, P. and Joret, J.C. 2005. Biodegradable organic matter in drinking water treatment and distribution. American Water Works Association, Denver, CO, USA.

Racke, K.D. 2003. What do we know about the fate of pesticides in tropical ecosystems? In: *Environmental Fate and Effects of Pesticides.* (Coats, J.R. and Yamamoto, H., eds.). American Chemical Society, Washington, DC, USA, pp. 96-123.

Ritter, L., Solomon, K.R., Forget, J., Stemeroff, M., and O'Leary, C. 1995. A Review of Selected Persistent Organic Pollutants DDT, Aldrin, Dieldrin, Endrin, Chlordane, Heptachlor, Hexachlorobenzene, Mirex, Toxaphene, Polychlorinated biphenyls, Dioxins and Furnas. PCS/95:39. International Programme on Chemical Safety (IPCS) within the framework of the Inter-Organization Programme for the Sound Management of Chemicals (IOMC). 167 pages.

Reynoldson, T.B. and Metcalfe-Smith, J.L. 1992. An overview of the assessment of aquatic ecosystem health using benthic invertebrates. *J. Aquatic Ecosystem Health* 1: 295-308.

Roberts, T.R. 1999. *Insecticides and Fungicides* (Metabolic Pathways). Royal Society of Chemistry, Cambridge, England.

Rosen, J.D. 1987. Application of New Mass Spectrometry Techniques in Pesticide Chemistry. Wiley, New York, NY, USA.

Rosh, J.D. 1987. *Applications of New Mass Spectrometry Techniques in Pesticide Chemistry.* Wiley, New York, NY, USA.

Schreiber, L. 2005. Polar paths of diffusion across plant cuticles: New evidence for an old hypothesis. Ann. Bot. 95: 1069-1073.

Scott, J.C. 1999. Cytochromes and insecticide resistance. Insect Biochem. *Mol. Biol.* 29:757-777.

Shimazu, M., Chen, W. and Mulchandani, A. 2004. Biological detoxification of organophosphate and detoxification pesticides. In: *Pesticide Contamination*, (Gan, J., Zhu, P.C., Aust, S.D. and Lewley, A.T., eds.). American Chemical Society, Washington, DC, USA. pp. 25-36.

Soberon, M., Pardo-López, L., López, I., Gómez, I., Tabashnik, B.E. and Bravo, A. 2007. Engieering modified Bt toxins to counter insect resistance. *Science* 318: 1640-1642.

Solomon, G., Kirsch, J. and Oganseitan, O.A. 2000. Pesticides and human health. *www.psrla.org/pesticides/pesticides and hratth kit.pdf.*

Stephenson, G.A. and Solomon, K.R. 1993. *Pesticides and the Environment.* Department of Environmental Biology, University of Guelph, Guelph, Ontario, Canada.

Sternersen, J. 2004. *Chemical Pesticides. Mode of Action and Toxicology.* CRC Press, Boca Raton, FL, USA.

Tasleva, M. 1995. *Anticoagulant Rodentcidies* (Environmental Health Criteria). World Health Organization, Geneva, Switzerland.

Thiev, Has-Peter and Zeumer, H. 1987. *Manual of Pesticide Residue Analysis.* VCH, New York, NY, USA.

U.S. Government Office Pesticides, Insecticides, Fungicides and Rodenticides. (SU Doc GP 3.22/2: 227/1990).

Van Eerd, L.L., Hoagland, R.E., Zablotowicz, R.M. and Hall, C.J. 2003. Pesticide metabolism in plants and microorganisms. *Weed Sci.* 51: 472-495.

Veleminsky, J. and Gichner, T. 2006. Scope 49-Methods to assess adverse effects of pesticides on non-target organisms. *http://www. icsu.scope.org/downloadpubs/scope49/chapter14 .html.*

Ware, G.W. 1999. *The Pesticide Book.* Ball Publishing. Batavia, IL, USA.

Werck-Reichart, D. and Feyereisen, R. 2000. Cytochromes P450: A success story. Genome Biol. 1 reviews 3003. 1- reviews 3003.9.

W.H.O. 1990. *Public Health Impact of Pesticides Used in Agriculture.* WHO, Geneva, Switzerland.

World Health Organization. 1995. B*iomodiolone Health and Safety Guide.* Geneva, Switzerland.

W.H.O. *The W.H.O. Recommended Classification of Pesticides by Hazard Guidelines to Classification.* 2004. W.H.O. International Programme on Chemical Safety. Geneva, Switzerland.

Wilkinson, C.F. and Denison, M.S. 1982. Pesticide interactions with biotransformation systems In: *Effects of Chronic Exposures to Pesticides on Animal Systems.* (Chambers, J.E. and Yarbrough, J.D., eds.). Raven Press, New York, NY, USA. pp. 1-24.

Yu, Ming-Ho. 2005. *Environmental Toxicology Biological and Health Effects of Pollutants.* CRC Press, Boca Raton, FL, USA.

Zablotowicz, R.M, Hoagland, R.E. and Hall, J.C. 2005. Metabolism of pesticides by plants and prokayotes. In: *Environmental Fate and Safety Management of Agrochemicals.* (Clark, J.M. and Ohkowa, H., eds.). Oxford Univ. Press, Oxford, England. pp. 168-180.

Ziegler, D.M. 1990. Flavin-containing monooxygenases: enzymes adapted for multisubstrate specificity. *Trends Pharmacol. Sci.* 11: 321-324.

CHAPTER 8

Radioactive Pollution

What is Radioactivity?

Isotopes (Browne and Firestone, 1986) are atoms (Kathren, 1984) that contain the same number of protons (possess the same atomic number) but a different number of neutrons (have different atomic weights). Figure 8.1 depicts the structure of an atom. Many naturally-occurring elements exist as a mixture of isotopes, e.g., ^{12}C and ^{14}C (Browne and Firestone, 1986). Radioactivity results from an unstable combination of protons and neutrons in the nucleus (Fig. 8.1) and an attempt to achieve a more stable combination (Zeidan and Dashek, 1996). This usually occurs via the emission of an α or β particle (Table 8.1; Fig 8.2).

Sources of Radioactivity

There are more than 60 naturally-occurring radioactive elements (radionuclides), which can occur in air, water, soil and rocks. Radionuclides have been classified as primordial, cosmogenic or human-produced (Table 8.2). The primordial radionuclides are the remains from creation of the World and Universe. In contrast, cosmogenic and human-produced radionuclides result from cosmic ray interaction and human actions, respectively (*http://www.physics.isu.edu/rad-nf/natural.htm*). Artificial radionuclides in the environment result from the atmospheric testing of nuclear weapons, the exploitation of nuclear fission for the production of energy and possible nuclear power plant accidents (Eisebud and Gessel, 1997; Tykva and Berg, 2004).

Characteristics of Radioactivity

Specific activity refers to the quantity of radioactivity (Table 8.3) per second and unit of substance. It is normally expressed as curies per gram (Ci/g), millicuries per milligram (mCi/mg), millicuries per millimole (mCi/mmole), disintegrations per minute per millimole (DPM) or counts per minute (CPM). The activity of each radionuclide is characterized by a constant, the half-life ($t^{1/2}$) which is the time taken for half of the original atoms to decay (Table 8.4).

Measurement of Radioactivity

Gamma-Ray Detection

Gamma radiation can be detected and counted by commercial gamma counters, commonly known as gamma well counters. Gamma emitting radiosotopes are expecially valuable in radioimmunoassays.

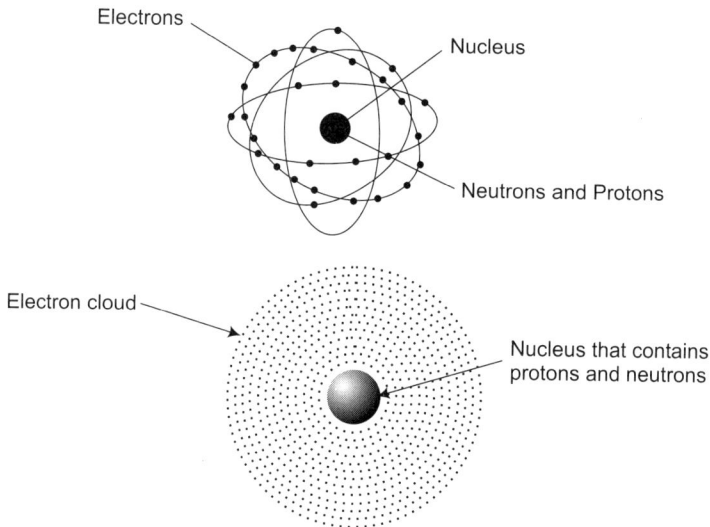

Particle	Charge	Symbol	Actual Mass	Relative Mass compared to Proton
Electron	$- (e^-)$	$^{0}_{-1}e$	9.109×10^{-28} g	1/1,837
Proton	$+ (e^+)$	$^{1}_{1}H$	1.673×10^{-24} g	1
Neutron	$0 (e^0)$	$^{1}_{0}n$	1.675×10^{-24} g	1

Fig. 8.1 Diagrams of the structure of the atom. Electron cloud diagram from Brewer (2006).

TABLE 8.1 Some Types of Radioactive Decay (see Fig. 8.2)[a]

Type	Example	Penetrating Power
Alpha (Helium Nuclei)	238U 234Th 4 92 90 2He	Stopped by the skin but very damaging due to ionization
Beta (High speed electron)		Penetrates human tissue to 1 cm.
Gamma (High energy proton)	238U 234Th 4He + $2^{0}_{0}Y$ 92 90 2	Highly penetrating but not very ionizing
Positron Decay	$^{23}_{12}Mg \rightarrow ^{23}_{11}Na + ^{0}_{1}\lambda + ^{0}_{0}\nu$	Positron (particle with same mass as electron). Positron decay occurs when there are too many protons in the nucleus.
Electron Capture	$^{7}_{4}Be + ^{0}_{1}\lambda \rightarrow ^{7}_{3}Li + \nu$	Orbital electron captured by the nucleus results in a proton being converted to a neutron and a neutrino V

[a] An extensive listing of radioactive decay occurs in Table 4.3 of Magill and Galy (2004).

TABLE 8.2 Classification of Radionuclides

Primordial Nuclides

Nuclide	Symbol	Half-Life	Natural Activity
Uranium 235	^{235}U	7.04×10^8 yr	0.72% of all natural uranium
Uranium 238	^{238}U	4.47×10^9 yr	99.2790% of all natural uranium; 0.5 to 4.7 ppm total uranium in the common rock types
Thorium 232	^{232}TH	1.41×10^{10} yr	1.6 to 20 ppm in the common rock types with crustal average of 10.7 ppm
Radium 226	^{226}Ra	1.60×10^3 yr	0.42 pCi/g (16 Bq/kg) in limestone and 1.3 pCi/g (48 Bq/kg) in igneous rock
Radon 222	^{222}Rn	3.82 days	Noble Gas; annual average air concentrations range in the US from 0.016 pCi/L (0.6Bq/m^3) to 0.75 pCi/L (28 Bq/m^3)
Potassium	^{40}K	1.28×10^9 yr	Soil – 1-30 pCi/g (0.037-1.1 Bq/g)

Cosmogenic Nuclides

Nuclide	Symbol	Half-life	Source	Natural Activity
Carbon 14	^{14}C	5730 yr	Cosmic-ray interactions, $^{14}N(n,p)^{14}C$	6 pCi/g (0.22 Bq/g) in organic material
Tritium 3	^3H	12.3 yr	Comic-ray interactions with N and O; spallation from cosmic-rays, ^6Li (n,alpha) ^3H	0.032 pCi/kg (1.2 x 10^{-3} Bq/kg)
Beryllium 7	^7Be	53.28 days	Cosmic-ray interactions with N and O	0.27 pCi/kg (0.01 Bq/kg)

Human Produced Nuclides

Nuclide	Symbol	Half-Life	Source
Tritium	^3H	12.3 yr	Produced from weapons testing and fission reactors; reprocessing facilities, nuclear weapons manufacturing
Iodine 131	^{131}I	8.04 days	Fission product produced from weapons testing and fission reactors, used in medical treatment of thyroid problems
Iodine 129	^{129}I	1.57×10^7 yr	Fission product produced from weapons testing and fission reactors
Cesium 137	^{137}Cs	30.17 yr	Fission product produced from weapons testing and fission reactors
Strontium 90	^{90}Sr	28.78 yr	Fission product produced from weapons testing and fission reactors
Technetium 99	^{99}Tc	2.11×10^5 yr	Decay product of ^{99}Mo, used in medical diagnosis
Plutonium 239	^{239}Pu	2.14×10^4 yr	Produced by neutron bombardment of ^{238}U ($^{238}U + n \rightarrow ^{239}U > Np + \beta \rightarrow ^{239}Pu + \beta$)

http://www.physics.isu.edu/radinf/natural.htm
http://www.umich.edu/~radinfo/introduction/natural.htm

Liquid Scintillation Counting

The sample usually contains [^3H] or [^{14}C] either dissolved or suspended in a solvent with one or more fluorescent substances. The sample is 'counted' in liquid scintillation instruments which 'hold' quench curves in memory to convert cpm to dpm (Zeidan and Dashek, 1996).

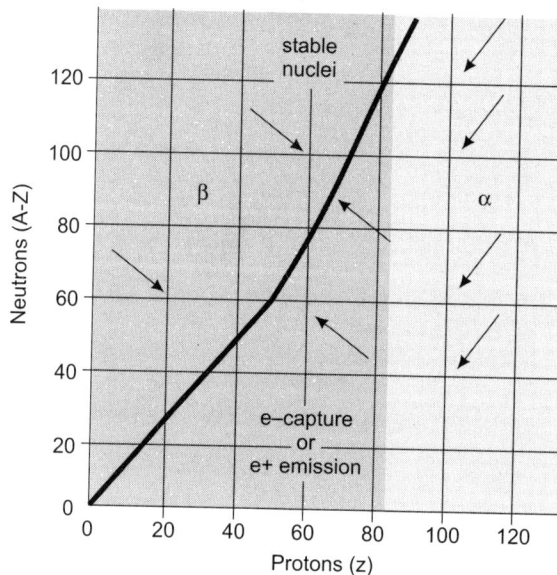

Types of decay

Alpha decay
$$^{238}U \rightarrow {}^{234}TH + {}^{4}He \; ({}^{234}Th + {}^{2}e^{-} + 4He^{2+})$$

Beta decay
$$^{14}C \rightarrow {}^{14}N + b^{-} + \bar{n}$$
$$^{22}Na \rightarrow {}^{22}Ne + e^{-} b^{+} n + Q - 1022 \text{ keV}$$
$$^{22}Na \rightarrow {}^{22}Ne + n + Q(\text{energy})$$

Gamma decay
$$^{137m}Ba \rightarrow {}^{137}Ba + Y$$

Fig. 8.2 Types of radioactive decay. From: *http://207.10.97.102/chemzone/lessons/11nuclear/nuclear.htm*. Other diagrams of alpha, beta and gamma decay can by observed at *http://library.thinkquest.org/3471/radiation_types_body.html*.

Gas Flow Counting

If a gas is contained in a chamber in which there are two charged electrodes, a β-particle can pass through this gas and can dislodge an orbital electron from one of the atoms. This results in an ion pair, i.e., the dislodged electron and the remaining positively charged ion. The β-particle possesses sufficient energy to ionize several atoms successively. If the gas is contained in a chamber with two charged electrodes, the secondary electrons and positive ions will be attached to the anode and cathode, respectively. This can be detected as a tiny pulse of charge (Zeidan and Dashek, 1996). In addition to the above commonly used methods, there is a variety of other methods (Table 8.5) for determining radioactivity (Mann et al., 1980; Tsoulfandis and Tsoulfandis, 1985; Knoll, 1999; L'Annanziata, 2003).

TABLE 8.3 Units of Radioactivity

Unit or Quantity	Symbol	Application
Becquerel	Bq	SI quantity of radioactivity
		Bq = 1 disintegration/sec
		Bq = 2.7 x 10^{11} Ci
Curie	Ci	Quantity of radioactivity
		1 Ci = 3.7 x 10^{10} dps
		1 Ci = 3.7 x 10^{10} Bq
Gray	Gy	SI unit of absorbed dose
		1 Gy = 100 rad = 1 J/kg
Rad	Rad	Unit of absorbed dose
		1 rad = 0.01 Gy = 100 erg/g
Rem	Rem	Unit of dose equivalent rad x Q x other modifying factor

TABLE 8.4 Half-lives of Some Common Radionuclides[a]

Nuclide	t = ½
^{14}C	5.73 x 10^4 y
$^{137}C_3$	30 y
^{131}I	8 d
^{55}Fe	657 d
^{32}P	14 d
^{239}Pu	2.4 x 10^5 y
^{24}Na	15 h
^{90}Sr	29 y
^3H	12 y
^{235}U	7.1 x 10^8 y
^{65}Zn	245 d

[a]Whole body

Radioactivity as a Pollutant

Chapter 4, which discusses energy, describes nuclear reactors (Lamarsh and Baratta, 2001) and draws a distinction between them and nuclear weapons. However, both can release radioactivity (Eisebud and Gesel, 1997). This occurs via either a meltdown or inappropriate waste storage and disposal of the reactor waste or a blast of the weapon. Unfortunately, a meltdown of the Chernobyl reactor oc-

TABLE 8.5 Summary of Additional Methods for Determining Radioactivity

Method	Isotope Detected
Cherenkov Liquid	Sr-90
Scintillation Counting	Pb-210
Alpha Spectroscopy	Po-210
	Th-228
	Th-232
	U-238
	Pu-239
	Pu-240
Radon Transfer System	Ra-226

http://www.cartage.org/b/en/themes/sciences/chemistry/ analyticalchemistry/methodsInstr and Povinec, P. and Baxter, M. 2007. Analysis of Environmental Radionuclides, Elsevier, Cambridge, England.

curred (Milhaud, 1991; Tsyb, 1997; Zamostian et al., 2002). The main radioactivity that would be released in a nuclear reactor meltdown is tritium which possesses a long half-life (t ½ = 12.3 yrs) and can enter food chains. However, the [^{3}H] β is of low energy and ingestion of foods would be required to yield deleterious effects. Other isotopes from a nuclear power plant disaster are much more dangerous than tritium, possess a long half-life and can accumulate within living organisms (World Health Organization, 1995). These byproducts are stored in steel drums which must be re-packaged from time to time as the drums can degrade (Table 8.6). Unfortunately, the drums have been disposed of in landfills (Hopkins, 2005), underground repositories (Milner, 1985; Chapman and McKinley, 1987) and oceans (Frankena and Frankena, 1986; Van Dyke, 1988; Ringus, 2000).

As the number of U.S. nuclear generating units has increased from 1957–1992 (Table 8.7), the U.S. nuclear industry employs a triple line of defense to prevent nuclear power plant disasters. These are: 1. incorporation of safety factors into the basic design, construction and operation of the

TABLE 8.6 Accumulated Volume and Radioactivity of High-level Nuclear Waste, by Source, 1980-1992

Year	DOE/Defense		Commercial	
	Volume (thousand cubic meters)	Radioactivity (million curies)	Volume (thousand cubic meters)	Radioactivity (million curies)
1980	295	1,310	2.2	33.4
1981	305	1,577	2.2	32.7
1982	340	1,317	2.2	31.9
1983	351	1,248	2.2	31.2
1984	361	1,397	2.2	30.5
1985	355	1,465	2.2	29.8
1986	364	1,417	2.2	29.1
1987	379	1,277	2.2	28.4
1988	383	1,174	2.1	27.9
1989	379	1,081	2.4	27.3
1990	397	1,015	1.2	26.7
1991	395	971	1.7	26.2
1992	397	988	1.2	25.6

Source: U.S. Department of Energy. Integrated Data Base for 1992: Spent Fuel and Radioactive Waste Inventories, Projections, and Characteristics, Table 2.1. 9. 47 – 48. (Oak Ridge, TN: DOE/ORNL, 1992).

TABLE 8.7 U.S. Nuclear Energy Capacity and Production, 1957 – 1992

Year	Nuclear generating units	Net generation of electricity	Year	Nuclear generating units	Net generation of electricity
	number	Billion kilowatt hours		number	Billion kilowatt hours
1957	1	<0.05	1975	54	172.5
1958	1	0.2	1976	61	191.1
1959	1	0.2	1977	65	250.9
1960	3	0.5	1978	70	276.4
1961	3	1.7	1979	68	255.2
1962	5	2.3	1980	70	251.1
1963	6	3.2	1981	74	272.7
1964	6	3.3	1982	77	282.8
1965	6	3.7	1983	80	293.7
1966	8	5.5	1984	86	327.6
1967	10	7.7	1985	95	383.7
1968	11	12.5	1986	100	414.0
1969	14	13.9	1987	107	455.3
1970	18	21.8	1988	108	527.0
1971	21	38.1	1989	110	529.4
1972	29	54.1	1990	111	576.8
1973	39	83.5	1991	111	612.6
1974	48	114.0	1992	109	618.8

Source: U.S. Department of Energy, Energy Information Administration. Annual Energy Review 1992, Table 9.2, p. 241. DOE/EIA-0384 (92). Washington, DC: DOE/EIA, 1993).

reactor, 2. assumptions that the first line of defense may fail and 3. assumptions that the first and second lines of defense could also fail. The possibility of an explosion is remote as the hydrogen isotopes supplied to the reactor are continuously consumed preventing accumulation. A thorough discussion of the triple line of defense occurs in Turk and Turk (1988). Table 8.8 presents a summary of nuclear power plant accidents

TABLE 8.8 Summary of Radiation Accidents that Resulted in Environmental Contamination

Accident	Effects
1. Thermonuclear deterioration of March 1, 1954 Bikini Atol in the Mid Pacific Ocean Fallout	Effect on skin – itching and burning sensation 1 – 2 days followed by most groups on neighboring atolls Hematological effects – Diminution in white blood cell count Thyroid abnormalities
2. Windscale Reactor No. 1 in October, 1957 Coastal Strip in NW England Release of Fission Products	Dose of radioiodine to individuals from inhalation or direct exposure was compared to the dose received from dairy products
3. Oak Ridge plutonium release of 1959 Chemical explosion releasing plutonium to nearby streets and building surfaces Plutonium forced into a chemical processing plant	None was injured
4. Army stationary low power reactor in 1961 Explosion of the national reactor testing in Idaho	Radiation levels outside the reactor building were minimal
5. Houston Incident of March, 1957 Opening of a sealed can containing 10 pellets of ^{192}Ir	Minor radiation burns
Fermi Reactor fuel meltdown Monroe, MI	No exposure to the employees or the public resulted
6. Abortive Reentry of the SNAP 9-A April 21, 1964 failed navigational satellite	^{238}Pu content of ground-level air was monitored and it was concluded that the 50 year dose culminated the pulmonary lymph node between 1961 and 1968 was 36 Millirem
7. Lubmin Nuclear Power Plant, 1976, East Germany	Radioactive core almost melted down
Three Mile Island March 28, 1979 Harrisonburg, PA, USA	One of two reactors lost coolant thereby causing overheating and partial meltdown of the uranium core, some radioactive gases and water released
8. Chernobyl April 26, 1986 (see text discussion)	
9. Uranium – Processing Nuclear Fuel Plant Sept. 30, 1999 Tokaimua, Japan	Uncontrolled chain reaction in a uranium-processing nuclear power plant released high levels of radioactive gas, killed two and injured one
10. Explosion on a Russian Nuclear Powered Submarine August 12, 2000, Barents Sea	All 118 on board are killed, submarine recovered and defuelled
11. Nuclear Power Plant Accident, Sept, 2000 Bulgaria	Two to five people contaminated
12. Millstone Power Plant March 7, 2003 Radioactive Leakage Connecticut, USA	Emergency Declared
13. Paks Nuclear Power Plant April 10, 2003 Hungary	Leakage of radioactivity into the environment
Nuclear Power Plant in Japan August 9, 2004 Mihama, Japan	Non-radioactive steam leaked from killing four and burning seven
14. Nuclear Power Plant Explosion 2005, St. Petersburg,	3 injured, no reported radioactive leakage
15. Russia	

[a]Adapted from Eisebud and Gessel (1997) and http://www.nuclearfiles.org.hitimeline/nwq/oo/2000s.htm

TABLE 8.9	Summary of Radiation Exposures from Everyday Risks
Exposure	Millirems Annually/Individual
Television	Less than 10
Nuclear Energy	Less than 10
Water Supply	Less than 10
Mining and Farming	Less than 10
Burning Fuels	Less than 10
Atomic Weapons	Less than 10
Nuclear Medicine	15
Cosmic Rays	27
Soil and Rocks	28
Diagnostic X-rays	40
Food and Water	40

as well as other incidents which have released environmental contamination. In addition, graphs presenting trends for nuclear power plant incidents and failures can be found at *http://www.atom.meti.go. jp/atom-db/h/trouble.*

While possible nuclear power plant accidents can be very serious (Table 8.8), there are other types of radiation accidents which could produce adverse environmental radiation. These accidents can result from mishaps involving radiation devices, nuclear medicine radioisotopes (teletherapy), industrial radiography, nuclear powered submarines and uranium processing plants (*http://www.nuclearfiles.org* and *http://www.infoplease.com*).

What are the biological effects of radioactivity?

While Table 8.4 presents the exposure to radioactivity from everyday risks (Mettler and Mosely, 1985), the biological effects of various doses of radioactivity are depicted in Table 8.10. A useful exercise is to relate these effects to those resulting from atomic blasts (RERF, 1987; UNSCEAR, 2000, Israel, 2002) and the Chernobyl nuclear power

disaster (Agence pour l'energie nucle'aire, 2003; International Federation of Red Cross and Red Crescent Societies, 2002; IPHECA, 1994; Medvedev, 1992; World Health Organization, 1995). The biological effects prompt a need for radiation protection (Noz and Maguire, 1995) which includes instrumentation, absorbed dose, distance shielding as well as radionuclides and the law.

Research Activities

Some research activities regarding radioactivity are presented in Table 8.11. It is apparent that these activities are quite diverse ranging from medical to environmental concerns.

Storage and Disposal of Radioactive Waste

The United States Environmental Protection Agency lists five characteristics of radioactive waste (rad waste). There are: 1. spent nuclear fuel from nuclear reactors and high-level waste originating from spent nuclear fuel reprocessing, 2. transmutation waste deriving chiefly from defense programs, 3. uranium mill tailing originating from the mining and milling of uranium ore, 4. low-level rad waste and 5. naturally occurring and accelerator-generated rad waste. Low-level rad waste (LLW) includes: contaminated clothing, exhausted resins or other non-usable or non-recyclable materials. In contrast, high-level waste consists of fuel assemblies, rods and waste separated from reactor spent fuel (American Nuclear Society, 2003). Willard and Miller (1990) have published a handbook of rad waste.

Before radioactive waste can be either stored or disposed of (Fig. 8.3) (Gershey et al., 1990; Wasserman, 1985), it is usually analyzed, i.e., detecting the occurrence of

TABLE 8.10 Pathophysiological Events[a]

Dose Range (cGy)	Prodromal Effects	Manifest-Illness Effects	Survival
75 – 150	Mild	Slight decrease in blood cell count	Virtually certain
150 – 300	Mild to Moderate	Beginning symptoms of bone marrow damage	Probable (>90%)
300 – 530	Moderate	Moderate to severe bone marrow damage	Possible – bottom third of range: $LD_{5/60}$ middle third: $LD_{10/60}$ top third: $LD_{50/60}$
530 – 830	Severe	Severe bone marrow damage	Death within 3.5 to 6 weeks Bottom half: $LD_{90/60}$ Top half: $GD_{99/60}$
830 – 1100	Severe	Bone marrow pancytopenia and moderate intestinal damage	Death within 2 to 3 weeks
1100 – 1500	Severe	Combined gastrointestinal and bone marrow damage, hypotension	Death within 1 to 2.5 weeks
1500 – 3000	Severe	Gastrointestinal damage upper half of range: early transient incapacitation (ETI), gastrointestinal death	Death within 5 to 12 days
3000 – 4500	Severe	Gastrointestinal and Cardiovascular damage	Death within 2 to 5 days

[a] From: Bier, V. 1990. Toxic Effects of Exposure to Low Levels of Ionizing Radiation. National Academy of Sciences. Radiation protection information can be found in Noz and Maguire (1995). Radiation hazards to fish, wildlife and invertebrates can be found in Eisler (1994).

Exposure (rem)	Effect on Health	Time to Onset
	radiation burns more severe as exposure increases.	
5 – 10	Changes in blood chemistry	
50	Nausea	hours
55	Fatigue	
70	Vomiting	
75	Hair loss	2 – 3 weeks
90 diarrhea		
100 hemorrhage		
400	Death from fatal doses	within 2 months
1,000	Destruction of intestinal lining internal bleeding death	
2000	Damage to central nervous system loss of consciousness Death minutes hours to days	

radionuclides as well as their amounts and levels. Their presence can be detected by alpha or gamma spectroscopies, neutron interrogation and liquid scintillation counting. Then, radioactive waste can be stored or disposed of (Eisebud and Gessel, 1997). Some wastes are stored to allow them to decay (DIS) to a reduced level of radioactivity, to hold for processing or to retain them for approved disposal (Radwaste.org). Disposal may require transportation via ship or airplane under IAEA regulation of the safe transport of radioactive materials.

In Sweden, spent nuclear fuel may be disposed in Swedish crystalline bedrock if formal legal consent is obtained.

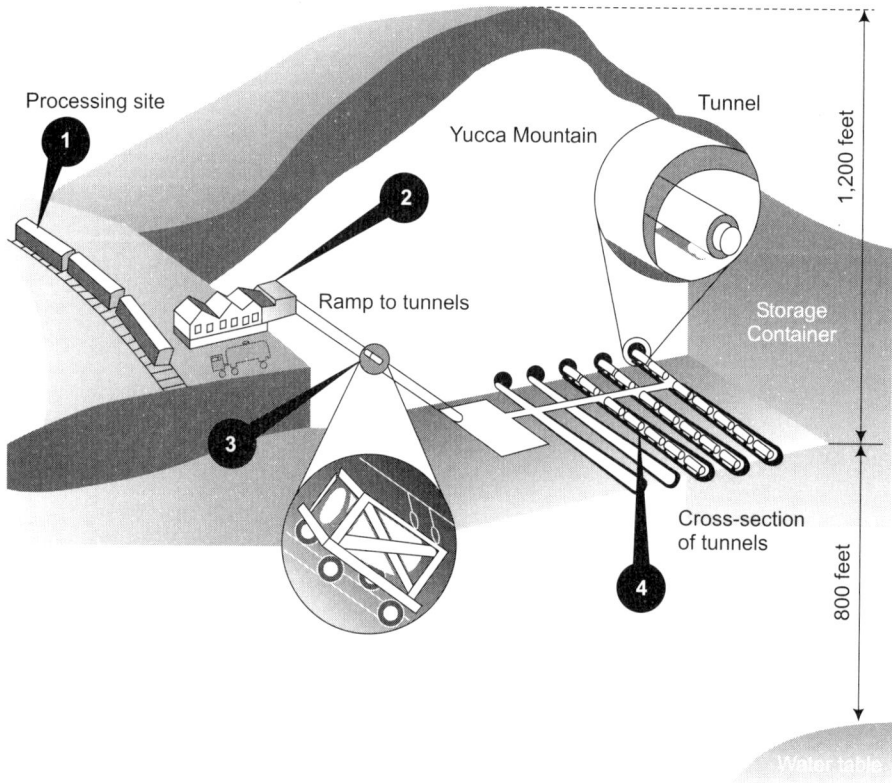

Fig. 8.3 Proposed Yucca Mountain Radioactive Waste Depository.
From: *http://www.nrc.gov/waste/hlw-disposal/design.html.*

The proposed disposal by Svensk Karnbranslehantering employs a copper cannister with cast iron encapsulating spent fuel. These would be buried at approximately 500m. The cannisters would be surrounded by a betonite backfill. Another proposed Swedish method is deep sea disposal of radioactivity. The radwaste would be packaged in glass or corrosion-resistant containers. Then, burial in deep ocean sediments is accomplished by either penetrators (a device dropped form a ship to the ocean floor) or drilling placement employing oil-industry-based technology). Whereas the penetration can embed itself 50 m into the

sediment, drilling of holes to 800 m or perhaps, 2000-3000 m below the sea-bed is foreseen for the drilling method. Then , the holes would be filled with 900 cubic m of radwaste and 'topped-off' with a 300m long concrete plug (*http:www.ssi.se/ssi-vapporter/pdf/ssi-rapp-2005-02.pdf*).

A method of converting long-lived radionuclides is transmutation (Khankhasayen et al., 1996), a process of altering the atom's nucleus. During the process, a long-lived radioactive nucleus is changed into one with a shorter half-life. Next, radioactive nuclei are separated from non-radioactive nuclei. Then, the former are

TABLE 8.11 Summary of Research Activities Regarding Radioactive Pollution

Activity	Purpose	Reference
Environmental Transfer Models	Modeling of radionuclide behavior	Momoshima (2002)
Development of Analytical Instruments	Measurement of low level Characteristics of elements	Momoshima (2002)
Problems of Spent Nuclear Fuel Inventory	Storage of radioactive waste	http://www.fiet/structure/300/300.htm
Migration and Accumulation of Radionuclides in the Atmosphere, Water Systems and Soil	Investigate long distance transport Involving physical and chemical properties of radionuclide carriers and their transformation in the middle and lower troposphere	http://www.fiet/structure/300/300.htm
Global Geophysical Problems Related to Atmospheric Radioactivity	Role of inert gases, El-Nino Phenomenon, Total ozone amount variations	http://www.fiet/structure/300/300.htm
Development of radioactive tracer methods	Application in geophysics and physics	http://www.fiet/structure/300/300.htm
Evaluation of self-cleaning from radionuclide	Self-cleaning of waste systems and watersheds	http://www.fiet/structure/300/300.htm
Investigation of unstable and/or radioactive iodine transport in human body and thyroid dose assessment	Reduce iodine deficiency	

TABLE 8.12 Some Recently Proposed Methods for Treating Low Level Radioactive Waste[a]

Method	Brief Summary of the Process
Vitrification	Low-level rad waste is heated together with glass making material. The mixture melts and the melted mixture is used to make glass beads or discs.
Plasma Arc Furnace	High temperature plasmas can be employed to 'cut' and melt waste material. Plasma is any high temperature gas that has an electrical charge.
Molten Metal Treatment	Low level waste is passed through a bath of metal at 1500 - 1575°. Many of the radioactive materials are captured in a froth that rises to the metal's top.

[a]From: Fentiman, A.W. 2005. http://www/ag.ohio-state.edu/~rer/rerhtml/rer-50.html

bombarded with high-speed subatomic particles generated by either accelerators or nuclear reaction. When the particles are captured by radioactive nuclei, the latter are changed. This method does not appear to be economically feasible since it requires either expensive accelerators or nuclear reactors. To date, the method has only been performed in laboratories and with small quantities of radioactive material.

In recent times, there have been attempts (Table 8.12; Fig.8.4) to dispose of nuclear wastes by means other than in landfills, respositories and oceans (Van Dyke, 1988; Livingston, 2004; Ringus, 2000). However, no satisfactory procedure exists for the permanent disposal of radioactive waste as there are social issues (Walker et al., 1983).

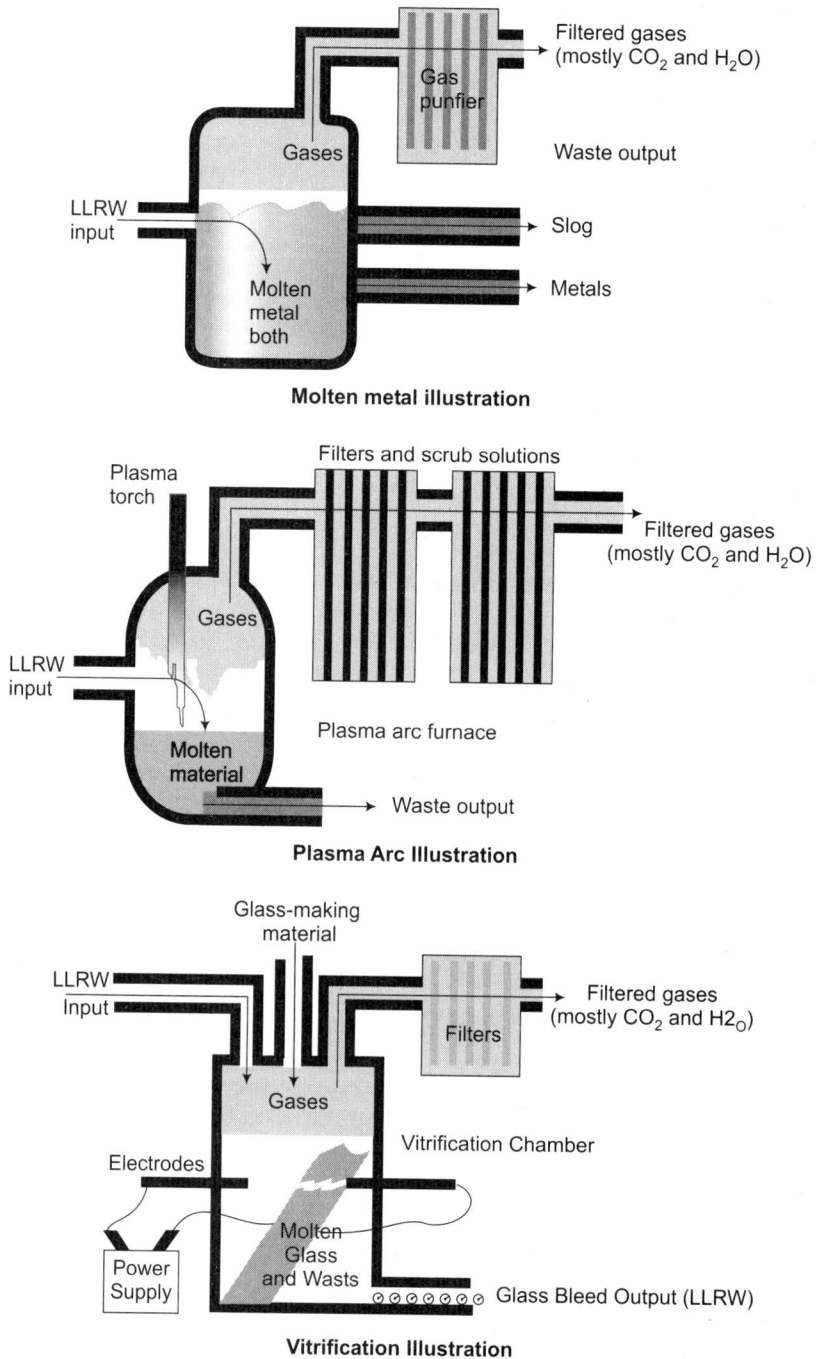

Molten metal illustration

Plasma Arc Illustration

Vitrification Illustration

Fig. 8.4 Recently proposed methods for the disposal of radioactive waste. From: Fentiman, A.W. 2005. http://www.ag.ohio-state.edu/~rer/rerhtml/rer-50.html.

References

Agence pour l'energie nucle'aire. 2003. Chernobyl: Assessment of Radiological and Health Impacts. *http://www.nea.tr/html/rp/chernobyl/welcome.html.*

American Nuclear Society. 2003. *10th International High-Level Radioactive Waste Management Conference.* Las Vegas, Nevada, USA.

Bier, V. 1990. *Health Effects of Exposure to Low Levels of Ionizing Radiation.* National Academy Press, Washington, DC., USA.

Browne, E. and Firestone, R. B. 1986. *Table of Radioactive Isotopes.* Wiley, NewYork, NY, USA.

Chapman, N. A. and McKinley, I.G. 1987. *The Geological Disposal of Nuclear Waste.* Wiley, New York, NY, USA.

Eisebud, M. and Gessel, T.F. 1997. *Environmental Radioactivity From Natural, Industrial and Military Sources.* Elsevier, St. Louis, MO, USA.

Eisler, R. 1994. *Radiation Hazards to Fish Wildlife and Invertebrates A Synoptic Review.* US Dept. of Interior, Washington, DC, USA.

Frankena, F. and Frankena J.K. 1986. *Radioactive Waste Disposal in the Ocean.* Vance Bibliographies, Monticello, IL, USA.

Gershey, E. L.,Wilkerson, A., Party, E. and Klein, R.C. 1990. *Low-Level Radioactive Waste From Cradle to Grave.* VanNostrand Reinhold, New York, NY, USA.

Hopkins, P.W.O. 2005. *Geology and Mineralology of Radioactive Waste Repositories.* Springer Verlag, New York, NY, USA.

Hooshyar, M.A., Reichstein, I and Malik, F.B. 2005. *Nuclear Fission and Cluster Radioactivity: An Energy-Density Functional Approach.* Springer. New York, NY, USA.

International Federation of Red Cross and Red Crescent Societies. 2002. Chernobyl: A chronic disaster. *http://www.ifrc.org/publicat/wdr2000/wdrch5.asp.*

IPHECA. 1994. *International Program on the Health Effects of the Chernobyl,* 1994 Report of International Consultation on "Chernobyl Accident Recovery Workers Project", St. Petersburg, Russia.

Israel, Y.A. 2002. *Radioactive Fallout after Nuclear Explosions and Accidents.* Elsevier, Cambridge, England.

Kaku, M., Trainer, J. and Trainer Thompson, J. 1990. *Nuclear Power Both Sides – The Best Argument For and Against the Most Controversial Technology.* Norton and Co., New York, NY, USA.

Kathren, R.L. 1984. *Radioactivity in the Environment.* Harwood Academic Publishers. New York, NY, USA.

Khankhasayen,M.Kh., Kurmanov, N.Zh. and Phedl, H.S. 1996. *Nuclear Methods for Transmutation of Nuclear Waste. Problems, Perspectives Cooperative Research* Proceedings of the International Workshop. Dubna, Russia.

Knoll, G. 1999. *Radiation Detection and Measurement.* Wiley, New York, NY, USA.

Lamarsh, J. R. and Baratta, A. 2001. *Introduction to Nuclear Engineering.* Prentice Hall, Upper Saddle River, NJ, USA.

L'Annanziata, M.F. 2003. *Handbook of Radioactivity Analysis.* Elsevier, Cambridge, England.

Livingston, H.D. 2004. *Marine Radioactivity.* Elsevier, Cambridge, England.

Magill, J. and Galy, J. 2004. *Radioactivity Radionuclides Radiation.* Springer Verlag, New York, NY, USA.

Mann, W. B., Ayers, R.L. and Garfinkel, S.B. 1980. *Radioactivity and Its Measurement.* MacMillan, New York, NY, USA.

Medvedev, Z.A. 1992. *The Legacy of Chernobyl.* W.W. Norton and Co., New York, NY,USA.

Mettler, F.A. and Mosely, R.D. 1985. *Medical Effects of Ionizing Radiation.* Grume and Stratton, Orlando, FL, USA.

Milhaud, G. 1991. The lesson of the Chernobyl disaster. *Biomedpharmacother* 45:219-220.

Milner, A. G. 1985. *Geology and Radwaste.* Academic Press, New York, NY.

Momoshima, V. 2000. Present research on the environmental radioactivity and its perspectives to Japan. *J. Nucl. Radiochem. Sci.* 1:43-45.

Noz, M.E. and Maguire, G.Q. 1995. *Radiation Protection*. Health Sciences, Singapore, Malayasia.

Ringus, L. 2000. *Radioactive Waste Disposal at Sea: Public Ideas, Transnational Policy, Entreprenews, and Environmental Regimes (Global Environment Accord Strategies for Sustainability and Institutional Innovation)*. MIT Press, Cambridge, MA, USA.

Scott, E.M. 2003. *Modelling Radioactivity in the Environment*. Elsevier, Burlington, MA, USA.

Tsoulfanidis, N. and Tsoulfanidis, N. 1985. *Measurement and Detection of Radiation*. Taylor and Francis, London, England.

Tsyb, A.F. 1997. The Chernobyl disaster: global impact. International Conference on Human Detoxification. Stockholm, Sweden.

Turk, J. and Turk, A. 1988. *Environmental Science*. Saunders College Publishing, College Station, Texas, USA.

Twardowska, I., Allen, H.E., Kettrup, A.F. and Lacy, W.J. 2004. *Solid Waste: Assessment, Monitoring and Remediation*. Pergamon, Cambridge, England.

Tykva, R. and Berg, D. 2004. *Man-Made and Natural Radioactivity in Environmental Pollution and Radiochronology*. Springer, New York, NY, USA.

United Nations Scientific Committee on the Effects of Atomic Radiation (UNSCEAR). 2000 Report. *Sources and Effects of Ionizing Radiation*. United Nations, New York, NY, USA.

US – Japan Joint Reassessment of Atomic Bomb Radiation Dosimetry in Hiroshima and Nagasaki: Final Report: DS86: 1987. Radiation Effects Research Foundation (RERF). Hiroshima, Japan.

Van Dyke, J.M. 1988. Ocean disposal of nuclear waste. *Marine Policy* 12:82-95.

Walker, A., Gould, L.C. and Woodhouse, E.J. 1983. *Too Hot To Handle? : Social and Policy Issues in the Management of Radioactive Waste*: Yale University Press, New Haven, CT, USA.

Wasserman, U. 1985. Disposal of radioactive waste. *J. World Trade* 19:425-428.

Willard, M.E. and Miller, R.M. 1990. *Environmental Hazards: Radioactive Materials and Waste, A Reference Handbook*. Santa Barbara, CA. American Nuclear Society. (2001). 9th International High-Level Radioactive Waste Management Conference. Las Vegas, Nevada, USA.

World Health Organization. 1995. *Health Consequences of the Chernobyl Accident* (summary report) Geneva, Switzerland. pp. 21-24.

Zamostian, P., Moysich, K.B., Mahoney, M.C., McCarthy, P., Bondar, A., Noschenko, A.G. and Michale, A.M. 2002. Influence of various factors on individual radiation exposure from the Chernobyl disaster. *http://www.ehjournal.net*

Zeidan, H. and Dashek, W.V. 1996. *Experimental Approaches to Biochemistry*. McGraw-Hill/Wm. C. Brown, Dubuque IW, USA.

CHAPTER 9

Thermal Pollution

What is Thermal Pollution and What are its Sources?

Thermal pollution is the introduction of waste heat into water such as streams, rivers and lakes (Neves and Lourenco, 1996). Often, this introduction results in a diminution of dissolved oxygen (Science Junction, 2004) resulting in major environmental problems for the water's organisms, (Iversen, 1995). The sources of thermal pollution are the discharge of hot water from nuclear power plants and other industrial facilities (*http://www.powerscorecard.org /issue-detail and Gonyeau, J. 2001. http://www.nucleartourists.com/systems/ct.cfm?issue_id =6*). An example of a source of water thermal pollution is depicted in Fig. 9.1

The once-thorough cooling of a steam turbine employs nearby water to cool the condenser water. In this system, utilised river or lake water is returned to the stream at an elevated temperature. While hydroelectric and thermoelectric power plants

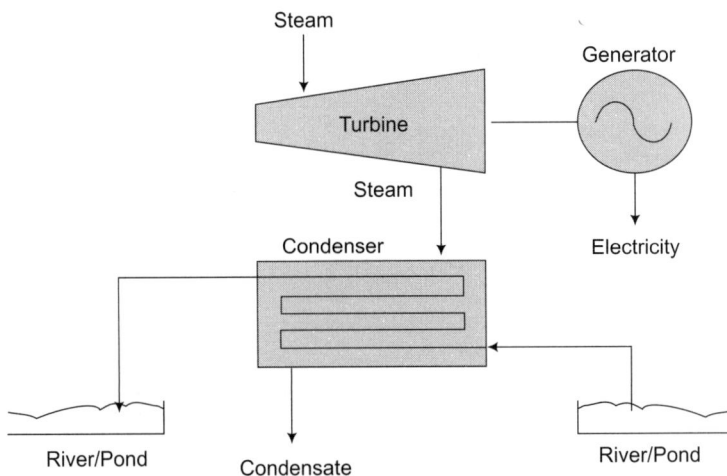

Fig. 9.1 An example of thermal pollution through a steam turbine.
From: *http://www.nrel.gov/docs/fy04osti/33905.pdf.*

TABLE 9.1 Effects of Hydropower Facilities on Water Use, Consumption and Quality[a]

Effect	Result
Diversion of water out of the river	Removes water from healthy in-stream ecosystems
De-watering of streches below dams	Organismal kills
Fluctuation in water flow from peaking operations	Create a "tidal effect" disrupting riparian community
Slowing of water flows allows silt to collect on river and resevoir bottoms	Hinders natural downstream migration of certain fishes

[a]Adopted from *http://www.powerscorecard.org/issue_detail.cfm?issue_id=5.*

consume water for cooling, gas turbine facilities and wind turbines and most solar photovoltaic systems do not (Table 9.1).

Biological Effects of Thermal Pollution

Perhaps the chief effect of thermal pollution affecting aquatic organisms (Kennedy and Mihursky, 1999) is the lowering of dissolved oxygen content via a decrease in the solubility (Fig. 9.2) of the gas in water (Krenkel and Parker, 1979). The negative effects (Table 9.2) of this decrease can be enhancements in photosynthesis and plant decay, a rise in respiration of certain organisms, shifts in types of species, elevated sensitivity to toxic wastes, parasites and diseases and increases in pathogenic organisms (Langford and Langford, 1990; USEPA, 2000). The effects are noticeable in young fish (Hart and Reynolds, 2002) which become susceptible to disease (*http://www. darisonk12.mi.us/academic/global/thermlpol. htm*). In addition, the growth of aquatic algae is affected by water temperature (Fig. 9.3) and coral reefs can be stressed especially from cooling water discharges from desalination and power plants adding to the thermal load. On a positive note, thermal pollution has been employed for agriculture.

Finally, the shutting down of power plants for maintenance can cause thermal shocks from abrupt temperature changes.

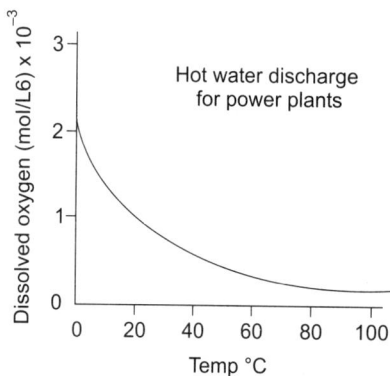

Fig. 9.2 Relationship between oxygen solubility and temperature.

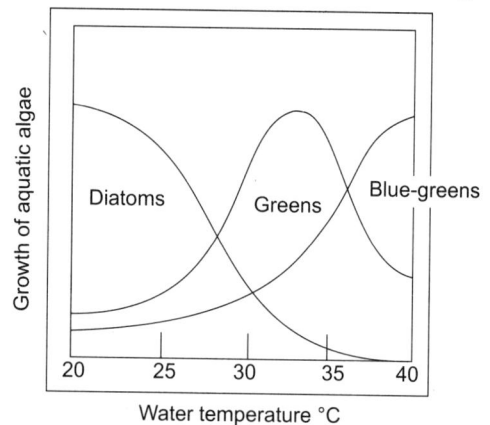

Fig. 9.3 Temperature shifts of algae populations From: Krenkel, P.A. and Novotry, V. Water Quality Management 2000. Academic Press, New York, NY, USA.

TABLE 9.2 Effects of Thermal Pollution[a]

Effects	Result	Species	Effects
Lower dissolved O_2 content by decreasing solubility of O_2 in water	Death of young fish	Pathogenic Organisms	Enhance growth causing disease in fish
Causes aquatic organisms to increase their respiration rates and consume O_2 more quickly	Enhances susceptibility to disease, parasites and toxic chemicals	Aquatic species[b] clams	Shift of clams to yield insect larval forms. Shift of clam population to insect larval forms near discharge.
Promotes faster metabolic rate in fish	Decrease in life expectancy of insects causing a shortage of fish food	Catfish	Catfish near hot discharge plane appear to be in inferior physical condition
		Brook trout	Swimming is affected by temperature

[a]Adapted from: http://www.darisonk12.mi.us/academic/global/thermpol.htm.
[b]Aquatic ecosystems subjected to not only the effects of an elevated average temperature but also to thermal shocks of unusual rapid temperature change.

This is especially important to organisms which may have acclimated to heated water discharge, e.g., certain insect larvae.

How can Thermal Pollution be Controlled?

The control mechanisms (Table 9.3) are diverse and dependent upon the causes of thermal pollution (*http://www.darisonk12.mi.us/academic/global/thermpol.htm*). With regard to deforestation and soil erosion, plowing of fields in a circular motion, discing fields rather than plowing them, using ground cover species and restriction of tree removal from shorelines have been proposed. As for hot water discharges from power plants (Fig. 9.4) the following have been suggested: 1. Use wind, solar and hydroelectric power as alternatives to nuclear power, 2. Operate power plants at a higher efficiency employing magnetohydrodynamics (Turk et al., 1984) and 3. Find alternative uses of hot water discharge from plants. For the latter, Turk et al. suggest the measures in Table 9.3. Finally, the question of whether thermal pollution can influence climate change needs to be explored.

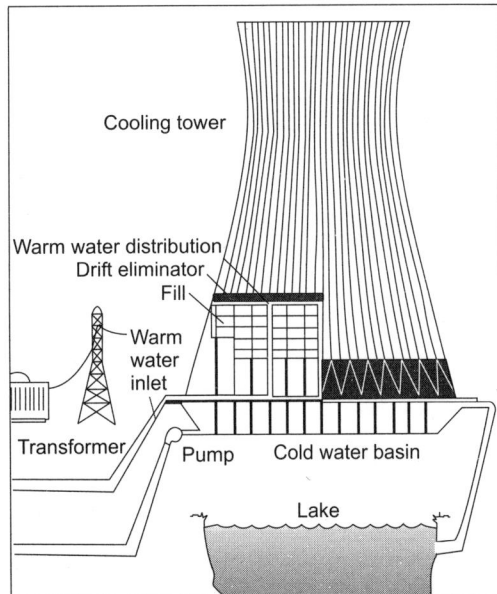

Fig. 9.4 Schematic of cooling tower. From: *http://teaching-ust.hk/~ceng51bites/CENG131 Lecture5.pdf*

Table 9.4 presents some analytical techniques and research efforts concerning thermal pollution.

TABLE 9.3 Pollution Control Mechanisms

Thermal pollution control, *http://www.darisonk12.mi.us/academic/global/thermpol.htm*, involves:
  Prevention of soil erosion
  Plowing of fields in a circular motion
  Discing fields instead of plowing them
  Using ground cover species
Energy chips
  Convert excess heat into energy
Less nuclear power
  Use wind power, solar power. hydroelectric power or hydrogen fuel (photovoltaic cells) cells as alternatives
End shoreline deforestation
  Restrict the removal of trees that protect the shoreline from the sun's heat.
Desalinization of seawater
  Employ barometric columns to evaporate water discharge
Operate power plants at higher efficiency
  Use the principal of magnetohydodynamics (Turk et al., 1984) to produce hot electrified air
Alternatives uses of hot water discharge from plants (Turk et al., 1984)
  Improve growth conditions in greenhouses
  Utilization of hot water for irrigation open field crops
  Enhance growth rate of certain fish by rearing them under controlled temperatures
  Speed sewage decomposition
  Dispose of heat in the air
Finally, thermal pollution is monitored by the Federal Water Pollution Control Administration. The question of whether Thermal Pollution influences climate change is open.

TABLE 9.4 Some Analytical Techniques and Research Activities for Thermal Pollution

Analytical Techniques	References
Dissolved Oxygen	Keith (1996), Virginia Dept. of Environmental Quality (2003), Washington State Dept. of Ecology (1994)
Application of Satellite Remote Sensing to Coastal Management	Sanderson (2001)
Invertebrate Species Counts Trent Biotic Index Biological Monitoring Working Party	NERC'S Center for Ecology and Hydrology
River Invertebrate Prediction and Classification System	http://www.seagrant.umn.edu/groundwater/pdfs/gwsourcew.pdf
Research Efforts	Miloh and Kit (2005)
Examination of Cooling-water Discharge From Power Plants on Thermal Pollution of Rivers and lakes	http://www.eng.tauoc.it/units/advanced/env/environment.html
Cleaner Ecosystem Model Employed for Reservoirs, Well-Stratifies Lakes, Shallow Lakes and High Mountain Lakes Lakes	Center for Ecological Monitoring Report (1978)
Develop Aerial Reconnaissance and Infrared Scanners and Temperature Measurement of Cooling Water Discharge from Power Plants	Scholt (1979) AA (Calspan Corp., Buffalo, NY, USA.)

[a]An overview of methods for stream ecology can be found in Hauer and Lamberti (1996).

References

Hart, P.J.B. and Reynolds J.D. 2002. *Handbook of Fish Biology and Fisheries*. Blackwell Pub., Malden. MA., USA.

Hauer, F.R. and Lamberti G.A. 1996. *Methods in Stream Ecology*. Academic Press, San Diego, CA., USA.

Iversen, E.S. 1995. *Living Marine Resources Their Utilization and Management*. Springer, New York, NY, USA.

Keith, L.H. 1996. *Compilation of EPA's Sampling and Analysis Methods*. CRC Press, Boca Raton, FL, USA.

Kennedy, V. and Mihursky, J., eds. 1999. *The Effects of Temperature on Invertebrates and Fish. A Selected Bibliography*. Maryland Sea Grant. University of Maryland, College Park, MD, USA.

Krenkel, P.A. and Parker, F.L. 1979. *Physical and Engineering Aspects of Thermal Pollution*. Vanderbuilt University Press, Nashville, TN, USA.

Langford, T.E. and Langford, T.E. 1990. 'Ecological Effects of Thermal Discharges'. Elsevier, New York, NY, USA.

Neves, R. and Lourenco, S. 1996. Thermal Pollution. *http://cape.Canterbury.ac.nz/archive/THERMAAL/ttel.htm*.

Sanderson, P.G. 2001. The application of satellites remote sensing to costal management in Singapore. *AMBIO* 30:43-48.

Science Junction – Water What Ifs – Dissolved Oxygen Lessons. 2004. www.ncsu.edu/sciencejunction/depot/experiments/waterlessons/do./uk.

Turk, J., Turk, A. and Arms, K. 1984. *Environmental Science*. W.B. Saunders Co., Philadelphia, PA, USA.

U.S. Environmental Protection Agency. 2000. Ambient aquatic life water quality criteria for dissolved oxygen. (saltwater): Cape Cod to Cape Hatteras. EPA. Quality Office of Water. Washington, DC, USA.

Virginia Dept. of Environmental Quality. 2003. Virginia Citizen Water Quality Monitoring Program Methods Manual. *http://www.deq.stateva.u/cmonitor*.

Washington State Department of Ecology. *Water Quality Program: A Citizens Guide to Understanding and Monitoring Lakes and Streams. 1994. Chapter 4 From the Fields to the Lab. How to Measure Dissolved Oxygen*. Olympia, WA, USA.

CHAPTER 10

Noise Pollution

WHAT CONSTITUTES NOISE POLLUTION AND WHAT ARE ITS SOURCES?

Noise pollution (Tripathy, 1999) is difficult to define as some noises are offensive to certain individuals and not to others. In addition, noise is transient and, thus, difficult to monitor its cumulative effects. However, it is possible to measure sound intensity (the decibel or dB) and relate dBs to both human and animal health. dB is the magnitude of the fluctuation in air pressure caused by sound waves. Rather than being arithmetic, the decibel scale is logarithmic. The dB scale is usually weighted to dis-

criminate against the lower frequencies. Table 10.1 presents noise nomenclature and Table 10.2 depicts A-weighted intensities for certain human activities. Full, audible frequency range for young healthy ears extends from 20 Hz (cycles per second) to 20,000 Hz (Fig. 10.1). Human hearing is not sensitive to sounds in the 500 to 8,000 Hz range (Luttman and Haggart, 1983). Of interest is the observation that night time noise yields greater annoyance than do the same levels at day.

A non-quantitative way of defining noise pollution is if it promotes annoyance, sleeplessness, fright or any other stress reaction (*http://www.quiet.org/faq.html*).

TABLE 10.1 Some Important Noise Nomenclature[a]

H_z = frequency as measured in cycles per second.

Coulness Level = a figure expressed in a unit called phon or the sound pressure level weighted decibels of an equally loud pure tone.

dB = decibel, a measure on a logarithmic scale of a quantity such as sound pressure, sound power or intensity with respect to a standard reference value.

L eq = continuous sound level measured in dB. It is recommended for use as a common measure of noise exposure.

A = weighted sound level, a convenient and fairly accurate index of speech interference. Some recommendation noise levels are: working environment - <75 dB (A), for indoor good speech intelligibility – 45 dB (A), background sleep noise – 35 dB (A) and outdoor noise 45 dB (A). a.) a.) From: *http://www.defra.gov.uk/environment/noise/health/page10.htm*.

TABLE 10.2	Sources of Noise and Their Levels[a]
Source	dB
Aircraft takeoff 100 ft.	130 – 150
Air conditioner at 20 ft.	60
Alarm clocks at 2 ft.	80
Car passing at 15 ft.	70
Chainsaw	110
Firecracker	150
Freeway traffic	80
Garbage truck	100
Gunshot blast	140 – 170
Light traffic	50
Noisy restaurant	70
Normal conversation	55 – 60
Quiet library	30
Rocket launching pad	180
Rock band	120
Snowmobile	100
Subway	90
Threshold of pain	120 – 130
Thunderclap	130
Vacuum cleaner	70

[a]70 dB is the point at which noise begins to harm hearing.

A thorough review of community noise has been published by Berglund and Lindvall (1995). This publication, which was prepared for the World Health Organization (WHO) contains sections on 1. the physiology of hearing and related mechanisms, 2. pychoacoustics and 3. the mental and behavioural effects of community noise. This document also suggests guidelines for levels of community noise in different environments (Fig. 10.2).

WHAT ARE THE BIOLOGICAL EFFECTS OF NOISE?

The biological effects of noise have been thoroughly reviewed by Tobias et al. (1980), Vallot (1993), Kryter (1994); Ising and Kruppa (1993) and Berglund and Lindvall (1990). These effects can include: hearing loss, annoyance responses, sleep disturbances, cardiovascular problems, psychoendocrine and immunological effects (Table 10.3). A few examples of excess

TABLE 10.3	Observed Effects on Human Health[a]	
Effects		Reference
Alteration in Blood Pressure		VanKempen et al. (2002)
Annoyance reaction		Neus and Boikat (2000); Babisch (2003)
Cardiovascular		Hygge et al. (2002)
Cognitive ability		
Colds		
Digestive system problems		
General fatigue		
Greater sensitivity to minor accidents		
Hearing impairment		Luttman and Hoggart (1983)
Increased headaches		
Increased mental hospital admission rates		
Increased reliance on sedatives and sleeping pills		
Interference with speech		
Sleep disturbances		
Work performance effects		

[a]The reader is referred to the following for in depth coverage: Babisch (2000, 2002, 2003, 2005); Bistrup et al. (2001); Fay (1991); Hygge et al. (2002); Neus and Boikat (2000); Passchier – Vermeer (2003) and Stansfield et al. (2000).

noise in animals include biophony distur-bances in amphibia and insects, enzyme stress levels in certain mammals, reduced bird density near roadways, bird abandon-ment of nests and young, and disruption and killing of whales, dolphins and other

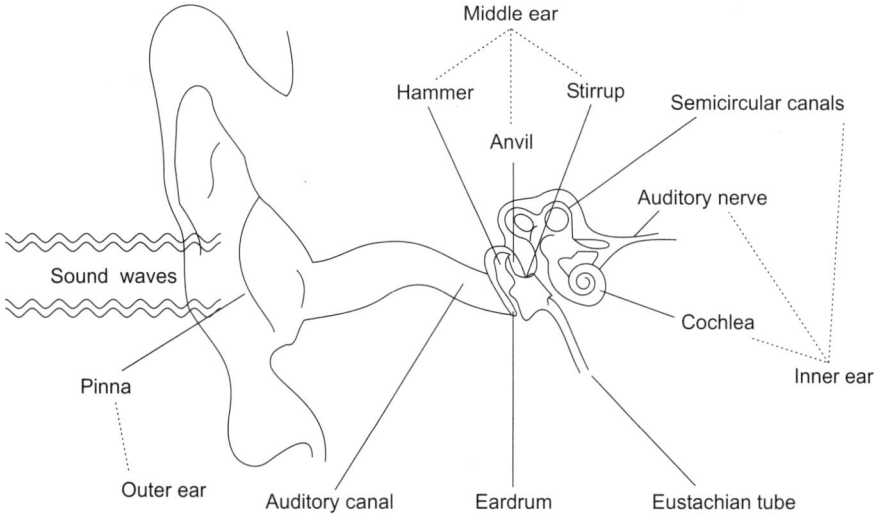

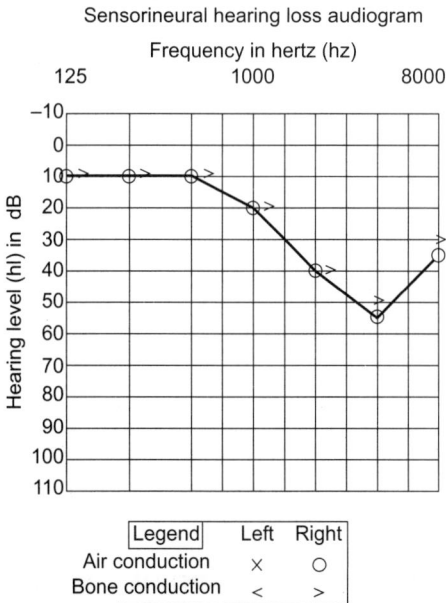

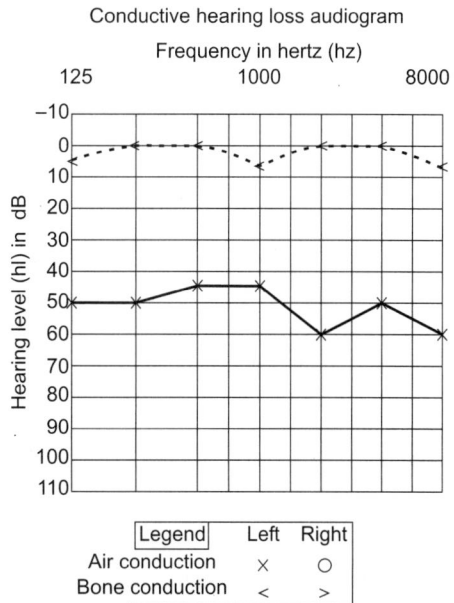

Fig. 10.1 Structure of the ear(top) and audiograms (bottom)
There are many varieties of these audiograms on Websites. Certain of these derive from Berlin, C.I. 1996. 'Hair Cells and Hearing Aids.' Singular. Publishing, San Diego, CA, USA.
Air Conduction   x = Left Ear, o = Right Ear   Bone Conduction > = Left Ear, < = Right Ear

OSHA Exposure Limits

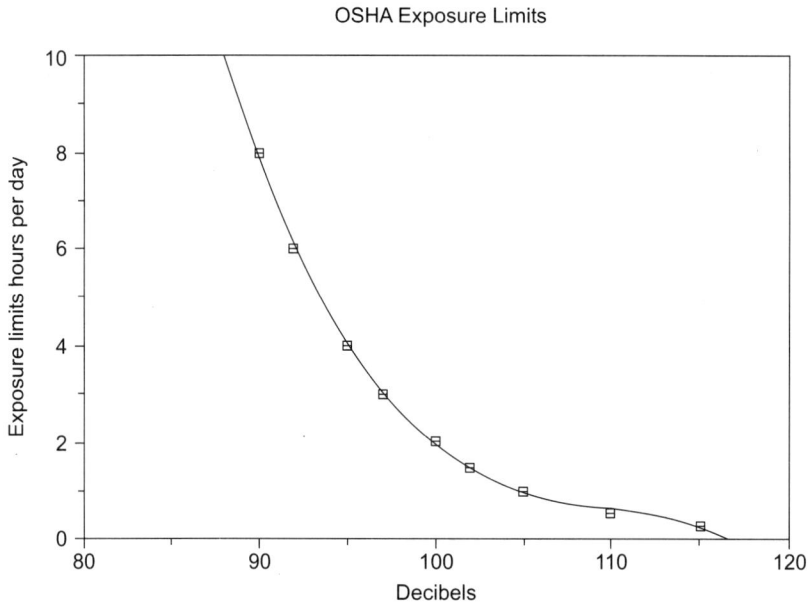

Duration of exposure (hrs/day)	Sound level dB(A)	
	NIOSH/ACGIH	OSHA
16	80	85
8	85	90
4	90	95
2	95	100
1	100	105
1/2	105	110
1/4	110	115*
1/8	115*	- - - -
		**

* No exposure to continuous or intermittent noise in excess of 115 dB(A)
** Exposure to impulsive or impact noise

Fig. 10.2 Comparison of OSHA and NIOSH Guidelines
From: *http://www.cdc.gov/niosh/hhe/reports/pdfs/1993-0863-2378.pdf.*

marine life. A thorough discussion of the effects of noise on terrestrial and marine wildlife can be found in Radle (1998).

The WHO noise guidelines are presented in Table 10.4. The symbols and units Laeq and dBA are defined in the caption to Table 10.1 and in Principal Descriptors of Environmental Sound (*http://www.non oise. org/library/envarticle/index.htm.*)

TABLE 10.4 The WHO Guideline Values (WHO, 2000, 2001, 2004)

Effects to be avoided	Effects criterion	1980	1995
Speech interference	100% intelligibility	45 LAeq	35 dBA
	reasonable		45 dBA
	intelligibility		
	loud speech		
	understood		55 dBA
Noise induced hearing loss	negligible risk	75 LAeq, 8hr	75 LAeq, 8hr
	increasing risk	140 dB	130-150 dB (peak)
Sleep disturbances	electrophysiological	35 LAeq	30 LAeq
	effects		
Cardiovascular disease		more research needed	more research needed
Performance effects	cognitive tasks		no specific criteria
	startle effects		no specific criteria
	reading skills in		no specific criteria
	children		
Thresholds of reported	moderate annoyance		50 LAeq
Annoyance	serious annoyance	50 LAeq	55 LAeq
Social behavior	reduced helping		80 dBA
	behavior		

TABLE 10.5 Some Analytical Noise Research Methods and Research Efforts

Analytical Method	Reference
Field Methods for Measuring Hearing Protection Performance	Franks et al. (2003)
Large Headphones that Enclose the External Ear	
Use of Speakers in a Small Testing Booth	
Loudness Matching	
Standard Laboratory Protocol for Estimation of the Field Attenuation of Hearing Protection Devices	Murphy et al. (2003)
Research Effort	Agency
Produce Strategic Noise Maps	Utrech the Netherlands
Provide Noise Control Standards and Information to Facilitate Decision-Making for Noise Action and Planning	http://www.nl.aeat.com
Contribute to a Better Understanding of the Effectiveness of Different Noise Control Policy Options	http://www.nl.aeat.com.
DEFRA Noise Mapping England Project	http://www.noisemapping.org.
Noise Reduction characterization of Nonlinear Hearing Protection Devices	US NIOSH
Field Evaluation of Hearing Protection Devices	http://www.idc.gov/noise.topics.noiseresearch/hearingprotectionresearch.html.

TABLE 10.6 Australian and USA Noise Control Measures[a,b]

Measure	Function
Noise Control Act of 1972	Act deals with the prevention of minimization of noise and vibration, Empowers the EPA, the waterways authority, local government and the police; EPA regulatory role in rail traffic noise
Environmental Planning and Assessment Act of 1979	Provides responsibility and opportunity for controlling environmental noise through the planning process
Motor Vehicles and Motor Vehicle Accessories Regulation 1995	Describes noise levels for classes of motor vehicles and restricts allowable noise levels for vehicles
Noise Abatement Program	A noise complaint register which provides noise abatement measures to reduce noise in sensitive locations, i.e., residences and schools
National Road Transport Commission	Developing new Australian design rules to limit noise from exhaust brakes
Commonwealth Government	Reduce or mitigate some of the noise at the Sydney airport
US Noise Laws and Regulations	
Noise Control Act of 1972	Federal Government Must Set and Enforce Uniform Noise Standards for Aircraft and Airports, Interstate Motor Carriers and Railroads, Workplace Activities, Medium and Heavy Duty Trucks, Motorcycles and Mopeds, Portable Compressors, as well as Federal Highways and Housing Projects
Aircraft Noise Abatement Act of 1968	Requires Federal Aviation Administration (FAA) to Develop and Enforce Safe Standards for Aircraft Noise
Airport and Airway Improvement act of 1982	Establish the Airport Improvement Program to Provide Assistance for Airport Construction Projects and Award Grants for Noise Mitigation
Federal-Aid Highway Act of 1970	Required Federal Highway Administration to Develop Standards for Highway Noise Levels Which are Compatible with Different Land Uses.
Aviation Environmental Analysis	Update and Improve Integrated Noise Model
Aircraft Noise Reduction and Control	Advances in Source Noise Reduction

[a]Adapted from: *http://www.epa.nsw.gov.ar/sol/97/chl/15-5htm*
[b]The reader is referred to *http://www.inceusa.org/pub.ncejindex.asp* for a review of analytical techniques.

WHAT ARE THE MEASURES TO CONTROL NOISE?

Noise control (Kamboj, 2002; Wang et al., 2004; Singal, 2005) has involved research (Table 10.5) as well as hearings and legislation (Behar et al., 2000). Some examples of EPA noise control measures in both Austraila and the USA are shown in Table 10.6. The USA EPA is responsible for Noise Control Collection, most of which was generated by the now unfunded office of Noise Abatement and Control for Congressional public Hearings. The 2001 website *http://www.epa.gov/historycollection/aid21.htm* provides a summary of the collection list. Noteworthy is the Noise Control Act of 1972 and a number of reports from congressional hearing in the 1970's (Table 10.6). It appears that the Australian EPA has kept more current than the US EPA (2004) as the US Federal Government has shifted its regulatory activities to individual states. Finally, the following are relevant to noise and society (Jones and Chapman, 1984; Tempst, 1985; Lara Saqnz and Steples, 1986; Bell, 1993; Bies, 2003).

WHAT ARE THE NOISE RESEARCH EFFORTS?

The research efforts involve developing analytical methods for noise pollution, diverse noise mapping as well as control and noise standards (Table 10.5). Many of the efforts are concerned with hearing protection Fig. 10.1.

References

Babisch, W. 2000. Traffic noise and cardiovascular disease; epidemiological review and synthesis. *Noise Health* 2:9-32.

Babisch, W. 2002. The noise/stress concept, risk assessment and research needs. *Noise Health* 4:1-11.

Babisch, W. 2003. Stress hormones in the research on cardiovascular effects of noise. *Noise Health* 5:1-11.

Babisch, W. 2005. Noise and Health. *Environ. Health Perspect.* 113:A14-A15.

Behar, A., Chasin, M. and Cheesman, M. 2000. *Noise Control: A Primer.* Singular Publishing/Thomson Learning. San Diego, CA, USA.

Bell, L.H. 1993. *Industrial Noise Control.* Marcel Dekker, London, UK.

Berglund, B. and Lindvall, T. (eds.). 1990. *Noise as a Public Health Problem.* Vol. 4. New Advances in Noise Research Part I. Stockholm, Sweden.

Berglund, B. and Lindvall, T. (eds.). 1995. *Community Noise. Archives of the Center for Sensory Research* 2:1-195.

Bies, D. 2003. *Engineering Noise Control: Theory and Practice.* Spon Press, Abingdon, UK.

Bronzaft, A.L. 1998. Effects of Noise. Encyclopedia of Environmental Science and Engineering '. Gordon and Breach (Pfallflin, J.R. and Ziegler, E.N., eds.). New York, NY, USA.

Bistrup, M.L., Hygge, S., Keiding, L. and Passchier-Vermeer, W. 2001. Health Effects of Noise on Children and Perception of Risk of Noise. National Institute of Public Health, Copenhagen, Denmark.

Fay, T.H. 1991. *Noise and Health.* Academy of Medicine, New York, NY, USA.

Franks, J.R., Murphy, W.J., Harris, D.A., Johnson, J.L. and Shaw, P.B. 2003. Alternative field methods for measuring hearing protector performance. *American Industrial Hygiene Association Journal.* 64:504-509.

Hygge, S., Evans, G.W. and Bullinger, M. 2002. A prospective study of some effects of aircraft noise on cognitive performance in school children. *Psychol. Sci.* 13:469-474.

Ising, H and Kruppa, B., (eds.). 1993. *Noise and Disease.* Gustav Fischer, Stuttgart, Germany.

Jones, D.M. and Chapman, A.J., (eds.). 1984. *Noise and Society.* Wiley, Chichester, UK.

Kamboj, N.S. 2002. *Control of Noise Pollution.* Deep and Deep Publications, New Delhi, India.

Kryter, K.D. 1994. *The Effects of Noise on Man.* Academic Press, San Diego CA, USA.

Lara Saqnz, A. and Steples, R.W.B. 1986. *Noise Pollution.* Wiley, Chichester, UK.

Luttman, M.E. and Haggart, M.P. 1983. *Hearing Science and Hearing Disorders.* Academic Press, London, England.

Murphy, W.J, Franks, J.R., Berger, E.H., Casali, J.G., Dixon-Ernst, C., Krieg, E.F., Mozo, B.T., Royster, J.D., Royster, L.H., Simon, S.D. and Stephenson, C. 2003. Development of a new standard laboratory protocol for estimation of the filed attenuation of hearing protection devices: Sample size necessary to provide acceptable reproducibility. *J. Acoust. Soc. Am.* 115:311-323.

Neus, H. and Boikat, U. 2000. Evaluation of traffic noise-related cardiovascular risk. *Noise Health* 2:65-77.

NPC Resources: Environmental Noise. 2006. *http://www.nonoise.org/library/envnoiseindex.htm.*

NSW EPA. 2004. *Noise Control Measures. http://www.epa.nsw.gov.ar/sol/97/chl/15-5htm*

Passchier-Vermeer, W. 2000. Noise and Health of Children. TNO report PG/VGZ/2000.042. Leiden: Netherlands Organization for Applied Scientific Research (TNO).

Passchier-Vermeer, W. 2003. Relationship between environmental noise and health. *J. Aviation Environ. Health Res. (suppl)*: 35-44

Radle, L. 1998. The effects of noise on wildlife: Literature review. World Forum For Acoustic Ecology. University of Oregon, Eugene, OR, USA.

Singal, S.P. 2005. *Noise Pollution and Control Strategy*. Alpha Science International, Oxford, UK.

Stansfield, S., Haines, M. and Brown, B. 2000. Noise and health in the urban environment. Rev. Environ Health 15:43-82.

Tempest, W. 1985. *The Noise Handbook*. Academic Press, London, UK.

Tobias, J.V., Jansen, G. and Ward, W.D., (eds.). 1980. *Noise as a Public Health Problem*. Rockville, ML, USA.

Tripathy, D.P. 1999. *Noise Pollution*. APH Publishers, New Delhi, India,

U.S. Dept. of Health and Human Services. 1998. *Noise Occupational Exposure*. Public Health Service, CDC, NIOSH. Cincinnati, Ohio.

Vallot, M. 1993. *Noise as a Public Health Problem*. Arcueil Cedex France.

Van Kempen, EEMM, Kruize, H., Boshuizen, H.C., Ameling, C.B., Staatsen, B.A.M. and de Hollander, A.E.M. 2002. The association between noise exposure and blood pressure and ischemic heart disease: a meta-analysis. *Environ. Health Perspect.* 110: 307-317.

Wang, L.K., Pereia, N.C. and Hung, Ying-Tse. 2004. *Advanced Air and Noise Pollution Control* Vol. 2. Humana Press, Totowa, NJ, USA.

WHO. 2000. Guidelines for Community Noise. Geneva: World Health Organization. *http://www.who.int/docstore/peh/noise/guidelines2.html.*

WHO. 2001. Occupational and Community Noise. Fact Sheet No. 258. Geneva: World Health Organization. *http://www/who.int/inf-fs/en/fact258.html.*

WHO Regional Office for Europe. 2004. Noise and Health Home. Bonn, Germany: WHO European Centre for Environment and Health. *http://www.euro.who.int.noise*

CHAPTER 11

Agricultural Pollution

What are the Major Agricultural Pollutants?

Agriculture includes horticulture, fruit growing, seed growing, dairy farming, livestock breeding and keeping, the use of land as grazing land (Smolen and Shanholtz, 1980), meadow land, market grounds, nursery grounds, and the utilization of land for woodlands (*http://www.environment-agency. gov.uk/netregs/processes/34246212lag=_edregion=*). Agroecology is the science of sustainable agriculture (Alteri, 1995; Gliessman et al., 1997).

The major agricultural pollutants (Table 11.1) are animal wastes (Vance, 1984; *http:// www.scotland.gov.uk/librarys/environment/ prepf-00.asp;2002)*, sediments (Millman and Meade, 1983; Ongley et al., 1992), fertilisers (Mergel, 1992; Johnston and Steen, 1997; Carpenter et al., 1998; Driscoll et al., 2002), pesticides (WHO, 1990)(see chapter 7), pathogens (Watershed Science Institute, 2000) and salts (*http://www.ers.usda.gov/ briefing/conservationandenvironment/questions/conservwg2.html*).

A significant health and economic problem is agricultural runoff which can pollute waterways. This runoff involves the hydrologic cycle (Fig. 11.1a, b). The key elements of water movement in this cycle are:

Water Moves
1. downward as precipitation
   ↓
   into the soil
   ↓
   through the unsaturated zone as infiltration
   ↓
   through the shallow and deep aquifiers as recharge
2. laterally as surface runoff
   ↓
   lakes, rivers, streams, wetlands and underground as groundwater
3. upward as evapotranspiration
   ↓
   from groundwater, lakes, plants, soil, streams and wetlands
   ↓
   discharge to surface waters
4. laterally as atmospheric moisture
   ↓
   cloud formation via condensation

TABLE 11.1 Summary of Major Agricultural Pollutants

Pollutant	Mechanisms of Runoff
Nutrient Pollution By Nitrogen and Phosphorus Applied to Cropland in Manufactured Fertilizers and Animal Manures to Enhance Crop Yields (European Fertilizer Manufactures Association, 2003)	Mechanism by Which Nutrients Enter Water *Runoff* Nutrients move from croplands to surface water when dissolved in runoff water or absorbed to soil particles
Nitorgen (Carpenter et al., 1998, Merrington et al., 2002; Driscoll et al., 2002) Phosphorus (Johnston and Steen, 1997)	*Runin* Transports chemicals to ground water through sinkholes, porous or fractured bedrock on poorly constructed wells.
	Leaching Movement of pollutants through the soil by percolating rain, melting snow or irrigation water.
	Atmospheric Deposition - Rain
Sedimentation and Turbidity (Ostry, 1982)	Tillage and cultivation leaving soil without vegetative cover resulting in soil erosion.
Minerals	A portion of irrigation water provided to farmlands runs off the land into ditches and flows back to a receiving body of water. Irrigation return flows can carry dissolved silts.
	Sewage sludge applied to farmlands can contain minerals (Davis, 1996).
Pathogen Damage	Bacteria are the second worst cause of river impairment and the major cause of estuary impairment.
Irrigation Water Applied to Cropland	Portion returns to ditches and is carried back to receiving waters. Irrigation can carry dissolved salts, nutrients and pesticides into surface or ground water (Reiff, 1987).
Pathogens	Animal waste contains pathogens that pose threats to human health; these pathogens cause disease through direct contact with contaminated water, consumption of contaminated drinking water or contaminated shellfish.

Glossary of Terms for Hydrologic Cycle

Condensation – A process which is the reverse of evaporation. Water changes from the vapor state into the liquid or solid states.

Groundwater – Water that is in the zone of saturation of the ground.

Infiltration – The flow of a fluid into a substance through pores or small openings.

Percolation – Flow through a porous substance.

Interception – The process and the amount of rain or snow stored on leaves and branches and eventually evaporated back to the air.

Throughfall – In a vegetated area, the precipitation that falls directly to the ground or the rainwater or snowmelt that drops from twigs or leaves.

Watershed – The divide separating one drainage base from another.

Zone of saturation – The zone in which the functional permeable rocks are saturated with water under hydrostatic pressure.

From: *http://www.water.usgs.gov/wsc/glossary.html.*

The hydrologic cycle in tropical rainforests differs (Fig. 11.2a) from that in agricultural systems in canopy characteristics (Fig. 11.2a). While rainforest canopies

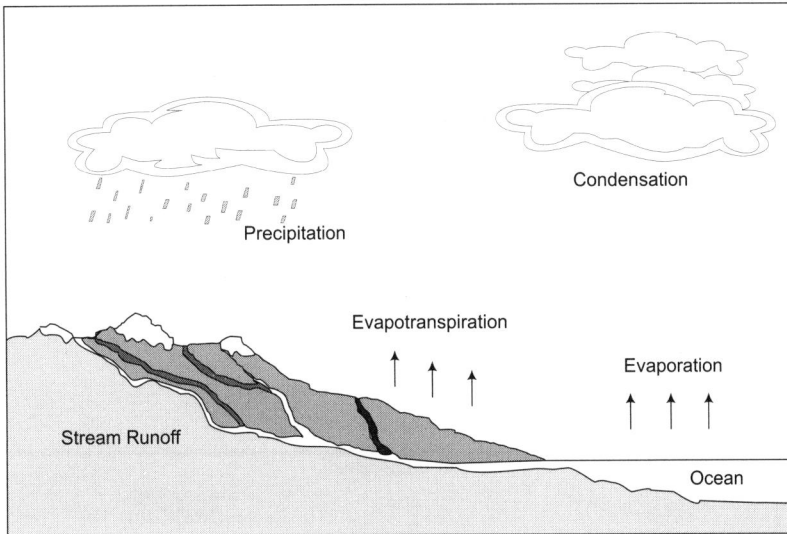

a

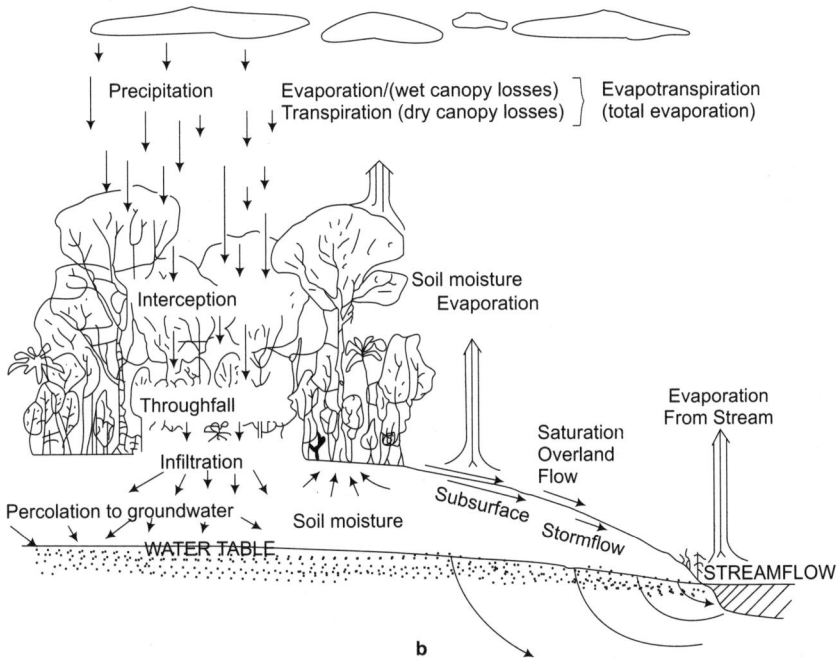

b

Fig. 11.1a, b Hydrologic cycle in non-tropical and tropical (bottom) environments.
a = *http://www.p2pays.org/ref/32/31342.pdf.* Another thorough diagram occurs at *http://ga.water.usgs.gov/edu/watercyclesummarytext.html.*
b = Gilmour, D.A. 1975. Catchment water balance studies in the wet tropical coast of North Queensland. Thesis/Dissertation. Dept. of Geography, James Cook Univ.,North Queensland Townsville
There are many depictions of the hydrologic cycle on websites. Those of *M.Ritter@uswp.edu* are very pleasing.

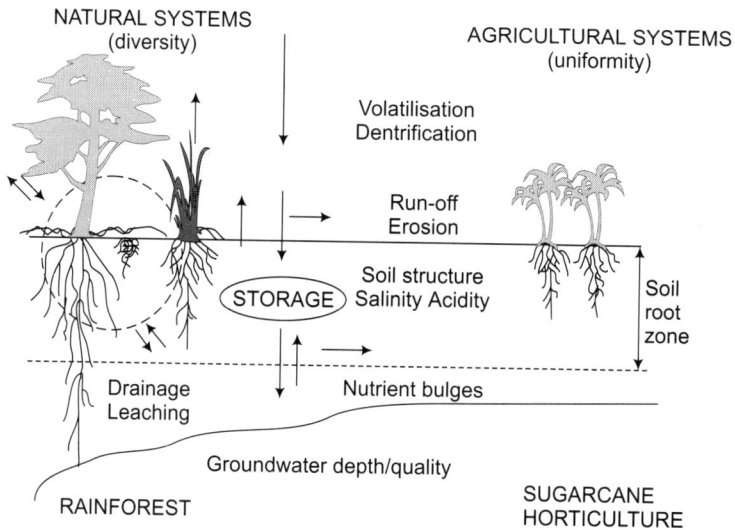

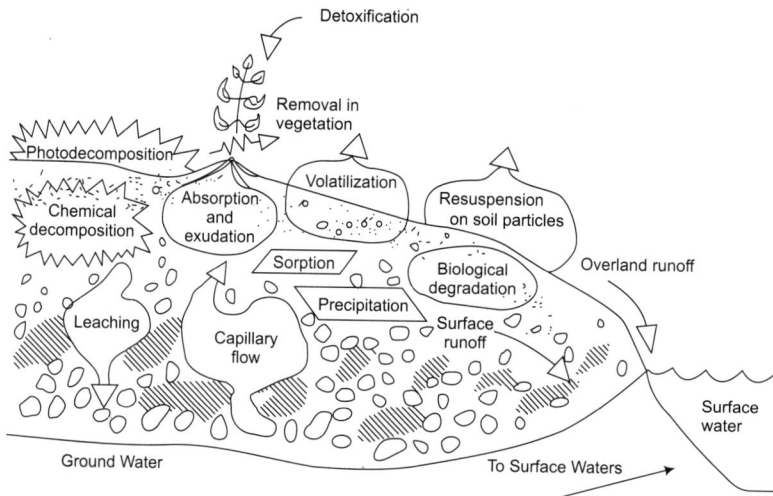

Fig.11.2a, b Comparison of water balance in natural (rainforest) (a) and agricultural (b) systems. From: *http://www.lwa.gov.ar/downloads/publications-pdf/pr980319.pdf.*

and root systems are diverse, those in agricultural systems appear to be more uniform resulting in greater runoff and/or deep drainage (Fig. 11.2b).

Once agricultural wastes enter soils (Forbes and Watson, 1991), they undergo a number of fates (Fig. 11.3) which depend upon the soil type (Table 11.2). "Soil is a natural body comprising of solids (minerals and organic matter), liquids and gases that occur on the land surface, occupies space." "The upper limit of soil is the boundary be-

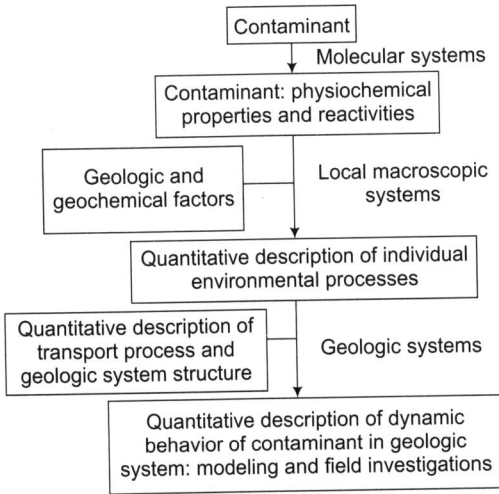

Fig. 11.3 Processes influencing behaviour and fate of waste in soils. From: http://faculty.evansville.edu/ch81/biol100sp3e/Bio100_ch2_ppt.pdf.

tween soil and air, shallow water, live plants, or plant materials that have not begun to decompose. The lower boundary that separates soil from non-soil is most difficult to define." (http://soils.usda.gov/education/facts/soil.html).

What are the Effects of Agricultural Pollutants in Biological Systems?

These are complex and presented in Table 11.3. While many agricultural pollutants can affect aquatic systems resulting from agricultural runoff, some (e.g., pesticides, see chapter 7), can affect terrestrial organisms (Powas and McSarley, 1999).

What are the Research Methods and Efforts Regarding Agricultural Pollutants?

With regard to agricultural research efforts, the USDA-ARS describes 17 research projects (www.ars.usda.gov/researchproje cts.htm). Some of the efforts are concerned with developing methods for detecting agricul-

TABLE 11.2 The Twelve Orders of the Earths Soils[a, b, c, d]

Soil Classification	Characteristics
Alifisol	Moderately leached soils with a subsurface zone of clay; agriculturally useful; forms in humid climates
Andisol	Forms in volcanic ash; retains water
Aridisol	$CaCO_3$ - Containing soils of dry regions; contains few organic particles
Entisols	Soils with little or no morphological development; formed in all climates
Gelisol	Soils with permafrost within 2m of the surface; formed in high mountain tops and polar regions
Histosol	Organic soils, formed in wet places, e.g., bogs and swamps
Inceptisol	Soils with weakly (thin) developed subsurface horizons; formed in humid areas
Mollisol	Grassland soils which are fertile; formed in prairie regions
Oxisols	Chemically weathered soils; formed in tropical regions
Spodosol	Acidic soils with a subsurface accumulation of metal – humus complexes, formed in cool, humid pine forests
Utilisol	Strongly leached soils with a subsurface zone of clay; formed in warm, humid regions
Vertisol	Soils with clay and elevated shrink/swell capacity; formed in relatively dry, warm regions

[a] Adopted from: http://soils.org.uidaho.edu/soilorders/orders.htm and soils.usda.gov/education

[b] Of these soils Mollisol is the most prevalent occupying 21.5% of the Earth's soil area.

[c] Sims et al. (1984) review in place treatment techniques for contaminated soils.

[d] Pierzynski et al. (2004) present a thorough review of soils and environmental quality.

TABLE 11.3 Effects of Agricultural Pollutants on Biological Systems[a]

System	Pollutant	Effect
Aquatic and Other Systems Consuming Aquatic Life (Hecna, 1995; Marco et al., 1999; Rouse et al., 1999)	Ammonium nitrate fertiliser nitrates from animal manures Phosphates from fertilizers via soil erosion (Tiessen, 1995; Johnston and Stein, 1997).	Eutrophication (Correll, 1998; Bricker et al., 1999) (Fig. 11.4) Algal blooms (Anderson et al., 1998) Death of fish and other aquatic organisms (Rouse et al., 1999); Nitrates in surface and groundwaters can result in drinking water contamination.
	Sedimentation and turbidity result of soil erosion (Sheet, Well and Gully); Sediment particles can carry phosphorous and pesticides (Dillaha et al., 1987)	Suspended soils reduce sunlight available to aquatic plants; cover fish spawning regions and food supplies) smother coral reefs, clog filtering of filter feeders, clog fish gills.
	Animal wastes manures Includes fecal and urinary wastes of livestock and poultry, process water and feed, bedding, litter and soil; can cause contamination of both surface and groundwater	Fish kills from manure run-off into surface waters because of ammonia or dissolved O_2 depletion; Shellfish and beach closure from high fecal coliform counts in manure runoff
	Salts Produced by natural weathering processes of soil and geological material	High salt concentrations in streams (Rhodes, 1993) can harm freshwater aquatic plants and anodromous fish.
	Feedlots Can contaminate surface water with metals contaminated in urine and feces; potential leaking to groundwater	Chronic public health problems from many pathogens

[a]Adapted from: *http://www.epa.gov/owow/nps/mmgi/Chapter2/ch2-1html* and *http://www.fao.org/docrep/w2598E/W2598e04.htm.*

Eutrophic Lake

Large nutrient supply → High productivity → High phytoplankton standing stock → Lake less transparent

Eutrophic	*Oligotrophic*
Nutrient rich	Nutrient poor
Shallower lake basin	Deep lake basin
Muddy (silt, clay) bottom	Sandy or gravel bottom
Cloudy, turbid water	Clear water
Dissolved solids high	Dissolved solids low
Plankton few but high numbers	Plankton scarce but many species
Oxygen high at surface but deficient under ice and thermocline	Oxygen high
Rooted veg. abundant	Rooted veg. scarce
Warm water	Cold water

Small nutrient supply → Low productivity → Low phytoplankton standing stock → Lake more transparent

Oligotrophic Lake

Fig. 11.4 Comparison of eutrophic and oligotrophic lakes. Eutrophication can occur from nitrates and phosphates in leached fertilizers as well as raw and treated sewage and run-off from poor land use. This type of water pollution occurs in areas of intensive cereal crops. Mesotrophic is intermediate between eutrophic and oligotrophic. Thermocline refers to an intermediate layer of lake stratification.

tural runoff pollution (Table 11.4). Rasmussen et al. (1998) review long-term agroecosystems experiments. Useful websites are presented in Table 11.5.

Can Agricultural Wastes Be Managed?

There are many Internet websites and various books (Sweeten, 1991); National Council for Science and the Environment, 1996; Loehr, 1997; Romstad et al., 1997; Theodore, 1999; Araji, 2002; Peart, 2004) concerning agricultural waste management. The basic components of an agricultural waste management system are presented in Fig. 11.5. This section considers the manage-ment of certain of the wastes (Joly, 1993). Economics are an integral part of a management system (Penson et al., 2002).

In addition, the agency considers "vegetated filter strips, field borders, sediment retention ponds and terraces" as strategies to control field runoff.

Control of Agricultural Runoff

Table 11.7 presents a variety of conservation practices (USDA-NRCS, 1997) which have been employed to minimize pollutants in agricultural runoff. The table summa-rises the expected beneficial detrimental effects of the conservation practices.

TABLE 11.4 Some Analytical and Non-Analytical Techniques for Agricultural Pollutants[a]		
Technique	*Usefulness*	*Reference*
Analytical Optical Spectroscopy Nucleic Acid Analysis	Measure bacteria and other pathogens	Pacific North-West Station Laboratory *http://www.pml.gov/new/back/ogfoodby.htm*
Fluorescence Spectroscopy	Make pathogenic bacteria visible	USDA Agricultural Research Service *www.ars.usda.gov/is/AR/archive/jano5/food/05.htm*
Immunomagnetic Beads Coated with Antibodies that Bind to Specific Bacteria	Detection of food contaminants	USDA Agricultural Research Service *www.ars.usda.gov/is/AR/archive/jano5/food/05.htm*
Liquid Chromatograph/tandem Mass Spectrometer	Analyzing animal-derived food for veterinary drug residues	USDA Agricultural Research Service *www.ars.usda.gov/is/AR/archive/jano5/food/05.htm*
Biosensor Immunoassay Using Surface Plasmon Resonance	Detect toxins in foods that cause gastroenteritis	USDA Agricultural Research Service *www.ars.usda.gov/is/AR/archive/jano5/food/05.htm*
Electroimmunoassay Technology	Food-Borne Pathogen Detection	IVD Technology Medical device link *www.plascomicrotek.com*
Non-agricultural Floating cattail mats	Removal of ammonium nitrates from retention ponds	Mikro-Tek Inc. (2003)
Vegetative Filter Strips	Agricultural runoff	Dillaha et al. (1987) Magette et al. (1987)
Model diffusion and dispersion discharge		Mills et al. (1985); Young et al. (1994)

[a] A thorough review of agricultural chemical analyses occurs in Faithful, N.T. 2002. *Methods in Agricultural Chemical Analyses: A Practical Approach*, CABI Publishing, Wallingford, Oxon, UK.

Anaerobic digester —— Biogas utilization —— Energy

Manure and
organic wastes

Liquids   Nutrient recovery- —— Reusable water
             treatment

Solid/liquid separation

Solids   Aerobic digester- —— Organic fertilizer
            nutrient enrichment

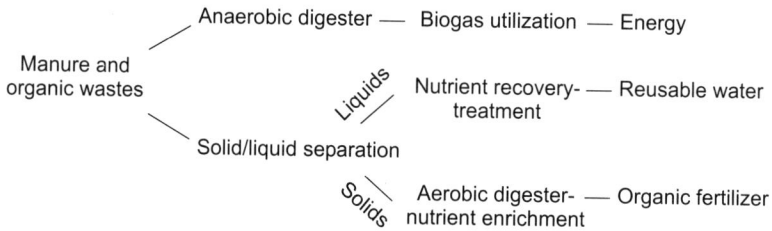

Fig. 11.5 Procedure to utilize agricultural wastes
Adapted from Integrated Waste Management Solution. Alberta Research Council, Alberta, Canada.

TABLE 11.5	Some Agricultural Websites
Topic	Websites
Pesticides	Agrifor.ac.uk/hb/7f6b30490fcab04d802843cleb04f369.html–88k
	www.eawag.ch/publicationssci-publ/publ.2001.html
	www.sepa.org.uk/publications/leaflets/agriculturalleaflets/pesticide.pdf.
Agricultural waste topics	http://www.science.gov/browse/w_105c.htm
Prevention of environmental pollution from agricultural activity	www.scotland.gov.uk/library3/pepfa-01-ag
Agricultural food and animal carcasses and animal manures	
Nitrogen and phosphorous pesticides, silos and sewage effluent	
Agricultural waste-sediments	www.wvc.wa.gov.air/public/waterfacts/3-pollution/table2.html.iak
Biological effects of agricultural runoff	Bibloalerts.com
Graphical results of 1999 pesticides in dietary, fruit and vegetables, cereal and cereal products, animal products and fish in the UK	www.pesticides.gov.uk/co.pdf
Presents management measures to protect costal waters from agricultural sources of nonpoint pollution.	http://www.epa.gov.owow.nps.mmgi.chapter2/ch2-1.html

TABLE 11.6	Manure Production by Livestock Type[a, b]
Livestock Type	Total Manure
Beef (High forage diet)	59.1
Dairy (Lactating cow)	80.0
Swine (Grower)	63.1
Chicken (layers)	60.5
Chicken (broilers)	80.0
Turkeys	43.6

[a] Adopted from: NRCS/RCA Issue Brief 7 Animal Manure Management 1995.

[b] lbs/day1,000 lb animal unit

What are some Commercially available Mechanisms to Contain Runoff?

Geosynthetics are synthetic materials employed in diverse applications to generate barriers for controlling erosion and provide containment (*http://www.landwater.com /features/vol149no4/vol149no4-2.html*).

Agricultural drainage wells (Fig. 11.8) are designed to collect excess water in relatively flat low areas where its removal

TABLE 11.7 USDA-NRCS Recommended Conservation Practices

Resource Concerns	Ground water					Surface Water								
Conservation Practices	Pesticides	Nutrients & Organic	Salinity	Heavy metals	Pathogens	Pesticides	Nutrients & organic	Salinity	Heavy metals	Pathogens	Temperature	Low dissolved oxygen	Suspended sediments & turbidity	Aquatic habitat suitability
Channel Vegetation	D						B		B	B	B		B	B
Conservation Cover		B	B	B	B	B	B	B	B	B	B	B	B	B
Constructed Wetland	D	D	D	D	D	B	B		B	B		B	B	B
Contour Buffer Strips	D	D	B	B		B	B		B		B	B	B	B
Critical Area Planting	B	B		B	B	B	B		B	B	B	B	B	B
Floodwater Diversion		B					D					D	D	
Forest Site Preparation														D
Grasses Waterway						B	D		B	B	B	B	B	B
Heavy Use Area Protection							B							
Hedgerow Planting							B						B	B
Irrigation System - Micro	B	B	D	B	B	B	B	B	B	B	B	B	B	
Irrigation System - Sprinkler	D	D	D	D	D	B	B	B	B	B			B	B
Manure Transfer		B	B				B			B		B		
Mulching	D	D	D	D	D	B	B		B	B	B	B	B	B
Nutrient Management		B	D	D	B	B	B	B	B	B		B	B	
Prescribed Grazing		B	B	B	D	B	B	B	B	B	B	B	B	B
Residue Management, Seasonal	D	D	D	D	B	B	B	B	B	B	B	B	B	B
Riparian Forest Buffer	B	B	B	B	D	B	B		B	B	B	B	B	B
Sediment Basin	D	D	B	D	D		B						B	B
Well Decommissioning	B	B	B	B	B									B
Wetland Restoration						B	B		B	B		B	B	B

B - Beneficial effects expected
D - Detrimental effects expected
Blank - Not Rated

From: Natural Resources Conservation Service. (1997)
Source: USDA – NRCS (1997)

Transport to

Treatment

(Examples) 1. Farm manure handling facilities including lagoons, composters, oxidation ditches, chemical additives, 2. microbial digestion, 3. use of vegetative filter strips, 4. biogas waste treatment

Storage

(Examples) 1. Banels containing dug-up soil, 2. injection wells and sinkholes, 3. ocean dumping

Environmental re-utilization (Examples) 1.) Alternative energy source 2.) replacing

Production

Collection

Storage   Transfer   Treatment

Utilization

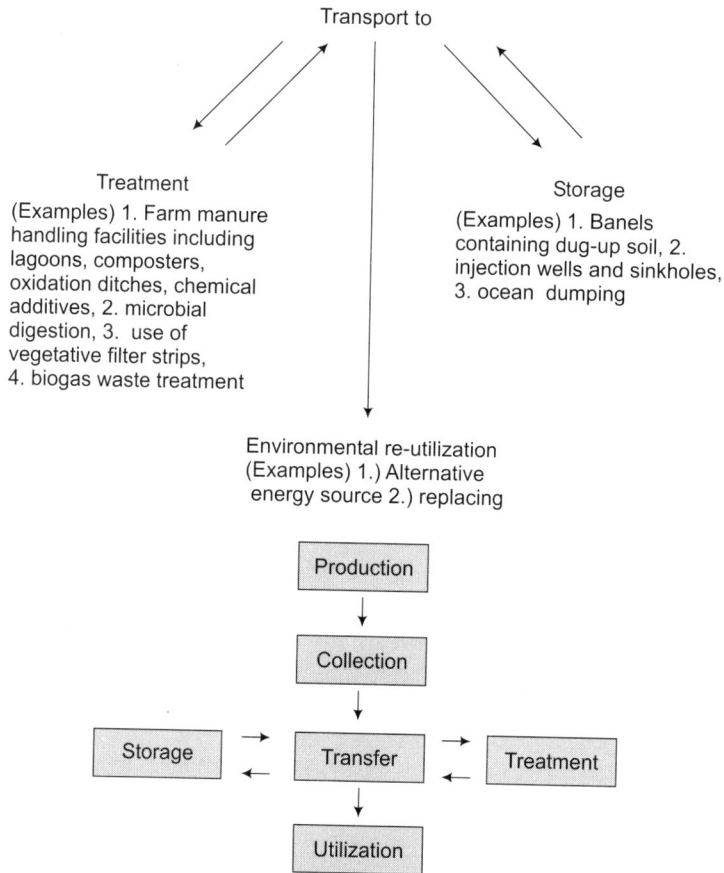

Fig. 11.6 Manure Management
USDA-NRCS. 1992. US Department of Agriculture – Natural Resource Conservation Service. Agricultural Waste Management Field Handbook. Publication No. 210. AWMFH, 4/92.
Touisseng (2001) developed a biochemical process (Fig. 11.9) which turns horse or cow manure into a liquid organic fertilizer (Fig. 11.10) and compost. The process has been modified for aquaculture.

through surface drainage or tile lines is impeded (*http://www.extension.umn.edu/distribution/cropsystems/dc6265.html*). These drainage systems consist of a collection cistern which empties into a bottom well. Unfortunately, contaminated water can enter groundwater which is used for drinking (Andreoli, 1993; Broden and Lovejoy, 1998).

Farm Manures

Farm handling systems (Fig. 11.6) for managing beef, poultry and swine wastes (Sweeten, 1991) are depicted in Figs. 11.7. These wastes contain feces and urine generated by the metabolism of feeds to derive energy. Other farm wastes include: leaking waters, flush water, rinsinates from wash-

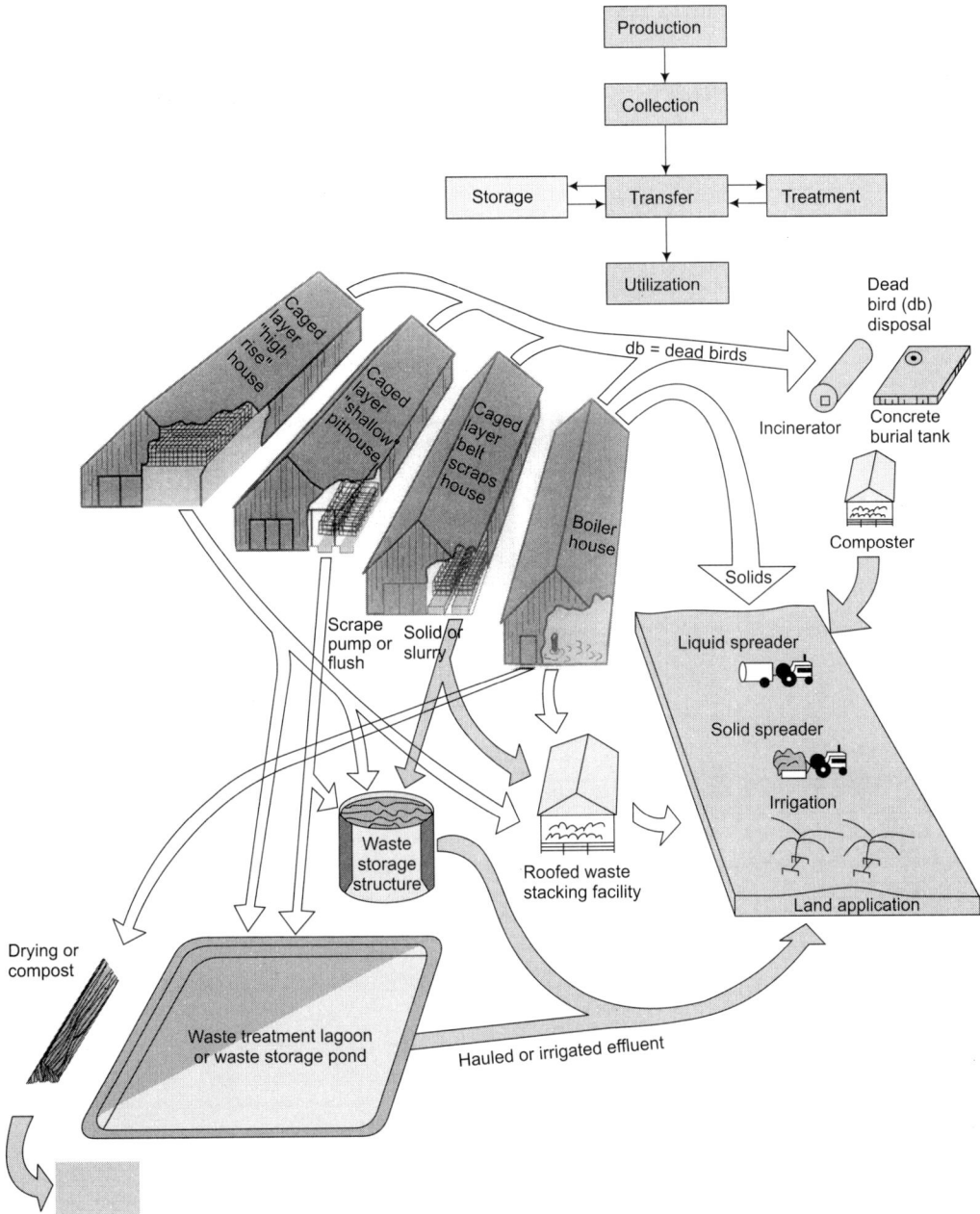

Production
Collection
Storage ← Transfer → Treatment
Utilization

Caged layer "high rise" house
Caged layer "shallow" pithouse
Caged layer belt scraps house
Boiler house

db = dead birds

Dead bird (db) disposal

Incinerator   Concrete burial tank

Composter

Scrape pump or flush
Solid or slurry

Solids

Liquid spreader
Solid spreader
Irrigation

Waste storage structure

Roofed waste stacking facility

Land application

Drying or compost

Waste treatment lagoon or waste storage pond

Hauled or irrigated effluent

a

(Contd.)

(Contd.)

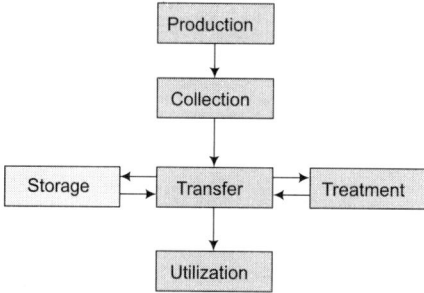

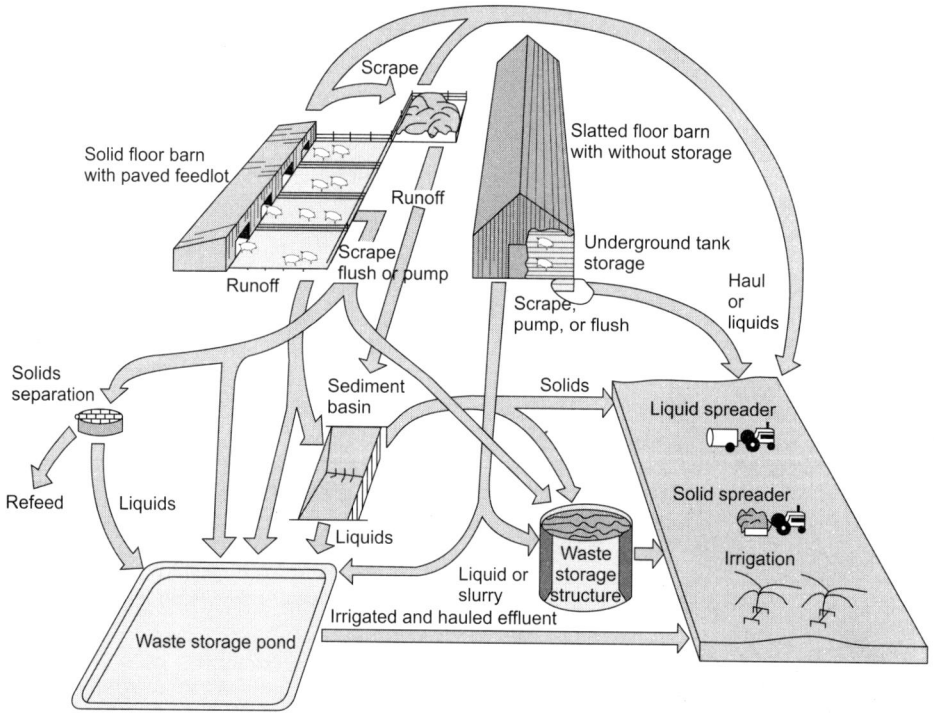

b

(Contd.)

(Contd.)

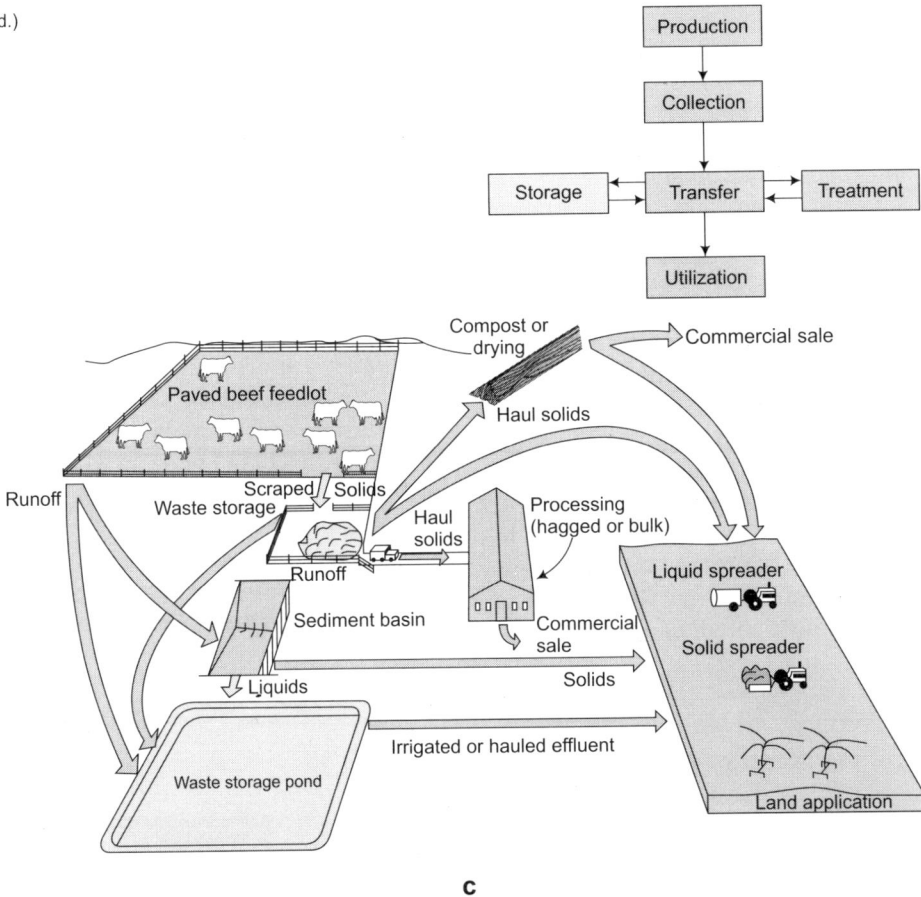

Fig. 11.7a,b,c Commercial waste handling operations for poultry (a), swine (b) and beef (c) operation. From: *ftp://ftp.wcc.nrcs.usda.gov.downloads/wastemgmt/AWMFH/awmfh-chap.*

Fig. 11.8 An agricultural drainage well. From: Lohmann, P., Iowa Department of Natural Resources. *http://www.igsb.uiowa.edu/inforsch/adv99/adv.htm.*

ing equipment as well as spilled feed and bedding. The management of certain of these wastes (Araji, 2002) involves recycling manure nutrients (Table 11.6) and organic matter (*http://www.mda.state.md.us/management/manual/manure*; Romstad et al., 1997; Chadwick and Chen, 2002) to reconnect farms to ecosystems (Jackson and Jackson, 2002).

Manure treatment employs lagoons, composters, oxidation ditches, solid separators and chemical additives (Fig. 11.6). The altered manure destined for land application is liquid, slurry or semi-solid. With regard to manure operations, the US-EPA has suggested that "animal feeding operations should be designed and operated to avoid waste discharge by having engineered runoff controls, waste storage, waste utilization and nutrient management.

Sediment

The US EPA has addressed management measures for erosion and sediment pollution. This Agency is concerned about minimizing soil detachment, erosion and sediment transport from the field. The EPA promotes products which "maintain crop residue in vegetative cover on the soil; improve soil properties; reduce effective water and/or wind velocities".

Finally, a more recent area of agricultural management involves the transformation of animals (Fig. 11.11) and plants (Fig. 11.12) (Lopez et al., 2005) to yield transgenic organisms (FAD, 2004). These are ones in which foreign DNA (a transgene) is incorporated into their genomes via recombinant DNA technologies early in development (*www.csa.com/hottop-*

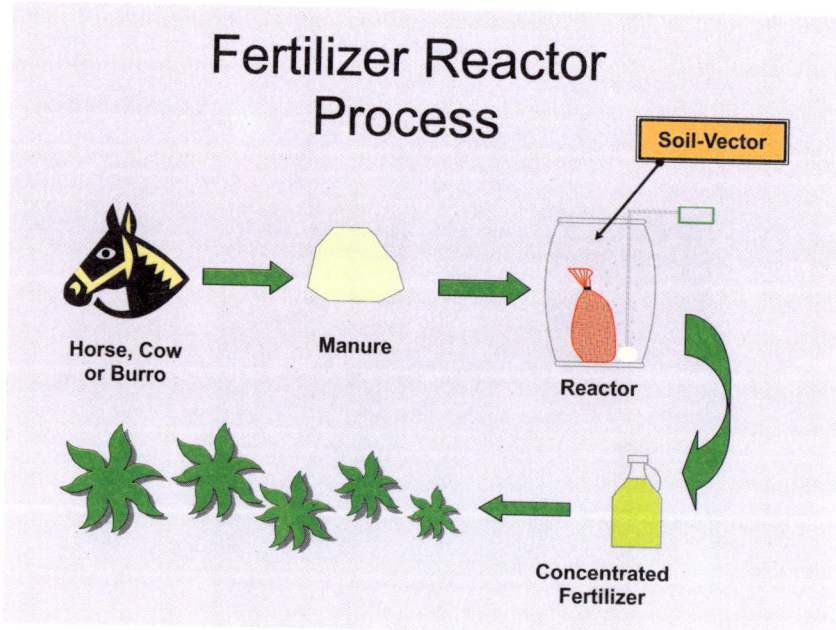

Fig. 11.9 Fertilizer reactor process for converting manure to a liquid fertilizer.
From: Toussieng, C. (2008). *chucktee@mac.com*.

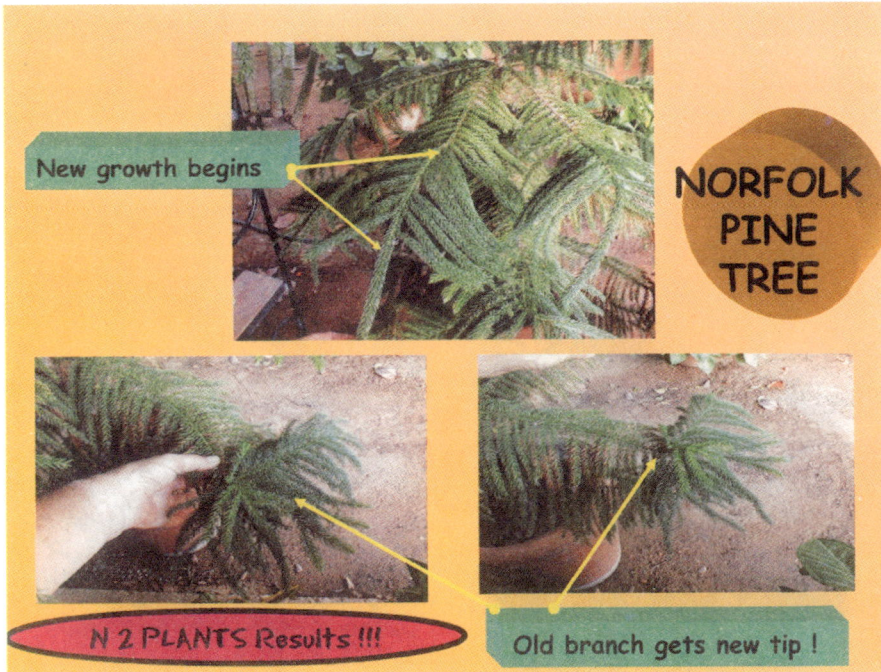

Fig. 11.10 Application of fertilizer from fertilizer reactor process to Norfolk pine tree. From: Troussieng. C. (2008). *chucktee@mac.com.*

ics/gmfood/gloss.php). There are many reviews concerning transgenic animals (Houdebine, 2003; Khachatourions et al., 2002; National Academies Press, 2002; Tifac, 2003; Pena, 2005) Chrispeels and Sadava (2006) and Murphy (2007) discuss the societal impact of transgenic plants.

References

Alteri, M.A. 1995. *Agroecology: The Science of Sustainable Agriculture.* Westview Press, Philadelphia, PA, USA.

Anderson, D.M., Cambella, A.D. and Hallengraeff, G.M. 1998. *Physiological Ecology of Harmful Algae Blooms.* Springer, Berlin, Germany.

Andreoli, C.V. 1993. The influence of agriculture on water quality. In: *Prevention of Water Pollution by Agriculture and Related Activities.* Proceedings of the FAO Expert Consultation, Santiago, Chile, 20-23 Oct. 1992. Water Report 1. FAO Rome, Italy pp. 53-63.

Araji, A.A. 2002. *Management of Agricultural Wastes As An Alternative Energy Source.* Dep't. of Agricultural Economics and Rural Sociology, College of Agriculture, University of Idaho, Moscow, Idaho, USA.

Broden, J.S. and Lovejoy, S.B. 1990. *Agriculture and Water Quality: International Perspectives,* Boulder, CO, USA.

Bricker, S.B. Clement, C.G., Pirhalla, D.E., Orlando, S.P. and Farrow, D.R.G. 1999. *National Estuarine Eutrophication Assessment Effects of Nutrient Enrichment in the Nations Estuaries.* NOAA, Silver Spring, MD, USA.

Broden, J.B. and Lovejoy, S.B. 1998. *Agriculture and Water Quality: International Perspectives,* Boulder, CO, USA.

Chadwick, D.R. and Chen, S. 2002. Manures. In: *Agriculture, Hydrology and Water Quality.* (Haygarth, P. and Jarvis, S., eds.). CAB International, pp. 58-82.

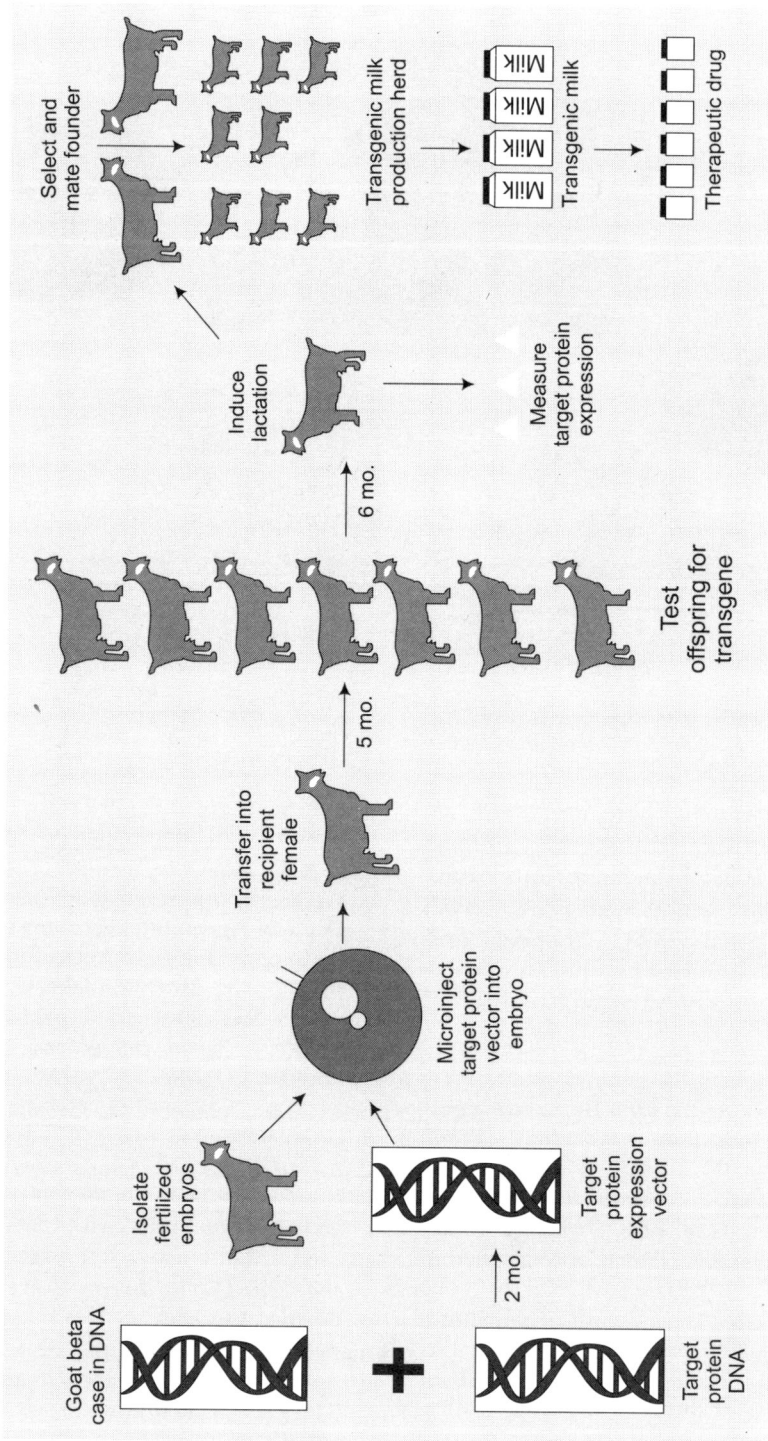

Fig. 11.11 Production of a transgenic animal. From: *http://www.transgenic.com/images/May2001Gavin.pdf.*

GENE
identified and
isolated

A

Agrobacterium

Gene inserted
into *ti* plasmid

Bacterium mixed
with plant cells

B

Gene Gun

Gene
replication

Gold particles
coated with DNA

ti plasmid moves
into plant cell and
inserts DNA into
plant chromosome

Cells shot with gene
gun and DNA
incorporated into
plant cell chromosome

C

**Screening for
cells with
transgene**

Cells screened
for transgene

transformed cells
selected with
selectable marker

transgenic plant
regenerated from single
transformed cell

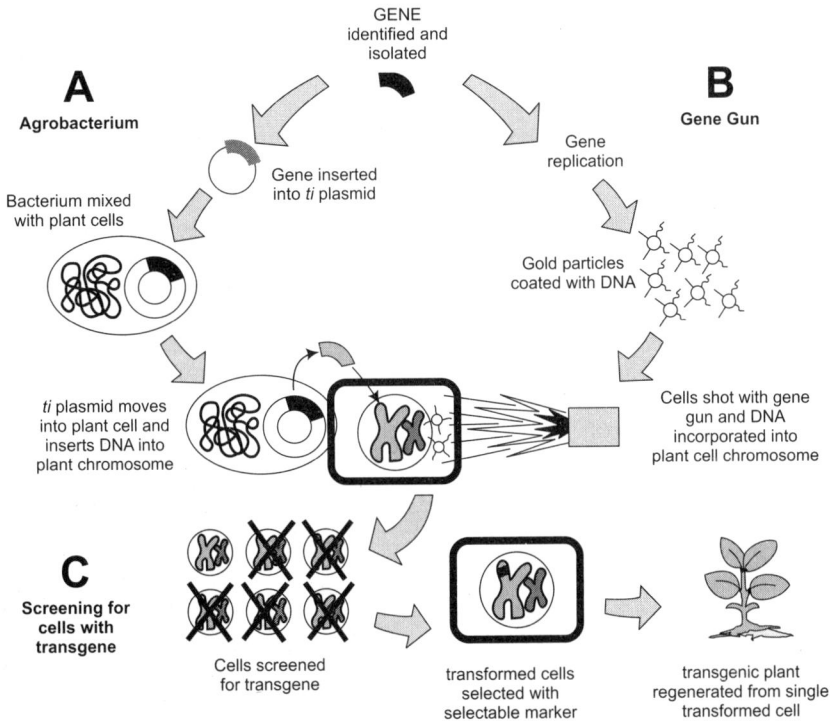

Fig. 11.12   Generation of a transgenic plant.
From: *http://www.ag.ndsu.edu/pubs/plantsci/crops/a1219-2.gif*. Peel, M.D., 2001. A Basic Primer on Biotechnology.
Another diagram comparing the 'Agrobacterium' and particle gun gene splicing methods is *http://www.ag.uiuc.edu/
~vista/html_pubs/biotech/genen.htm*.

Carpenter, S.R., Caraco, N.F., Correll, D.L. and Howard, R.W. 1998. Nonpoint pollution of surface waters with phosphorus and nitrogen. *Ecol. Appl.* 8:559-568.

Chadwick, D.R. and Chen, S. 2002. Manuers. In: *Agricultures, Hydrology* and *Water Quality*. (Haygarth, P.M. and Javis, S.C, eds.) CABI, Wallingford, UK.

Chrispeels, M.J. and Sadava, D.E. 2006. *Plants, Genes and Crop Biotechnology*. Academic Internet Publishers, Moorpark, CA, USA.

Christou, P. 1996. Transformation technology. *Trends Plant Sci.* 1:423-431.

Clark, J.M. and Ohkawa, H. 2005. *New Discoveries in Agrochemicals*. Oxford Univ. Press, Oxford, England.

Committee On Environmental Impacts Associated With Commercilization of Transgenic

Plants. 2002. *Environmental Effects of Transgenic Plants*. National Academy Press, Washington, DC, USA.

Correll, D.L. 1998. The role of phosphorus in the eutrophication of receiving waters: a review. *J. Environ. Qual.* 27:261-266.

Dasi, C. 2001. *Agricultural Use of Groundwater*. Kluwer Academic Publishers, Dordrecht, Netherlands.

Davis, R.D. 1996. The impact of EU and UK environmental pressures on the future of sludge treatment and disposal. *Water Environ. J.* 10: 65-69.

Dillaha, T.R., Reneau, R.B., Mostaghimi, S., Shanholtz, V.O. and Magette, W.V. 1987. Evaluating nutrient and sediment losses from agricultural lands: Vegetative filter strips: EPA, Region III Chesapeake Liaison Office, Annapolis, MD, USA.

Driscoll, C.T., Whitall, D., Aben, J., Boyer, M., Casbo, M., Cromer, C., Goodole, C., Groffman, P., Hopkinson, C., Lambert, K., Lawrence, G. and Ollinger, S. 2002. Nitrogen pollution in Northeastern United States: Sources, effects and management options. *Bioscience* 53: 357-374.

European Fertilizer Manufactures Association. 2003. N content of the various nitrogen fertilizers consumed in the ED. *http://www.efma.opg/statistics/section01.asp.*

Food and Organization of the United Nations (FAD), The state of food and agriculture 2003-2004: Agriculture biotechnology meeting the needs of the poor 2004. FAD, Rome, Italy.

Forbes, J.C. and Watson, R.D. 1991. *Plants in Agriculture.* Cambridge Univ. Press. Cambridge, England.

Fox, M.W. 1986. *Agricide: The Hidden Crisis That Affects Us All.* Shocker Books, New York, NY, USA.

Galin, E. and Breiman, A. 1997. *Transgenic Plants.* Imperial College Press, London, England.

Garder, B. and Rausser, G.C. 2001. *Handbook of Agricultural Economics: Agriculture Products.* Elseiver, St. Louis, MO, USA.

Gliessman, S.R., Engles, E. and Kreiger, R. 1997. *Agroecology: Ecological Processes in Sustainable Agriculture.* Lewis Publisher, Boca Raton, FL, USA.

Greene, S.A. and Pohanish, R.P. 2005. *Sittig's Handbook of Pesticides and Agricultural Chemicals.* William Andrew Pub., New York, NY, USA.

Harmful algal blooms: 2003. *http://www.nwfsc.noaa.gov/lad/blooms.htm.*

Haygarth, P. and Jarvis, S.C. 2002. *Agriculture, Hydrology and Water Quality.* CABI Publishing, Wallingford, Oxfordshire, England.

Hecna, S.J. 1995. Acute and chronic toxicity of ammonium nitrate fertilizer to amphibians from southern Ontario. Environ. Toxicol. Chem. 14:2131-2137.

Houdebine, L.M. 2003. *Animal Transgenesis.* Wiley, New York, NY, USA.

Jackson, D.L. and Jackson, L.L. 2002. *The Farm as a Natural Habitat. Reconnecting Food Systems with Ecosystems.* Island Press, Washington, DC, USA.

Johnston, A.E. and Steen, I. 1997. Understanding phosphorus and its use in agriculture. European Fertilizers Manufactures Association. Brussels, Belgium.

Joly, C. 1993. Plant nutrient management and the environment. In: 'Prevention of Water Pollution by Agriculture and Related Activities.' *Proceedings of the FAO expert Consultation, Santiago, Chile, 20-23 Oct. 1992. Water Report 1.* FAO, Rome, pp. 223-245.

Jones, J.G. 1993. *Agriculture and the Environment.* Horwood, New York, NY, USA.

Khachatourians, G.G., McHughen, A., Scorza, R., Nip, W.K., and Hu, Y.H. 2002. *Transgeneic Plants and Crops.* CRC Press, Boca Raton, FL, USA.

Loehr, R.C. 1997. *Pollution Control for Agriculture.* Academic Press, New York, NY, USA.

Lopez, O., Olmedor, G. and Guevora, F. 2005. *Introduction to Molecular Biotechnology for Food and Agricultural Sciences.* CRC Press, Boca Raton, FL, USA.

Magette, W., Brinsfield, R.B., Pallmer, R.E., Wood, J.D., Dillaha, T.A. and Reneau, R.B. 1987. *Vegetated filter strips for agricultural runoff treatments.* US EPA, Philadelphia, PA, USA.

Marco, A., Wuilchano, C. and Blauster, A.R. 1999. Sensitivity to nitrate and nitrite in pond breeding amphibians from the Pacific northwest, USA. *Environ. Toxicol. Chem.* 18:2836-2839.

Marvel, J.T. 1985. Biotechnology in crop improvement ACS Symposium Services 276: 477-510. American Chemical Society, Washington, DC, USA.

McLean, N. 2006. *Animals with Novel Genes.* Cambridge Univ. Press, Cambridge, England.

Mergel, K. 1992. Nitrogen agricultural productivity and environmental problems. In: (Mergel, K. and Pibena, D.L., eds.). *Nitrogen Metabolism of Plants.* Clarendon Press, Oxford, England pp. 1-15.

Merrington, G., Winder, L., Parkinson, R. and Redmon, M. 2002. *Agricultural Pollution: Environmental Problems and Practical Solutions.*

Spon Press, Abindgon Oxon, OX, UK.

Millman, J.D. and Meade, R.H. 1983. World-wide delivery of river sediment to the oceans. *J. Geology* 91:1-21.

Mills, W.B., Porcella, M.J., Ungs, S.A., Gheini, S.A., Summers, M., Lingsurg, G.L., Rupp, G.L., Bowie, G.L. and Haith, D.A. 1985. *Water Quality Assessment: A Screening Procedure for Toxic and Conventional Pollutants*, EPA-600/6-82-004a & b. Volumes I and II. US EPA, Washington, DC, USA.

Murphy, D. 2007. *Plant Breeding and Biotechology: Societal Content and the Future of Agriculture.* Cambridge Univ. Press, Canbridge, UK.

National Academies Press. 2002. *Animal Biotechnology:* Social Based Cancers. Washington DC, USA

National Council for Science and the Environment. 1996. Washington, DC, USA.

Newell, C.A. 2000. Plant transformation technology. Development and applications. *Mol. Biotechnol.* 16: 53-65.

Novarty, V. and Olens, H. 1994. *Water Quality: Prevention, Identification and Rearrangement of Diffuse Pollution.* Van Nostrand Reinhold, New York, NY, USA.

Ongley, E.D. 1994. Global water pollution: Challenges and opportunities. *Proceedings: Integrated Measures to Overcome Barriers to Minimizing Harmful Fluxes from Land to Water.* Publication No. 3, Stockholm, Sweden, pp. 23-30.

Ongley, E.D., Krishnappan, B.G., Droppo, I.G., Rao, S.S. and Maguire, R.J. 1992. Cohesive sediment transport: Emerging issues for toxic chemical management. Hydrobiologia 235/236:177-187.

Ostry, R.C. 1982. Relationship of water quality and pollutant loads to land uses in adjoining watersheds. *Water Res. Bull.* 18:88-104.

Peart, R.M. 2004. *Agricultural Systems Management: Optimizing Efficiency and Performance.* CRC Press, Boca Raton, FL, USA.

Pena, L. 2005. *Transgenic Plants: Methods and Protocols.* Humana Press, Totowa, NJ, USA.

Penson, J.B., Capps, O. and Rosson, C.P. 2002. 'Introduction to *Agricultural Economics'.* Prentice Hall, London, England.

Pepper, S.L. 1996. *Pollution Science.* Academic Press, New York, NY, USA.

Philip, R.B. 2001. *Ecosystems and Human Health. Toxicology and Environmental Standards.* Lewis Publishers, Boca Raton, FL, USA.

Pierzynski, G.M., Sims, J.T. and Vance, G.F. 2004. *Soils and Environmental Quality.* CRC Press, Boca Raton, FL, USA.

Powas, L. and McSarley, R. 1999. *Ecological Principles of Agriculture.* Thomson Delmar Learning. Florence, KY, USA.

Proger, M.J. 1973. Automated waste monitoring instrument for phosphorus contents. Office of Research and Monitoring, EPA. Washington, DC, USA.

Rasmussen, P.E., Goulding, K.W.T., Brown, J.R., Grace, P.R., Janzen, H.H. and Korschens, M. 1998. Long term agroecosystem experiments: Assessing agricultural sustainability and global change. *Science* 282: 893-896.

Recheigl, J.E. and MacMinnon, H.C. 1997. *Agricultural Uses of By- Products and Wastes.* American Chemical Society, Washington, DC, USA.

Reiff, F.M. 1987. Health aspects of waste-water reuse for irrigation of crops. In: *Proceedings of the Interregional Seminar on Non-conventional Water Resources Use in Developing Countries,* 22-28 April 1985, Series No. 22, United Nations, New York. pp. 245-259.

Reynolds, T.B. and Metcalfe-Smith, J.L. 1992. An overview of the assessment of aquatic ecosystem health using benthic invertebrates. *J. Aquatic Ecosystem Health* 1:295-308.

Rhodes, J.D. 1993. Reducing salinization of soil and water by improving irrigation and drainage management In: *Prevention of Water Pollution by Agriculture and Related Activities.* Proceedings of the FAO expert Consultation, Santiago, Chile, 20-23 Oct. 1992. Water Report 1. FAO, Rome. pp. 291-320.

Rickert, D. 1993. Water quality assessment to determine the nature and extent of water pollution by agricultural and related activities. In: ' Prevention of Water Pollution by Agricultural and Related Activities.' *Proceedings of the FAO Expert Consultation,*

Santiago, Chile, 20-23 Oct. 1992. Water Report 1. FAO Rome. pp. 171-194.

Romstad, E., Simonsen, J. and Vatn, A. 1997. *Controlling Mineral Emissions in European Agriculture Economics, Policies and the Environment.* CABI International, Wallingford, UK.

Rouse, J.D., Bishop, C.A. and Strager, J. 1999. Nitrogen pollution: An assessment of its threat to amphibian survival. *Environ. Health Perspect.* 107:799-803.

Sagardoy, J.A. 1993. An overview of pollution of water by agriculture. In: *Prevention of Water Pollution by Agriculture and Related Activities,* Proceedings of the FAO Expert Consultation, Santiago, Chile, 20-23 Oct. 1992. Water Report 1. FAO, Rome. pp. 19-26.

Shortte, J.S. and Abler, D.G. 2001. *Environmental Policies for Agricultural Pollution Control.* CABI Pub.,Wallingford, UK.

Shortte, J.S., Abler, D.G. and Ribaudo, M. 2001. Agriculture and Water Quality: The Issues. In: *Environmental Policies for Agricultural Pollution Control.* CABI Pub., Wallingford Oxon, UK. pp. 1-18.

Shugar, G.J., Drum, D.A., Lauber, J. and Bauman, S.L. 2000. *Environmental Field Testing and Analysis Ready. Reference Handbook.* McGraw Hill, New York, NY, USA.

Sims, R.C., Sorensen, D.L., Sims, J.L., McLean, J.E., Mahmoort, R. and Dupart, R.R. 1984. Review of in-place treatment techniques for contaminated soils. Vol. 2. Background information for *in-situ* treatments. US EPA, Cincinnati, Ohio, USA.

Skinner, D.Z., Muthukuighnam and Liag, G.H. 2004. Transformation: A Powerful Tool for Crop Improvement. In: *Genetically Modified Crops, Their Development, Uses and Risks.* (Liag, G.H. and Skinner, D.Z. eds.), Food Products Press, Binghamton, NY, USA pp. 1-16.

Smith, S.R. 1996. *Agricultural Recycling of Sewage Sludge and the Environment.* CABI, Wallingford, UK.

Smolen, M.D. and Shanholtz, V.O. 1980. *Agricultural land use: Effects on the chemical quality of runoff.* Water Resources Research Center Virginia Tech, Blacksburg, VA, USA.

St. Amand, P. 2004. Risks associated with genetically engineered crops. In: *Genetically Modified Crops, Their Development Uses and Risks.* (Liang, G. and Skinner, D., eds.). Haworth Press, Binghamton, NY, USA pp. 351-363.

Stephenson, G.A. and Solomon, K.R. 1993. *Pesticides and the Environment.* Department of Environmental Biology, University of Guelph, Guelph, Ontario, Canada.

Sutherland, W.J. 2000. *The Conservation Handbook Research, Management and Policy.* Blackwell. Oxford, England.

Sweeten, J.M. 1991. 'Animal Waste Management (Agricultural and Silvcultural Non-Point Source Management).' Texas State Soil and Water Conservation, *www.tsswcb.state.tx. us/.*

Theodore, L. 1999. *Pollution Prevention: The Waste Management Approach to the 21st Century.* CRC Press, Boca Raton, FL, USA.

Tiessen, H. (ed.). 1995. *Phosphorus in the Global Environment: Transfers, Cycles and Management.* Wiley, Chichester, England.

TIFAC. 2003. Transgenic Plants and Concerns. *http://www.tifac.orgin/offer/tibo/rep/st164.htm.*

U.S. EPA 2003. Polluted runoff (nonpoint source pollution). Techniques for tracking, evaluating, and reporting the implementation of nonpoint source control measures for agriculture. *http://www.epa.gov/owow/nps/c2nact. html.*

Vance, M.A. 1984. *Agricultural Waste Monograph.* Vance Bibliographies. Monticello, IL, USA.

WHO. 1990. 'Public Health Impact of Pesticides In Agriculture.' WHO, Geneva, Switzerland.

Wateshed Science Institute 2000. *http:// www.wcc.hrcs. usda.gov/watershed.*

Young, R.A., Onstad, C.A., Bosch, D.D. and Anderson, W.P. 1994. *Agricultural Nonpoint Source Pollution Model.* USDA-ARS, Morrin, MN, USA.

CHAPTER 12

Speciation and Biodiversity

WHAT IS A SPECIES?

Evolution is the unifying concept of biology (Gould, 2002) consisting of microevolution (alteration beneath the species level) and macroevolution (change resulting in the origin of higher taxa). The latter involves common ancestry decent with modification, speciation (King, 1995; Hey, 2001; Stamos, 2004), genealogical relatedness of all life, species transformation and large-scale functional and structural alterations of populations through time at or above the species level (Theobald, 2004). Common decent, which concerns the genetic origins of living organisms (Landweber and Dobson, 1999), centers about branching and divergence of species as exemplified by phylogenetic trees (Fig. 12.1) or phylogenies (Wu, 2001; Avise and Wollenberg, 1997; Felsestein, 2004; Theobald, 2004; Avise, 2004; Steinke et al., 2006 and Hughes et al., 2006). The construction of such trees is dependent upon various research methodologies (Table 12.1).

While a comprehensive discussion of evolution is beyond the scope of this book, speciation is briefly reviewed because of the contemporary exploitation of the environment resulting in endangered and threatened species.

TABLE 12.1	Summary of Some Pertinent Research Methods Regarding Speciation
Biological Method	*Reference*
Non-molecular	
Leaf morphology and light microenvironments	Westmoreland (1989)
Cladistics and phylogenetic reconstruction	Theobald (2004)
Maximum parsimony	
Maximum likelihood	
Distance matrix methods	
Composite phenograms from	Avise and Wollenberg (1997)
Quantitative values of organismal co-ancestry	
Consensus phylogenetic traits across allelic transmission routes	Avise and Wollenberg (1997)
Molecular	
Mitochondrial DNA sequence	Machado et al. (2002)
Applications of DNA and protein markers	Avise (2004)
Expressed Sequence (ESTs)	Steinke et al. (2006)
	Hughes et al . (2006)

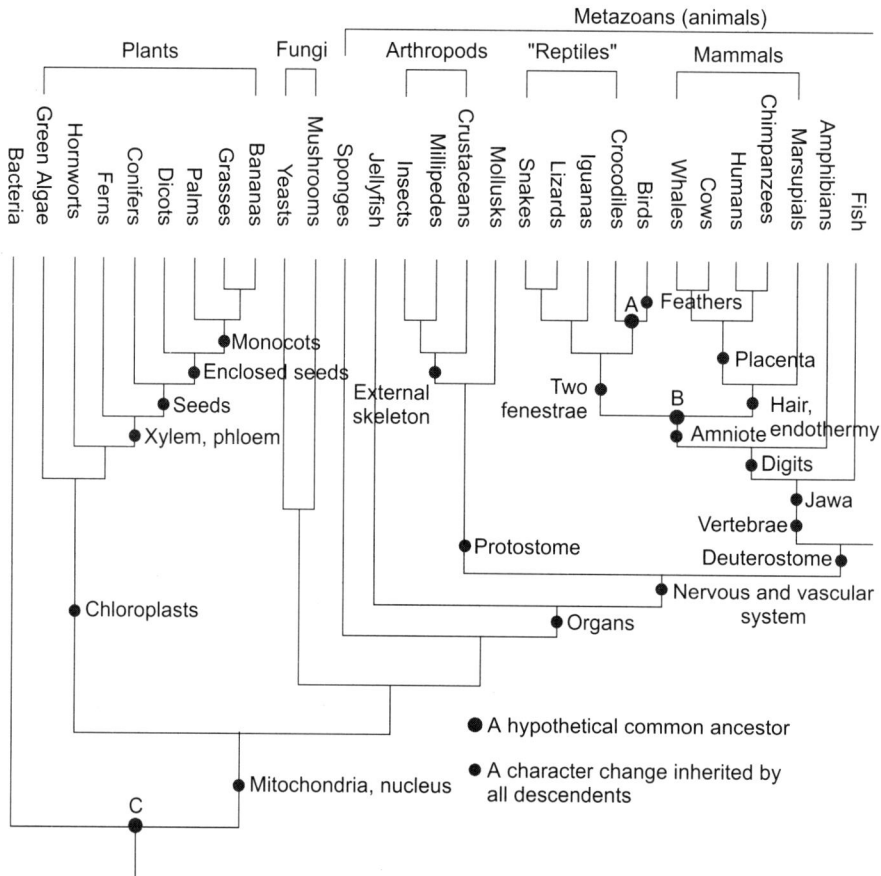

Fig. 12.1 Consensus phylogenetic tree of all life. From: Theobald, L. "29+ Evidences Macroevolution: The Scientific Case for Common Descent." The Talk Origins Archive. Vers. 283. 2004. 12 Jan, 2004. *http://www.talkorigins.org/faqus/condesc/.*

What is a species? The various definitions of species are:

1. Two organisms are members of distinct species if they do not produce fertile offspring when crossed. Thus, the species concept is based upon reproductive compatibility.

    However, geographical isolation is another important factor (Wheeler and Meier, 2000).

2. Species is a gene pool.

3. Populations that co-exist with one another in highly complex ecosystems are species.

4. Species is a taxonomic unit.

The types of species are presented in Table 12.2. Subspecies or varieties (cultivars in plants) are distinct populations.

What are the tests for speciation? There are three:

1. Do two organisms co-exist in the same area with no evidence of cross breeding?

TABLE 12.2 Summary of Categories of Species

Category	Definition
Exotic or introduced species	An organism that is not native to the locale where it is introduced; instead it was accidentally or deliberately transported to the new locale via human activity
Extinct	No reasonable doubt that the last individual died or is in immediate danger of extinction throughout all or a significant portion of its range.
Extinct in the wild	Species is known to survive only in captivity or well outside its range.
Critically endangered	Species is fairly or extremely high risk of extinction in the wild.
Endangered or vulnerable	Species is facing a very high risk of extinction in the wild.
Invasive species	Introduced by human action to a location where it is not native, but becomes capable of establishing a breeding population in the new locale without additional human interaction.
Mate-recognition	Group of organisms which recognize one another as potential mates.
Microspecies	Species that reproduce without either meiosis or mitosis; each generation is genetically identical to the previous.
Morphological	A population or group of populations which differ morphologically from other populations.
Phylogenetic/Evolutionary/ Darwinian	Group of organisms that shares an ancestry; a lineage witch maintains its integrity with respect to other lineages throughout both space and time.
Stable	Species is neither endangered nor threatened.

Adapted from: IUCN Red list and *http://www.wikipedia.org/wiki/species.*

2. When members of two groups are placed together and fail to mate or produce weak or sterile hybrids.

3. If two organisms are extremely different in appearance.

What are the causes of speciation?

The student is encouraged to read Howard and Berlocher (1998) regarding this aspect of speciation. The modes of speciation are depicted in Fig. 12.2.

1. *Phyletic speciation:* One species gradually becomes so altered that it is a new species. The flow or movement of alleles (Miglani, 2006) physically through space is significant.

2. *Divergent speciation:* Some populations of species evolve into two species or more descendant species which

become increasingly unlike. Reproductive isolation is a key factor in divergent speciation.

3. There are abiological and biological reproductive barriers (Table 12.3).

TABLE 12.3 Examples of Speciation from Abiological and Biological Reproductive Barriers for Divergent Speciation

Abiological	Biological
Allopatric Speciation (having different ranges)	*Sympatric Speciation* (having the same range)
Results from dividing the original species into two or more populations which cannot interbreed, e.g., habitat isolation or seasonal isolation (King, 1995).	Two groups become reproductively isolated even though they grow together, e.g., polyploidy (Coyne and Orr, 2004; Dieckmann et al., 2004).

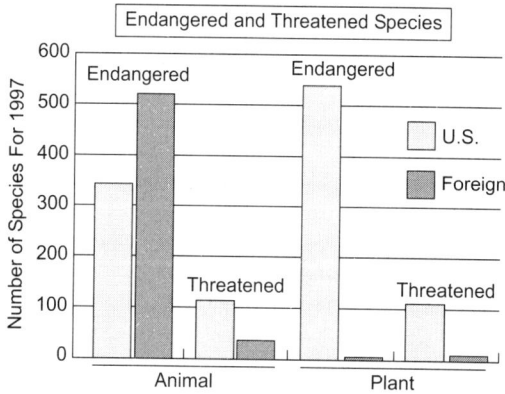

Fig. 12.2 Endangered and threatened species 1998. From: *http://endangered.fww.gov/wildlife.html.*

What Constitutes Endangered and Threatened Species?

A species is classified as endangered when it is in imminent danger of extinction throughout either a significant amount or all of its range. In contrast, a species is considered threatened when it may become endangered within the near future (Table 12.2). A list of threatened and endangered U.S. and non-U.S. species for 2007 is shown in Table 12.4. The World's endangered species (Fig. 12.3) have been catalogued by Baker (1990), Burton (1991), Beacham and Beetz (2001), Clifford (1998), Pollock (1993), Reading and Miller (2000), Turing (2002) and Vance (1989). In addition, lists of endangered species for each U.S. state can be found at *http://www.endangeredspecies. com/ map.htm.* Some of these endangered species are keystone species, i.e., ones in which other species are dependent upon. The video, *Natural Connections* (2002), presents some noteworthy examples where species extinction and threatened biodiversity (Wilson, 1992; Takacs, 1996) have been examined (Table 12.5).

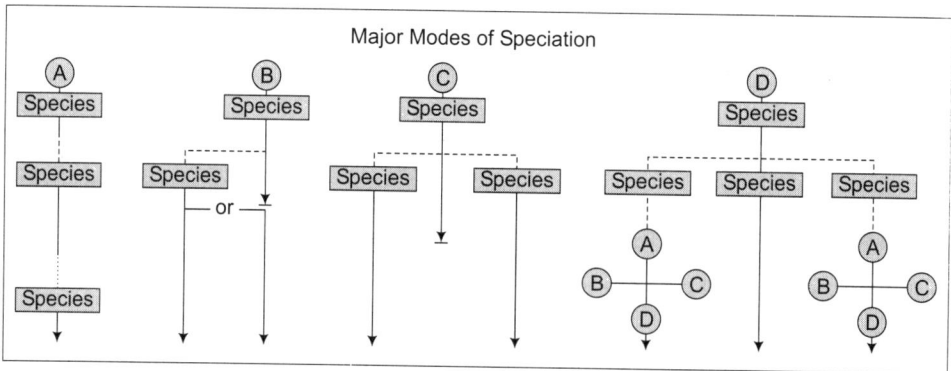

A. Anagenetic speciation. a species changes significantly over time that the changed form would not have been able to reproduce with original form. Dashed lines indicate periods of transition

B. Cladogenesis—allopatric of perapatric branching off of a daughter species. The (or) route indicates that the new species replaced its parent.

C. Allopatric or perapatric branching. Here, the parent species later goes extinct. Two species survive to the present (but there could be many).

D. Similar to (A) and (C) combined. But two species in this representation give birth to species swarms.

Fig. 12.3 Mechanisms of speciation

TABLE 12.4 USFWS Threatened and Endangered Species System (TESS)

Summary of Listed Species
Listed Populations[1] and Recovery Plans[2] as of 07/13/2007

Group	United States Endangered	Threatened	Total Listings	Foreign Endangered	Threatened	Total Listings	Total Listings (US and Foreign)	Listings with Recovery Plans[3]
Mammals	70	12	82	255	20	275	357	54
Birds	76	15	91	175	6	181	272	80
Reptiles	13	24	37	65	15	81	118	35
Amphibians	13	10	23	8	1	9	32	16
Fishes	74	64	138	11	1	12	150	98
Clams	62	8	70	2	0	2	72	69
Snails	25	11	36	1	0	1	37	30
Insects	47	10	57	4	0	4	61	33
Arachnids	12	0	12	0	0	0	12	6
Crustaceans	19	3	22	0	0	0	22	18
Animal Subtotal	*411*	*157*	*568*	*521*	*44*	*565*	*1133*	*439*
Flowering Plants	570	143	713	1	0	1	714	607
Conifers and Cycads	2	1	3	0	2	2	5	3
Ferns and Allies	24	2	26	0	0	0	26	26
Lichens	2	0	2	0	0	0	2	2
Plant Subtotal	598	146	744	1	2	3	747	638
Grand Total	*1009*	*303*	*1312*	*522*	*46*	*568*	*1880*	*1077*

[1] A Listing has an E or a T in the "status" column of the tables in 50 CFR 17.11(h) or 50 CFR 17.12(h) (the "List of Endangered and Threatened Wildlife and Plants").

32 animal species (16 in the U.S. and 16 Foreign) are counted more than once in the above table, primarily because these animals have distinct population segments (each with its own individual listing status).

[2] There are a total of 556 distinct approved recovery plans. Some recovery plans cover more than one species, few species have separate plans covering different parts of their ranges. This count includes only plans generate the USFWS (or jointly by the USFWS and NMFS), and only listed species that occur in the United States.

[3] This column includes only U.S. listings.

From: *http//ecos.fws.gov/tess_public/Boxscore.do*

TABLE 12.5 Some Examples Where Species Extinction and Threatened Biodiversity Have Been Investigated[a,b]

Ecosystem	Examples
Tropical Rain Forest	Costa Rican Rain Forest Insect Life Studies require a change in ethics and the way we interact with nature.
An Aquatic Ecosystem	Adams River in British Columbia Sockeye Salmon ;species other than man cannot compete with man-made pollution and water overuse.
Tidal Ecosystem	Keystone Species Concept; i.e., species critical to the balance of an ecosystem
Temperate Forest	Western Washington State USA, Modified harvesting technique to monitor biodiversity

[a] Adapted from *Natural Connections*. National Video Resources (2001). Conservation and biodiversity have been discussed by Dobson (1996).
[b] Bloomfield, M. 2007 discusses community action to protect biodiversity.

Why are Animals (Hodge, 1985) and Plants (Levin, 2000) Becoming Extinct?

The reasons vary with geologic time (Ehrlich and Ehrlich, 1981). At the time of writing this, the causes of the sixth great extinction (Kaufman and Mallory, 1993) appear to be those expressed by the acronym, HIPPO (Wilson, 2002).

H = Habitat destruction. One of the habitats which is currently being destroyed is the tropical rain forest, home of exotic species. Another habitat which has declined is the wetland, cleanser of the environment. Habitat conservation has been discussed by Bean et al. (1991).

I = Introduction of exotic species. Invasive species can displace native species by competing with or preying upon rare species.

P = Pollution. The pollutions affecting contemporary organisms are those in air, water and soil.

P = Population. The present World's human population of 6 billion is placing a strain on the World's resources.

O = Overconsumption. Contemporary man in industrialized countries equates quality of life with being wasteful. Some species have been harvested to extinction. Tudge (1992) describes how mass extinction can be stopped.

In addition, climate change is affecting species diversity (Lovejoy and Hannah, 2006; Sala and Knowlton, 2006).

What Can Be Done to Reduce the Loss of Species?

Individuals

Many individuals need to alter their life styles; accumulating possessions beyond what is required to sustain life is detrimental to the environment. In this connection, if the non-industrialized world develops in the same manner as the industrialized world, the world's ecosystems will be stressed further. This is a complex ethical problem, i.e., how can the industrialized world morally suppress the non-industrialized world's development?

Organizations

Besides individuals, organizations such as the Nature Conservancy and Sierra Club have been effective in saving endangered species.

Legislation

In the U.S., the Endangered Species Act was passed in 1973 to promote the preservation of endangered and threatened animal and plant species as well as their habitats (Yaffee, 1982; Burgess, 2001; Liebesma and Petersen, 2003). The Act is controversial as some see it as a means of preserving biodiversity while others view it as an impediment to employment and economic growth. The Act's authorization expired in 1992 and a new bill, the Endangered Species Improvement Act, is pending.

Research

There is a variety of contemporary research activities designed to prevent species loss (Table 12.6).

References

Avise, J.C. and Wollenberg, K. 1997. Phylogenetics and the origin of species. *PNAS* 94: 7748-7755.

Avise, J. 2004. *Molecular Markers, Natural History and Evolution*. 2nd Ed. Sinauer Associates, Suderland, MA, USA.

Baker, S. 1990. *Endangered Vertebrates: A Selected, Annotated Bibliography, 1981-1988*. Garland, New York, USA.

TABLE 12.6 Some Species Research Activities

Research Activity	Agency
Biodiversity Conservation[a] and Restoration	US Pacific Southwest Research Station (PSRS)
Determine ecological responses of aquatic and ecosystems to anthropogenic (human-induced) and natural environmental changes	PSRS
Develop Aquatic Species and Ecosystem Conservation and Restoration Techniques and Strategies	PSRS
Ecological Response of Terrestrial Species and Ecosystems to Management Activities and Natural Processes	PSRS
Habitat Relationships of Terrestrial Wildlife Species	PSRS
Impact of Mammalian Predators in Wildlife Species	US Northwest Fisheries Science Center (2001)
Understanding the Basic Ecology of Aquatic Organisms in Less Disturbed Habitats	PSRS

[a] Ziegler (2007)

Beacham, W and Beetz,K. 1998-2001. *Beacham's Guide to International Endangered Species.* Beacham Pub., Osprey, FL, USA.

Bean, M., Fitzgerald, S.G. and O'Connell, M. 1991. *Reconciling Conflicts Under the Endangered Species Act: The Habitat Conservation Planning Experience.* World Wildlife Fund, Washington DC, USA.

Bloomfield, M. 2007. *Biodiversity A Profile for Community Action.* Seattle Book Co., Seattle, WA, USA.

Burgess, B.B. 2001. *Fate of Wild: The Endangered Species Act and the Future of Diversity.* University of Georgia Press, Athens, GA, USA.

Burton, J. 1991. *The Atlas of Endangered Species.* Macmillan, New York, NY, USA.

Clifford, L. 1998. *Endangered Species: A Reference Handbook.* ABC-CCIO, Santa Barbara, CA, USA.

Coyne, J.A. and Orr, H.A. 2004. *Speciation.* Sinauer, Sunderland, MA, USA.

Dieckmann, U., Doebeli, M., Metz, J.A.J. and Tautz, D. 2004. Adaptive Speciation (Cambridge Studies in Adaptive Dynamics). Cambridge Univ. Press, Cambridge, UK.

Dobson, A.P. 1996. *Conservation and Biodiversity.* Scientific American, New York, NY, USA.

Ehrlich, P.R. and Ehrlich, A.H. 1981. *Extinction: The Causes and Consequences of the Disappearance of Species.* Random House, New York, NY, USA.

Felsestein,, J. 2004. *Inferring Phylogenetics.* Sinauer , Sunderland, MA, USA.

Gould, S.J. 2002. *The Structure of Evolutionary Theory.* Harvard Univ. Press, Cambridge, MA, USA.

Hey, J. 2001. *Genes, Categories and Species. The Evolutionary and Cognitive Cause of the Species Problem.* Oxford Univ., Press, Oxford, England.

Hodge, R.J. 1985. *Animal Extinction: What Everyone Should Know.* Smithsonian Institution Press, Washington, DC, USA.

Howard, D.J. and Berlocher, S.H.. 1998. *Endless Forms: Species and Specification.* Oxford Univ. Press, New York, NY, USA.

Hughes, J., Longhorn, S.J., Papadopoulou, A., Theodorides, K., deRiva, A., Mejia-Chang, M., Foster, P.G. and Vogler, A.P. 2006. Dense taxonomic EST Sampling and its applications for Molecular Systematics of the Coleoptera (beetles). *Mol. Biol. Evol.* 23: 268-278.

Kaufman, L. and Mallory, K. 1993. *The Last Extinction.* MIT Press, Cambridge, MA, USA.

King, M. 1995. *Species Evolution: The Role of Chromosome Change.* Cambridge Univ. Press, Cambridge, UK.

Landweber, L. and Dobson, A. 1999. *Genetics and the Extinction Species.* Princeton Univ. Press, Princeton, NJ, USA.

Levin, D.A. 2000. *The Origin, Expansion and Demise of Plant Species*. Oxford Univ. Press, Oxford, England.

Liebesma, L.R. and Petersen, R. 2003. *Endangered Species*. Environmental Law Institute, Washington, DC, USA.

Lovejoy, T.E. and Hannah, L. 2006. *Climate Change and Biodiversity*. Yale Univ. Press, New Haven, CT, USA.

Machado, C., Kliman, JR. M., Markert, M. and Hey, J. 2002. Inferring the history of speciation from meticulous DNA sequence data: the case of *Drosophila pseuoobscura* and its close relatives. Mol. Biol. Evol. 19: 472-488.

Miglani, G.S. 2006. Mendelion Genetics In: *Plant Cell Biology*. (Dashek, W.V. and Harrison, M., eds.). Science Publishers, Enfield, NH, USA, pp. 259-348.

National Video Resources 2002. Session 6 Biodiversity Natural Connections. New York, NY, USA.

Pollock, S. 1993. The Atlas of Endangered Species. Facts on File, New York, NY, USA.

Reading, R.P. and Miller, B. 2000. Endangered Animals: A Reference Guide to Conflicting Issues. Greenwood Press, West Port, CT, USA.

Sala, E. and Knowlton, N. 2006. Global marine biodiversity trends. *Annu. Rev. Env. Resour.* 31: 93-122.

Stamos, D.N. 2004. The Species Problem: *Biological Species Ontology and the Metaphysics of Biology*. Lexington Books, Lanham, MD, USA.

Steinke, D., Salzburger, W. and Meyer A.J. 2006. Novel relationships among ten fish model species revealed based on a phylogenomic analysis using ESTs. *J. Mol. Evol.* 62: 772-784.

Takacs, D. 1996. *The Idea of Biodiversity: Philosophies of Paradise*. John Hopkins University, Baltimore, MD, USA.

Theobald, D. 2004. 29+evidences for macroevolution. The scientific case for common descent. *http://www.talkorigins.org/faqs/com desc/default.html.*

Tudge, C. 1992. Last Animals at the Zoo. How Mass Extinction Can Be Stopped. Island Press, Washington, DC, USA.

Turing, J.P. 2002. *Endangered Species: A Bibliography With Indexes*. Nova Science Publishers, New York, NY, USA.

Vance, M.A. 1989. *Endangered Species: A Bibliography*. Vance Bibliographies. Monticello, IL, USA.

Westmoreland, D. 1989. Leaf morphology and light microenvironments: a field exercise *Amer. Biol. Teach.* S1: 303.

Wheeler, Q.D. and Meier, R. 2000. *Species Concept and Phylogenetic Theory: A Debate*. Columbia Univ. Press, New York, USA.

Wilson, E.O. 1992. *The Diversity of Life*. Harvard Univ. Press, Cambridge, MA, USA.

Wilson, E.O. 2002. *The Future of Life*. A.A. Knopf, New York, NY, USA.

Wu, G-D. 2001. The genic view of the process of speciation. *J. Evol. Biol.* 14: 851-865.

Yaffee, S.L. 1982. Prohibitive Policy Implementing the Federal Endangered Species Act. MIT Press, Cambridge, MA, USA.

Ziegler, D. 2007. *Understanding Biodiversity* Praeger, Westport, CT.

CHAPTER 13

Population Growth

INTRODUCTION

The student is encouraged to review basic population biology as an introduction to population growth. Some useful reference books include those of Lotka et al. (1998), Neal (2003), Renshaw (1993), Royama (1996), Silvertown (1993), Turchin (2003), Uirzo and Sarukan (1984), Weeks (2004), Ranta et al. (2005) Rockwood (2006) and Hastings (2007).

The Table 13.1/Fig. 13.1 illustrate the 0.04%/yr increase in the population growth rate from AD 1 to the 1600's. Between 1650 and 1970 the growth rate rose to 2.1% and then declined to 1.6%/yr (Cohen, 1995 a, b). Hardin (1993) suggests that population increases over time were associated with three revolutions, i.e., tool-making, agriculture and scientific/industrial.

Population growth has been more prolific in less developed countries, especially during the last 100 years (United Nations, 1996; 1999; 2003; 2005). In contrast, certain developed countries have experienced low birth rates (Day, 1992; Wattenberg, 2004). Human population growth in less-developed countries cannot continue to increase indefinitely because of both space and resource limitations (Herwig, 1995).

Eventually, the Earth will reach its human carrying capacity (Cohen, 1995 b). However, human choice is both dynamic and uncertain and, therefore, not adequately satisfied by an understanding of ecological carrying capacity. A concise history of the World population occurs in Livi-Bocci (1997) and Gilbert (2001) provides a handbook on World population.

What Factors Influence Human Population Growth?

1. *Food supplies* grow at an arithmetic rate in contrast to the geometric growth of populations (Fig. 13.2). Therefore, human population growth should surpass the food supply. However, the food supply has accelerated more rapidly than arithmetically because of improvements in agricultural technology (see Agricultural Pollution Chapter). In addition, geometric curves do not adequately explain human population growth. Some populations exhibit a sigmoidal growth curve (Fig. 13.2).

2. Birth rate, death rate and migrations are *vital* statistics studied by demographers (Namboodiri, 1991; Ness, 1999; Newell, 1990; Preston, 2001; Rowland,

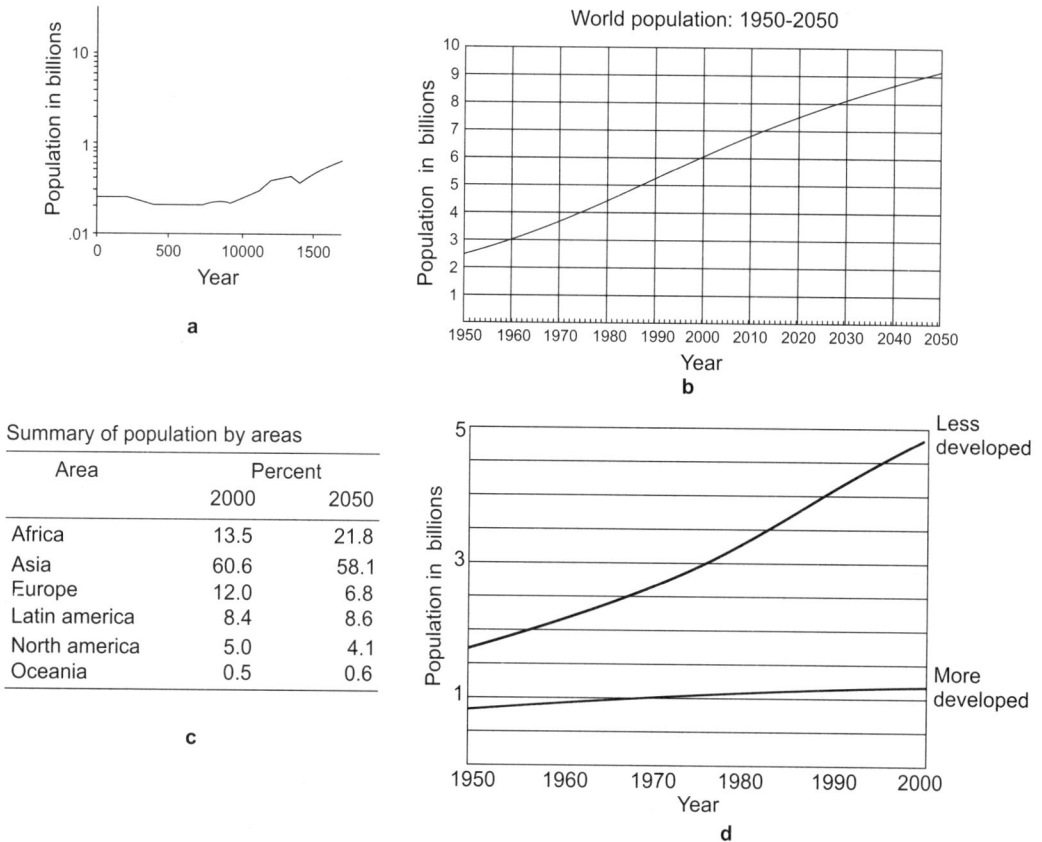

World population: 1950-2050

Summary of population by areas

Area	Percent	
	2000	2050
Africa	13.5	21.8
Asia	60.6	58.1
Europe	12.0	6.8
Latin america	8.4	8.6
North america	5.0	4.1
Oceania	0.5	0.6

c

Fig. 13.1 Time-Dependent Changes in Population by Areas and World Countries' Development
From: US Census Bureau, International Data Base, August 2006 Version.
http://www.census.gov/ipc/www/img/worldpop.gif.
In developing countries, there is a shift from rural to urban living (Grimm, N.B. et al., 2008. *Science* 319: 756-760.)

2004). Yearly birth and death rates are influenced by the age–sex composition of a population (Table 13.2). Furthermore, lifespans can increase as human populations age (Sanderson and Scherbov, 2005). Patterns of migrations often do not affect growth if a large percentage of the population migrates. Some measurements that demographers employ in population investigations are depicted in Table 13.3.

U.S. Department of Commerce, Bureau of the Census. U.S. Population Esitmates, by Age, Sex, Race, and Hispanic Origin. Current population Reports, Series p-25, No. 1045, Table 2, 9. 52 (Washington, DC: DOC/BOC, 1990) and Series P-25, No. 1095, Table 1, p. 2 (Washington, DC: DOC/BOC, 1993).

Population Growth Control

Depending upon the country, population growth trends range from 0.6% to 4.0% per

TABLE 13.2 Age Structure of the Population, 1940-1991
(population, in millions)

Year years	Under 5 years	5-14 years	15-24 years	25-34 years	35-44 years	45-54 years	55-64 years & over	65 years
				Age Class				
1940	10.58	22.36	24.03	21.45	183.42	15.56	10.69	9.03
1950	16.33	24.48	22.26	23.93	21.56	17.40	13.36	12.36
1960	20.34	35.74	24.58	22.92	24.22	20.58	15.63	16.68
1970	17.16	40.73	36.50	25.29	23.14	23.31	18.66	20.09
1980	16.46	34.85	42.80	37.63	25.87	22.70	21.76	25.70
1991	19.20	35.90	36.60	43.10	39.40	25.70	21.00	31.80

Source: U.S. Department of Commerce, Bureau of the Census. Historical Statistics of the United States: Colonial Times to 1970, Part I, Series 30-37, p. 10 (Washington, DC: DOC/BOC, 1976).

TABLE 13.3 Demographic Measurements[a]

Measurement	Mathematical Expression
Vital rate	Number of vital events in a population during a specified time
Crude birth rates	Size of the population Number of live births during a given year x1000
Crude death rates	Midyear population during a given year Number of deaths during a given year
Crude reproductive rate	Midyear population during a given year x1000 Number of live births – Number of deaths in a given year x1000
Survivorship curves	Midyear population during a given year Periodic measures of age and sex specific rates

[a]See Waller and Bouvier (1981), Wunsch and Termote (1978), Newell (1990), Namboodiri (1991), Lotka et. al. (1998) and Yaukey and Anderton (2001). The United Nations Population Division 2003 has made available Mortpak for Windows, a software package for demographic measurements.

annum for the 21st century. In the developed World, the replacement level has fallen below 2.1 children per woman. In marked contrast, a high birth rate exists in many poor countries creating 'a surplus population'. Therefore, curtailing population growth (Harkovy, 1995) in these countries is essential for World political stability, environmental integrity (Hunter, 2000) and quality of life (Bouvier and Bertrand, 1999).

What Factors (Natural and Man-made) can Control Population Growth?

Diseases – Throughout human history, there have been epidemics, e.g., the black plague and AIDS. In the contemporary developed World, obesity, cancer and heart attacks reduce the population.

Famine – Starvation can be caused by drought, flooding, weather conditions and political events (e.g., sanctions during war-time).

Anthropogenic Controls – Birth control is practiced in developed countries but not often in developing countries. Family planning varies from country to country for religious and cultural reasons.

Environmental Changes – These include water and air pollution which affect agricultural yield and aquatic life.

Economic Development – This usually results in a reduction in a nation's population growth. Some countries employ economic activities to reduce fertility rate, e.g., wars often discourage births via genocide.

What research methods and activities are being employed to examine population growth? These are presented in Tables 13.4 and 13.5. Individuals usually think of overpopulation as a human problem. However, we also need to consider wildlife overpopulation (Silby, 2003).

What are the environmental consequences of overpopulation?

There are many books and articles regarding overpopulation and the future. Some of these are Bongaarts (1994), Herwig (1995) and Shah (1998).

The environmental consequences of overpopulation are myriad including loss of habitats (rainforests and wetlands), biodiversity (Mckee, 2005) air and water pollution, global warming, increased demand for natural resources (Hunter, 2000) and threats to the World's food supply and security (Polumin, 1998).

Finally, Table 13.6 presents some significant websites and journals regarding overpopulation. These enhance the present text by providing additional detail.

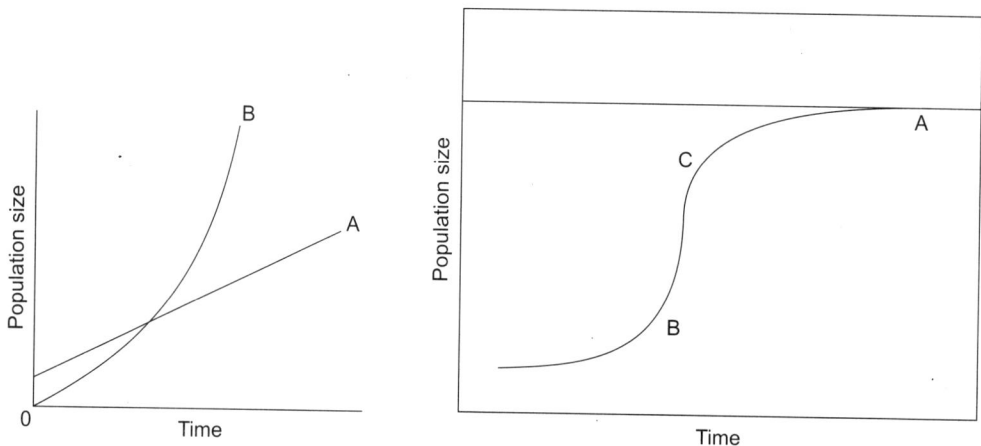

Fig. 13.2 Types of growth curves
Left: A = Arithmetic, B = Geometric
Right: Sigmoidal Growth Curve
    A = Plateau, B = Exponential, C = Beginnings of decline
http://www.mansfield.ohio-state.edu/~sabedon/biol.1540. Four Delta provides toolkits for modeling sigmoidial growth curves (*http://www.fourdelta.co.uh/sigmoid.html*)

TABLE 13.4 Some Research Methods for Investigating Population Growth

Method	Reference
Population Growth with Protists	http://mutualismswilliams.edu/biol203/labparamecium%20lab.htm
Remote Sensing Land-Use Change Detection	Qui, Woller and Briggs (2003)
Demographic Analyses	Krishnan, Tuan and Mahadevan (1992)
Fertility Levels	Bongaarts and Bulato (2000)
Census	www.guardian.co.uk/population/story

Bongaarts, J. and Bulato, R.A. (2000). *Beyond Six Billion Forecasting the World's Population.* National Academies Press, Washington, DC, USA.

Krishnan, P., Tuan, C.H. and Mahadevan, K. (1992). *Readings in Population Research: Policy, Methods and Perspectives.* Vedams, India.

Lemeshow, S and Levy (1999). *Sampling of Population Methods and Applications.* Wiley, New York, NY, USA.

Qiu, F., Woller, K.L. and Briggs, R. (2003). Modeling urban population growth from remote sensor and TIGER GIS record data. Photometric Engineering and Remote Sensing 69:1031 – 1042.

TABLE 13.5 Some Examples of Current Research Efforts Regarding Population Growth

Activity	Agency
Effect of population on climate change	Population Resource Center, Washington, DC
Water and population dynamics	World Conservation Union (IUCN)
Population – environment analysis and methods of population forecasting	World Population Program International Institute for Applied Systems Analysis
Advanced academic research on population and the environment	Population – Environment Research Network

TABLE 13.6 Summary of Selected Websites and Journals for Population Growth

Websites	Purpose
http://www.census.gov/ipc/www/:idbyr.html United States Census Bureau, Population Division	Create population pyramids which are graphical representations of the age and gender distribution of the country
http://www.prcdc.org/summaries/climateupdate02/climateupdate02.html Population Resource Center	Information on population and global warming, science and policy issues, effect of population on per capita emission trends, effect of climate change on population
http://www.socstats.soton.oc-uk/popsci/population	Website for teaching the population-related sections of a geography course
http://www.nwf.org National Wildlife Federation Population Growth	
http://www.populationaction.org Population Action International	
http://www.populationinstitute.org Population Institute	
http://www.Zpg.org Zero Population Growth	

http://www.npg.org
Negative Population Growth
http://www.popco.org
Population Coalition
http://www.undp.org/popin
United Nations Population Division
http://www.census.gov/ips
United States Census Bureau

Other Population Websites can occur at:
http://www.actionbioscience.org/environment/hinricnsen_robey.html

Journals
Demographic Research
Demography
European Journal of Population
Journal of Population Studies
Mathematical Population Studies
Population and Development Reviews
Population and Environment
Population Research and Policy Review
Population Review
Population Studies
Theoretical Population Biology

References

Bongaarts, J. 1994. Population policy options in the developing world. *Science* 263: 771 – 776.

Bouvier, L.F. and Bertrand, J.T. 1999. *World Population.* Seven Locks Press, Santa Anna, CA, USA.

Cohen, J. 1995a. *How Many People Can the Earth Support.* W.W. Norton, New York, NY, USA.

Cohen, J.E. 1995b. Population growth and earth's human carrying capacity. *Science* 269:341-346.

Day, L. 1992. *The Future of Low Birth-Rate Populations.* Routledge, London, England.

Gilbert, G. 2001. *World Population. A Reference Handbook.* ABC-CLIO, Santa Barbara, CA, USA.

Hardin, G. 1993. *Living Within Units.* Oxford Univ. Press, Oxford, England.

Harkovy, O. 1995. *Curbing Population Growth. An Insiders Perspective on the Population Movement.* Plenum Press, New York, NY, USA.

Hastings, A. 2007. *Population Biology: Concepts and Models.* Springer, New York, NY, USA.

Herwig, B. 1995. *World Population Projections for the 21st Century. Theoretical Interpretations and Qualitative Simulates.* St. Martin's Press, Frankfurt, Germany.

Hunter, L.M. 2000. *Environmental Implication of Population.* Rand Corporation, Santa Monica, CA, USA.

Lemeshaw, S. and Levy, P.S. 1999. *Sampling Population Methods and Applications.* Wiley, New York, NY, USA.

Livi-Bocci, M. 1997. *A Circle History of World Population 2nd Ed.* Blackwell, Holden, MA, USA.

Lotka, A.J., Smith, D.P. and Rossert, H. 1998. *Analytical Theory of Biological Populations.* Plenum Press, New York, NY, USA.

Mckee, J.K. 2005. Sparing Nature: *The Conflicts Between Human Population Growth and Earth's Biodiversity.* Rutges Univ. Press, Piscataway, NJ, USA.

Namboodiri, N.K. 1991. *Demographic Analysis: A Stochastic Approach.* Academic Press, San Diego, CA, USA.

Neal, D. 2003. *Introduction to Population Biology.* Cambridge Univ. Press, Cambridge, England.

Ness, I. 1999. *The Encyclopedia of Global Population and Demographics.* Sharpe Reference, Armonk, New York, NY, USA.

Newell, C. 1990. *Methods and Models in Demography.* Guilford Press, New York, NY, USA.

Polumin, N. 1998. *Population and Global Security.* Cambridge Univ. Press, Cambridge, England.

Preston, S.H. 2001. *Demography Measuring and Modelling Population Responses.* Blackwell; Malden, MA, USA.

Ranta, E., Kaitala, W. and Lundberg, P. 2005. *Ecology of Population.* Cambridge Univ. Press, Cambridge, England.

Renshaw, E. 1993. *Modelling Biological Population in Space and Time.* Cambridge Univ. Press, Cambridge, England.

Rockwood, L.L. 2006. *Introduction to Population Ecology.* Blackwell, Holden, MA, USA.

Rowland, D.T. 2004. *Demographic Methods and Concepts.* Oxford University Press, Oxford, England.

Royama, T. 1996. *Analytical Population Dynamics.* Kluwer Academic Publishers, New York, NY, USA.

Sanderson, W.C. and Scherbov, S. 2005. Average remaining lifetimes can increase as human populations age. *Nature* 435. 811-813.

Shah, A. 1998. *Ecology and the Crisis of Overpopulation: Future Prospects for Global Sustainability.* Edgar Allen, Cheltenham, UK.

Silby, R. 2003. *Wildlife Population Growth Rates.* Cambridge Univ. Press, Cambridge, England.

Silvertown, J.W. 1993. *Introduction to Plant Population Biology.* Blackwell Scientific Publications, Oxford, England.

Turchin, P. 2003. *Complex Population Dynamics a Theoretical Empirical Synthesis.* Princeton University Press, Princeton, NJ, USA.

Uirzo, R and Sarukan, J. 1984. *Perspectives on Plant Population Ecology.* Sinauer Associates, Surderland, MA, USA.

United Nations, Department of International Economic and Social Affairs, Long-Range World Population Projections: Two Centuries of Population Growth 1950-2150 (1996). United Nations, New York, NY, USA.

United Nations 1999. World Population Monitoring Population Growth, Structure and Distribution. *http://un.org*

United Nations. 2003. *World Population Prospects. The 2003 Revision: Data Table and Highlights.* United Nations, New York, NY, USA.

United Nations Population Division. 2005. *Population Challenges and Development Goals.* United Nations, New York, NY, USA.

Waller, R.H. and Bouvier, L.F. 1981. *Population Demography and Policy.* St. Martin's Press, New York, NY, USA.

Wattenberg, B.J. 2004. *Fewer: How the New Demography of Depopulation Will Shape Our Future.* Ivan R. Dee, Chicago, IL, USA.

Weeks, J.R. 2004. *Population: An Introduction to Concept and Issues.* Wadsworth, Philadelphia, PA, USA.

Wunsch, G.J. and Termote, M.C. 1978. *Introduction to Demographic Analysis.* Principles and Methods. Plenum Press, New York, NY, USA.

Yaukey, D. and Anderton, D.L. 2001. *Demography: The Study of Human Population.* Wavelend Press Inc., Long Grove, IL, USA.

CHAPTER 14

Environmental Law

AGREEMENTS, LAWS AND TREATIES

We might ask what are the limits of human existence (Kime, 1992)? The following are relevant to this significant question: 1.The degradation of the environment will most likely accelerate as the developing countries industrialize (Gibson and Halter, 1994) and the World's population expands, 2. Even remote World areas are affected by distant environmental damage, for example, the transport of certain air-pollutants can be long distance, 3. The development of new technologies often test environmental limits (Brune et al., 1997), 4. Protecting the World's environment requires substantial money (Farrow and Toman, 1999) and meaningful international co-operation. Given human history of wars, the latter seems very precarious, 5. The altruistic approach to protecting the environment is not very feasible as individuals who act without concern for maintaining environmental quality are not adequately punished. Furthermore, those who act prudently are often not rewarded, and 6. Many people fail to understand that man is dependent upon sustaining the World's fauna and flora. Thus, students need to take biology and bioethics courses during their educational training. If man will not behave in an altruistic manner, what can be done to protect the environment? Table 14.1 provides some examples of laws (Findley and Farber, 2008), agreements and treaties designed to protect the environment. The following environmental law and justice references are relevant to that protection (Beacham, 1993; Committee on Environmental Justice, 1999; Dobson, 1999; Ferry, 1997; Mintz,1994; Moskowitz, 1995; Patton-Hulce, 1995; Powell, 1999; Berry and Dennison, 2000; Portney and Stavins, 2000; Schoenbaum and Rosenberg, 2001; Benidickson, 2002; Kubasek and Silverman, 2007; Thornton and Beckwith, 2004; Speth and Haas, 2006).

ENVIRONMENTAL ORGANIZATIONS COUNTRIES POSSESSING

What are the existing organizations (Table 14.2) concerned with protecting the environment? Seredich (1991) catalogues these numerous agencies into non-governmental, federal and state organizations. This author lists the contact information, purposes, programs, accomplishments and publications of each of the mentioned organizations. In addition to Seredich's volume, Krupin (1994) has published an environmental

TABLE 14.1 Summary of US EPA Environmental Laws and International Agreements[a,b,c]

USA	Year
Chemical Safety Information, Site Security and Fuels Regulatory Relief Act	1991
Clean Air Act (CAA)	1970
Clean Water Act (CWA)	1977
Comprehensive Environmental Response, Compensation, and Liability Act (CERCLA or Superfund)	1980
Emergency Planning & Community Right-To-Know Act (EPCRA)	1986
Endangered Species Act (ESA)	1973
Federal Insecticide, Fungicide and Rodenticide Act (FIFRA)	1972
Food Quality Protection Act (FQPA)	1996
National Environmental Policy Act (NEPA)	1969
Noise Control Act	1972
Occupational Safety and Health Act (OSHA)	1970
Oil Pollution Act	1990
Pollution Prevention Act (PPA)	1990
Resource Conservation and Recovery Act (RCRA)	1976
Safe Drinking Water Act (SDWA) and 1996 Amendments	1974
Superfund Amendments and Reauthorization Act (SARA)	1986
The Freedom of Information Act (FOIA)	1966
Toxic Substances Control Act (TSCA)	1976

International (Hunter et al., 2002; Rajamani, 2006)

1. (Convention on Biological Diversity (*http://www.biodiv.org/convention/articles.asp.*)

2. Climate change United Nations Framework Convention on Climate Change and the Kyoto Protocol on Global Warming. *http//unfcc.int/essential-background/convention/items/2627.php.*

3. Desertification (United Nations Convention to Combat Desertification. *http://www/uncce.int/*).

4. Endangered species (Convention on International Trade in Endangered Species (CITES) *http://www.cites.org/end/disc/text.shtml.*

5. Hazardous materials and activities (Basel Convention (Table 14.4) on the Control of Transboundary Movements of Hazardous and Their Disposal (*http://www.basel.int/text/text.html.* and *http://www.ban.org/main/about-basel-conv.html*).

6. Marine pollution (Convention on the Prevention of Marine Pollution by Dumping of Wastes and Other Matter *http://www.londonconvention.org/main.htm.*)

7. Sustainable development (The Rio Declaration on Environment and Development *http://www.unep.org/documents/default.sap?documentid=78*).

8. Uses of the seas (United Nations Convention on Law of the Sea (UNCLOS) linked from *http://www.un.org/depts/los/index.htm*).

9. Environmental Protection Act (1990) – Great Britain. http://www2.environmentagency.gov.uk/epr/info.asp
"Waste management licensing system, a legal duty of care and registration system for business transporting waste."

[a] Adopted from *http://www.epa.gov/epahome/laws.htm* and *http://www/asil.org/resource/env1.htm.*

[b] See the following for US Environmental Policies (Beacham, 1993; Committee on Environmental Justice, 1999; Koren, 1991; Hoffman, 2000)

[c] The following are relevant to Global Environmental Policy (Beacham, 1993; Grant et al., 2001; Katz and Thorton, 1994; Miller and Miller, 1991; Rombke and Mottmann, 1995; Percival and Alevizatos, 1997; Taylor, 1998; Sands, 2003). An International Encyclopedia of laws was published by Kluwer, Dordrecht in 2000.

directory listing nationwide phone numbers of more than four thousand environmental firms, organizations, government agencies and private Institutions. With regard to numerous websites for international environmental organizations (Table 14.2), one can find websites for global issues, water, forests, mines and mining, animals, climate, economics and organizational/environmental support at *http://www.periclespress.com /webenvrn.html*. In addition, an environmental web directory can be located at *http://www.webdirectory.com*.

TABLE 14.2 Countries Possessing Environmental Organizations[a]

Intergovernmental	
Canada	
New Zealand	
United Kingdom	
United States	
Private Organizations	
Australia	New Zealand
Canada	South Africa
Germany	United Kingdom
Italy	United States

[a] See en.wikipedia.org/wiki/List-of-environmental organizations for specifics.

What can Individuals and Groups do to Promote Environmental Change?

Table 14.3 summarizes some environmental action methods. While certain of these are for individual action, others are for groups.

An example of international group action is Table 14.4. Finally, cost-benefit analysis (Lifoff and Swartzman, 1982; Farrow and Toman, 1999; Puttaswamaiah, 2002; Bateman et al., 2003) has been employed as a means of setting environmental standards at a lower cost to society (Table 14.5). While this analysis has many proponents, Heinzerling and Ackerman (2002,

TABLE 14.3 Ways By Which Individuals and Groups Can Promote Environmental Change[a]

Individuals	Groups
Career Choices	Waste Recycling
Involvement in Politics	Regulating World Population
Life Style Alterations	Pollution Control
Participation in Environmental Organizations	Research and Development of Alternative Energy Sources
	Develop Sustainable Agriculture Conservation of World's Resources
	Legislation and Fines

[a] Adapted in part from Turk and Turk (1997).

TABLE 14.4 Objectives of the Basel Convention[a]

Objectives

1. Co-operate on the exchange of information, technology transfer, and the harmonization of standards, codes and guidelines.
2. dispose of hazardous wastes within the country of generation, as much as possible.
3. ensure that hazardous waste generation is reduced to a minimum.
4. establish controls on exports and imports of hazardous waste
5. prohibit shipments of hazardous wastes to countries lacking the legal, administrative and technical capacity to manage and dispose of them in an environmentally sound manner

[a] adopted from *http://www/ec.gc.ca/tmb/eng/tmbbasel_e.htmp* and *http://www.basel.int/text/text.html*.

http://www.law.georget own.edu/gelpi/popers/Pricefml.pdf) have concluded that cost-benefit analysis should be rejected as a way of determining environmentally protective regulation. They elaborate three reasons for the rejection. Briefly condensed these are: 1. Cost-benefit will not yield more effective decisions since reducing health, life and the natural world monetary considerations is flawed., 2. Employment of discounting fi-

Approach evolves as attitudes and behaviours change over time	• Remove barriers • Give information • Provide facilities • Provide viable alternatives • Educate/train/provide skills • Provide capacity

Enable

• Tax system
• Expenditure – grants
• Reward schemes
• Recognition/ social pressure – league table
• Penalties, fines & enforcement action

Encourage → Catalyse ← Engage

Is the package enough to break a habit and kick start change?

• Community action
• Co-production
• Deliberative for a
• Personal contacts/ enthusiasts
• Media compaigns/ opinion formers
• Use networks

Exemplify

• Leading by example
• Achieving consistency in policies

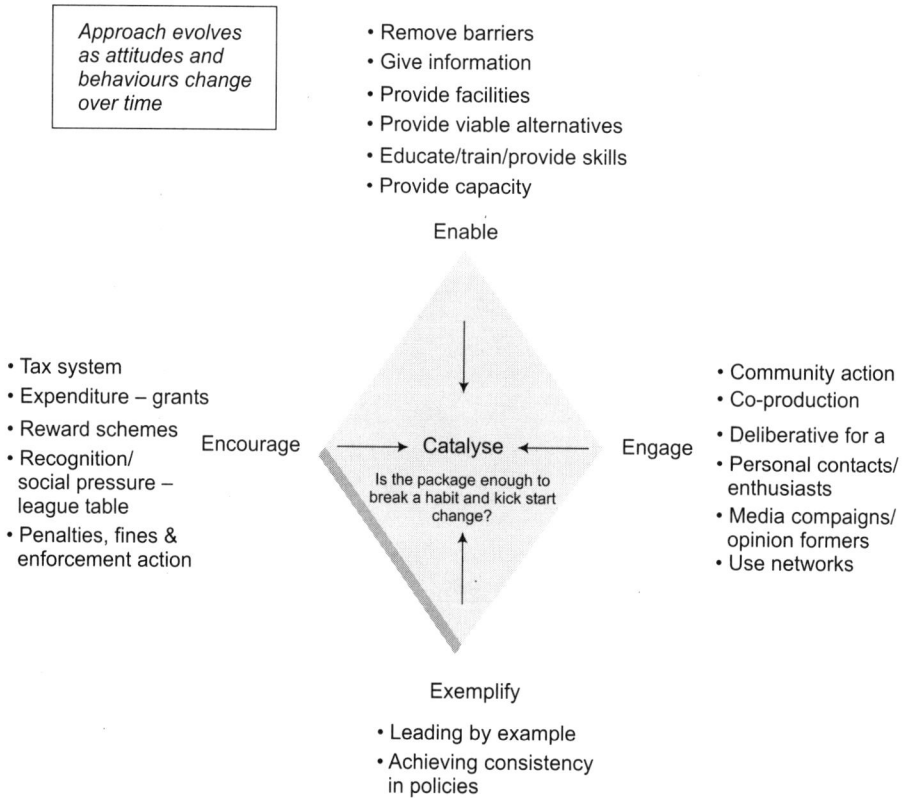

Fig. 14.1   United Kingdom governmental program to alter peoples' mindset regarding the Environment. From: *http://www.sustainable-development.gov.uk/publications/pdf/strategy/chap%202.pdf.*

TABLE 14.5   An Example of Cost-Benefit Analysis[a]

	Lives[1]	Benefit	Costs	Net Costs[2]	Cost-effectiveness[3]
EPA's model without accounting for latency	28	$170 million	$210 million	$40 million	$7.5 million per statistical life
EPA's model accounting for latency	28	$50 million	$210 million	$160 million	$26 million per statistical life
Our high estimate[4]	110	$200 million	$210 million	$10 million	$6.4 million per statistical life
Our best estimate[5]	11	$20 million	$210 million	$190 million	$65 million per statistical life
Our low estimate	5.5	$10 million	$210 million	$200 million	$130 million per statistical life

What the New Rule Brings
The effects of EPA's arsenic rule, with different assumptions

[1]Statistical lives saved shown here are not discounted for latency.
[2]Net costs are costs minus benefits.

[3]See article text for details.

[4]We obtain our upper-bound estimates by taking EPA's models, including "non-quantifiable" benefits and accounting for latency.

[5]Our best estimate includes "non-quantifiable" benefits, accounts for latency, and incorporates a sub linear dose-response function. See article text for details.

From: *http://www.cato.org/pubs/regulation/regv24n3/specialreport.pdf.*

Another example of cost-benefit analysis relative to the environment is that of the USA's Clean Air Act (*http://www.nescoline.org/nle/cisreports/risk/rsh-42.cfm*).

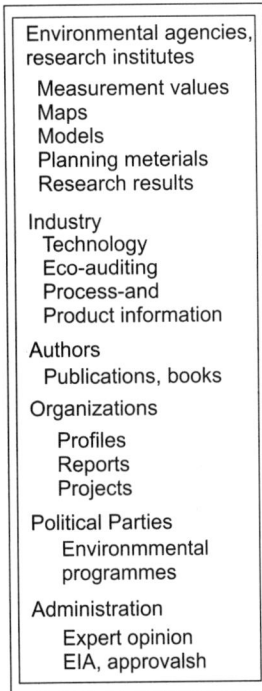

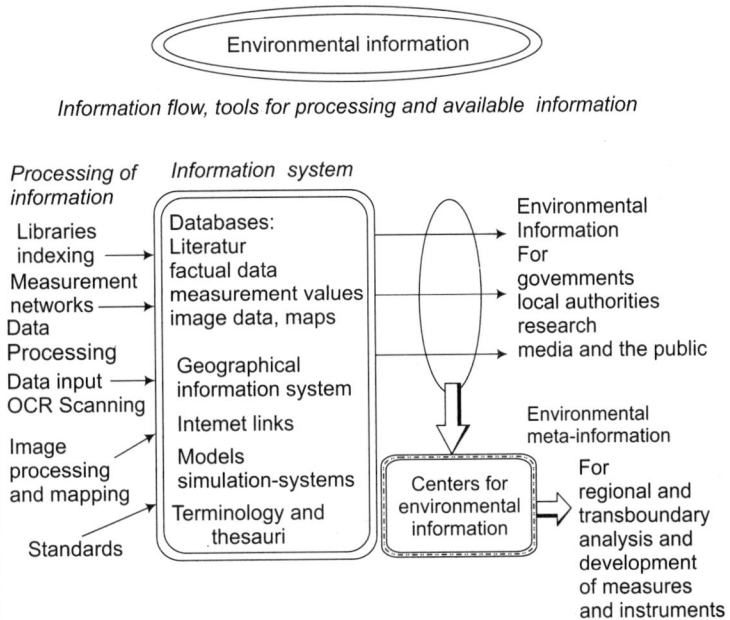

Fig. 14.2 Environmental Education
From: Pan-American User Forum on Environmental Telematics, March, 1999
http://www.p2parp.org/ref/04/03872.pdf.

nancial investments diminishes the significance of environmental regulation and 3. Cost-benefit does not address those who suffer from environmental problems. Morel and Linkov (2006) discuss environmental management. One example of environmental action is that proposed in England (Fig. 14.1). The goal is to assist people to alter their mindsets regarding the environment. Other educational/informational (Fig. 14.2) sources are the international proceedings of the working groups of the Intergovernmental Panels on climate change. The concerns of the 2007 report, the physical basis, are presented in Table 14.6.

TABLE 14.6 Concerns of IPCC Climate Change Feb. 2007. The Physical Science Basis[a]

Concerns

Historical Overview of Climate Change Science

Changes in Atmospheric Constituents and in Radiative Forcing

Observations: Surface and Atmospheric Climate Change

Observations: Changes in Snow, Ice and Frozen Ground

Observations: Oceanic Climate Change and Sea Level

Paleoclimate

Couplings Between Changes in the Climate System and Biogeochemistry

Climate Models and their Evolutions

Understanding and Attributing Climate Change

Global Climate Projections

Regional Climate Projections

[a] Schmittner et al. (2008) address certain of these concerns for a business as usual scenario until 4,000 AD.

References

Bateman, I.J., Lovett, A. and Brainard, J.S. 2003. *Applied Environmental Economics: AGIS Approach to Cost Benefit Analysis.* Cambridge University Press, Cambridge, England.

Beacham, W. 1993. *Beacham's Guide to Environmental Issues and Sources.* Beacham Publishing, Inc., Washington, DC. Vol. 1 Contains sections on Global Environmental Policy and U.S. Environmental Policy and Politics.

Benidickson, J. 2002. *Environmental Law.* Irwin Lab, Toronto, Canada.

Berry, J.F. and Dennsion, M.S. 2000. *Environmental Law and Compliance.* McGraw-Hill, New York, NY,USA.

Brune, D., Chapman, D.V., Gwyne, M.D. and Pacyra, J.M.. 1997. *The Global Environment, Science, Technology and Management.* Wiley and Sons, New York, NY, USA.

Committee on Environmental Justice – Institute of Medicine.1999. *Toward Environmental Justice: Research, Education and Health Policy Needs.* National Academy Press, Washington, DC, USA.

Dobson, A. 1999. *Justice and the Environment: Conceptions of Environmental Sustainability and Dimensions of Social Justice.* Oxford Press, Oxford, United Kingdom.

Farrow, S. and Toman, M. 1999. Using benefit – cost analysis to improve environmental regulations. *Environment* 41: 12-15.

Ferry, S. 1997. *Environmental Law: Examples and Explanations.* Aspen Publishers, New York, NY, USA.

Findley, R.W. and Farber, D.A. 2008. *Environmental Law in a Nutshell.* Thomson and West, Egon, MN, USA.

Gibson, J.E. and Halter, F.1994. Strengthening environmental law in developing countries. *Environment* 36: 40-44.

Grant, W., Matthews, D. and Newell, P. 2001. *The Effectiveness of European Union Environmental Policy.* Palgrave, New York, NY, USA.

Heinzerling, L. and Ackerman, F. 2002. Pricing the priceless cost-benefit analysis of environmental protection. *http://www.law. georgetown.edu/gelpi/papers/priceful. pdf.*

Heyes, A. 2001. *The Law and Economics of the Environment.* E. Elgar, Chettenham, UK.

Hoffman, A. 2000. Integrating environmental and social issues into corporate practice. *Environment* 42:22-33.

Hunter, D., Salzman, J. and Zaelke, D. 2002. *Hunter, Salzman and Zaelke's International Environmental Law and Policy.* Thompson, Florence, KY, USA.

Katz, M. and Thornton,D. 1994. *Environmental Management Tools on the Internet Accessing the World of Environmental Information.* St. Lucie Press. Delray Beach, FL, USA.

Kime, R.E. 1992. *Environment and Health.* Grolier Educational Group, Dushkin Pub. Group, Guilford, CT, USA.

Krupin, P.J. 1994. *Krupin's Toll Free Environmental Directory.* Direct Contact Publishing. Kennewick, Washington, USA.

Koren, H. 1991. *Handbook of Environmental Health and Safety Principles and Practices.* Lewis Publishers, Chelsea, MI, USA.

Kubasek,N. and Silverman, G. 2007. *Environmental Law.* Prentice Hall, Upper Saddle River, NJ, USA.

Law Firm of Mays and Valentine. 1992. *Virginia Environment Law Handbook.* Government Institute. Rockville, MD, 2nd Edition, USA.

Lifoff, R.A. and Swartzman, D. 1982. *Cost-Benefit Analysis and Environmental Regulations: Politics, Ethics and Methods.* World Wildlife Fund, Washington, DC, USA.

Miller, E.W. and Miller, D.M. 1991. *Contemporary World Issues Environmental Hazards. Toxic Wasteland Hazardous Material.* ABC-CLIO, Santa Barbara, CA, USA.

Mintz, J.A. 1994. *State and Local Government Environmental Liability.* Clark Boodman, Deerfield, IL, USA.

Morel, B. and Linkov, I. 2006. *Environmental Security and Environmental Management: The Role of Risk Assessment.* Springer, Amsterdam, The Netherlands.

Moskowitz, J.S. 1995. *Environmental Liability and Real Property Transactions.* Wiley Law Publications, New York, NY, USA.

Patton-Hulce, V.R. 1995. *Environment and the Law. A Dictionary.* ABC-CLIO, Inc. Santa Barbara, CA, USA.

Pearce, D., Atkinson, G. and Mourato, S. 2006. *Cost-benefit Analysis and the Environment.* OECD Publishing, Danvers, MA, USA.

Percival, R.V. and Alevizatos,D.C. 1997. *Law and The Environment, A Multidisciplinary Reader.*

Temple University Press, Philadelphia, PA,USA.

Portney, P. and Stavins, R. 2000. *Public Policies for Environmental Protection. Resources for the Future.* RFF Press, Washington, DC, USA.

Powell,C. 1999. *Introduction: Locating Culture, Identity and Human Rights.* HRL Rev.201.

Puttaswamaiah, K. 2002. *Cost-benefit Analysis Environmental and Ecological Perspectives.* Transaction Publishers, New Brunswick, NJ, USA.

Rajamani, L. 2006. *Differential Treatment in International Environmental Law.* Oxford Univ. Press, New York, NY, USA.

Rombke, J. and Mottmann, J.F.1995. *Applied Ecotoxicology, Chapter 11. Legal Foundations.* Lewis Publishers, Boca Raton, FL, USA.

Sands, P. 2003. *Principles of International Law.* Cambridge University Press, Cambridge, England.

Schmittner, A., Oschlies, A, Matthews, H.D. and Galbraith, E.D. 2008. Future changes in climate, ocean circulation, ecosystems, and biogeochemical cycling simulated for a business-as-usual CO_2 emission scenario until year 4000 AD. *Global Biogeo. Cycles* 22: GB 1013, doi: 10: 1029/2007 GB002953.

Schoenbaum, T.J. and Rosenberg,R.H. 2001. *Environmental Law.* Foundation Press, New York, NY, USA.

Seredich, J. 1991. *Your Resource Guide to Environmental Organization.* Smiling Dolphins Press, Irvine, CA, USA.

Speth, J.G. and Haas, P.M. 2006. *Global Environmental Governance.* Island Press, Washington, DC, USA.

Taylor,P. 1998. *An Ecological Approach to International Law, Responding To Challenges Of Climate Change.* Routledge, London, England.

Thornton, J. and Beckwith,S. 2004. *Environmental Law.* Sweet and Maxwell, London, UK.

Turk, J. and Turk,A. 1997. *Environmental Science.* Harcourt Brace, Orlando, FL, USA.

ENVIRONMENT, CONSERVATION & ECOLOGY JOURNALS

A&WMA Environmental Compliance News
Acta Oecologia
African Journal of Ecology
Agriculture, Ecosystems and Environment
Annual Review of Ecology and Systematics
Annual Review of Energy and the Environment
Applied and Environmental Microbiology
Aquatic Conservation
Aquatic Microbial Ecology
Archives of Environmental Health
Asia Environmental Review
Atmospheric Environment

Behavioral Ecology
Behavioral Ecology and Sociobiology
Biochemical Systematics and Ecology
Bulletin of Environmental Contamination and Toxicolgy
Bulletin of the Ecological Society of America

Canadian Environmental Protection
Chemoecology

Chemosphere – Global Change Science
China Environmental Review
Conservation Ecology
Critical Reviews in Environmental Science and Technology
Current Advances in Ecological & Environmental Sanitation

Dairy, Food and Environmental Sanitation
Dynamics of Atmospheres and Oceans

E: Environmental Magazine
Ecological Applications
Ecological Bulletins
Ecological Economics
Ecological Engineering
Ecological Entomology
Ecological Modelling
Ecological Monographs
Ecological Research
Ecologist
Ecology Law Quarterly
Ecology of Food and Nutrition
Ecology of Freshwater Fish
Ecosystem Health
Ecosystems
Ecotoxicology

Ecotoxicology and Environmental Safety
Energy & Fuels
Environment
Environment and Development Economics
Environment International
Environment Matters
Environmental and Ecological Statistics
Environmental and Experimental Botany
Environmental and Molecular Mutagenesis
Environmental Biology of Fishes
Environmental Conservation Journal
Environmental Engineering and Policy
Environmental Entomology
Environmental Ethics Journal
Environmental Geology
Environmental Health and Pollution Control
Environmental Health Perspectives
Environmental Impact Assessment Review
Environmental Law and Management
Environmental Lawyer
Environmental Management
Environmental Management and Health
Environmental Microbiology
Environmental Policy and Law
Environmental Pollution
Environmental Practice
Environmental Progress Environmental Quality Management
Environmental Research
Environmental Reviews
Environmental Science and Engineering
Environmental Science and Policy
Environmental Science and Technology
Environmental Technology
Environmental Toxicology and Chemistry
Environmental Toxicology and Pharmacology

Environmental Toxicology and Water Quality
Environmental Values
Environmentalist
Environmentrics
Erosion Control
Evolutionary Ecology
Evolutionary Ecology Research

FEMS Microbiology Ecology
Forest Ecology and Management
Functional Ecology

Geophysical Research Letters
Global Environmental Change
Global Environmental Politics

Human Ecological Risk Assessment

International Journal of Environmental Health Research
International Journal of Sustainable Development and World Ecology
Issues in Ecology

Journal of Agricultural and Environmental Ethics
Journal of Agricultural, Biological, and Environmental Statistics
Journal of Animal Ecology
Journal of Applied Ecology
Journal of Aquatic Ecosystem Stress and Recovery
Journal of Arid Environments
Journal of Atmospheric and Oceanic Technology
Journal of Chemical Ecology
Journal of Ecology
Journal of Ecosystems and Management

Journal of Environmental Assessment Policy and Management

Journal of Environmental Biology

Journal of Environmental Economics and Management

Journal of Environmental Education

Journal of Environmental Health

Journal of Environmental Health

Journal of Environmental Law

Journal of Environmental Management

Journal of Environmental Medicine

Journal of Environmental Planning and Management

Journal of Environmental Quality

Journal of Environmental Radioactivity

Journal of Environmental Science

Journal of Environmental Science and Health

Journal of Environmental Marine Biology and Ecology

Journal of Freshwater Ecology

Journal of Hazardous Materials

Journal of Industrial Ecology

Journal of Land Use & Environmental Law

Journal of Marine Environmental Engineering

Journal of the Atmospheric Sciences

Landscape Ecology

Marine Ecology

Marine Ecology Progress Series

Marine Environmental Research

Marine Pollution Bulletin

Microbial Ecology

Molecular Ecology

Ocean & Coastal Management

OIKOS

P2: Pollution Prevention Review

Plant Ecology

Pollution Ecology

Progress in Environmental Sciences

Radiation and Environmental Biophysics

Remote Sensing of Environment

Researches on Population Ecology

Restoration Ecology

Russian Journal of Ecology

Sciences and the Environment Bulletin

Science of the Total Environmental

Trends in Ecology and Evolution

Urban Ecosystems

Water Environment Research

Water, Air, & Soil Pollution

Wetlands

From: *http://www.medbioworld.com/* bio/journals/environ.html

SCIENTIFIC METHOD

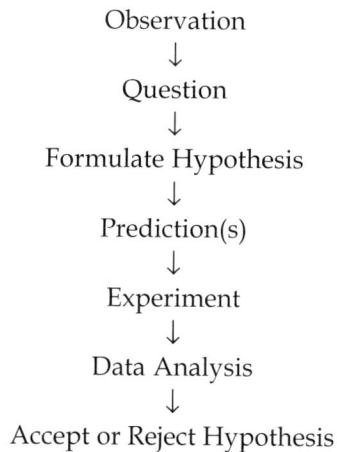

Observation
↓
Question
↓
Formulate Hypothesis
↓
Prediction(s)
↓
Experiment
↓
Data Analysis
↓
Accept or Reject Hypothesis

Summary of Important Terms[a]

Terms	Meaning
Hypothesis	A logical but unproven explanation for a set of facts; refers to knowledge state prior to experimentation
Model	When a hypothesis possesses at last limited validity
Theory	A hypothesis or a group of related hypothesis which has (have) been confirmed through repeated experimentation
Law	Laws are theories that have been repeatedly tested and not disproved

[a]Adapted from *http://teacher.nsrl.Rochester.edu/phy-labs…*
and *http://Pasadena.wr.usgs.gov/office/garderson…*

METRIC TERMS

Metric Terms Frequently Used in the Biological Sciences[a]

Terms	Definition Symbol
Length	
Meter	m
Centimeter	cm
Millimeter	mm
Micrometer	μm
Weight	
Kilogram	kg
Gram	g
Microgram	μg
Nanogram	ng
Picogram	pg
Solutions amount of a substance	
Molar	M
Millimolar	mM
Micromolar	μM
Nanomolar	nM
Normal	N
Volume	
Liter	L
Millilitre	ml
Temperature	
Kelvin Thermodynamic Temperature	K
Degree Celsius Temperature	°C

[a]milli = 10^{-3}, micro = 10^{-6}, Nano = 10^{-9}, pico = 10^{-12}
[s]The student is referred to the U.S. Metric Association, Northridge, CA for additional terms sometimes used by biologists

TEMPERATURE CONVERSION SCALE

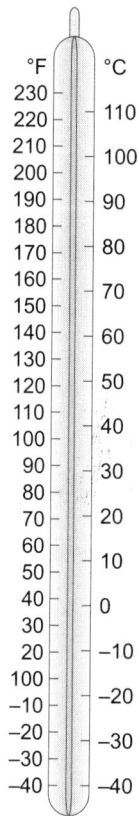

°F	°C
230	
220	110
210	100
200	
190	90
180	
170	80
160	70
150	
140	60
130	
120	50
110	40
100	
90	30
80	
70	20
60	10
50	
40	0
30	
20	−10
100	−20
−10	
−20	−30
−30	
−40	−40

To convert Fahrenheit to Celsius use the following formula: C = 5/9 °F − 32

To convert Celsius to Fahrenheit use the following formula: F = 9/5 °C + 32

STATISTICAL METHOLOLOGY

Statistical Software Packages[e]

Package	Usefulness
GLIM Version 3.77 [a]	ANOVA and linear and multiple regression
Minitab release 7.1 and 8.2 [b]	Comprehensive statistical package ANOVA, linear bivariate and multiple regression
SAS release 6.03 [a]	Comprehensive statistical package, ANOVA, linear and multiple regression, cluster analysis and others
SAS, Sigma plot and Deltagraph [a]	Non-linear curve fitting
SPSS/PC + version 3.0 [c]	Comprehensive statistical package, ANOVA linear and multiple regression, cluster analysis and other
Statgraphics version 2.1 [a]	Comprehensive statistical package, ANOVA bivariate linear and non-linear regression, multiple regression, cluster analysis and others
Delta Graph, Table Curve, Sigma Plot, Origin, Igor and Prism [d]	Curve fitting complex of functions
Matlab, Lab View [d]	Data analysis programs

Note: This table is a composite form multiple readings, e.g., *http://www.cas.lancs.ac.uk/glossary-/vbl/hyptest.html* and other web sites.

Summary of Type I and Types II Errors [a][b]

Statistical choice	Null Hypothesis Positive Ho	(Ho) Decision Negative Ho
Negate Ho	Type I Error [c]	Affirm
	P (type I error) = α	Decision
	More serious than a Type II Error	
Ho Not Negated Hi	Affirm	Type II Error c
	Decision	P(Type II error) = b Frequently due to a small sample size
		Error of the Second Kind

[a] Type I and II errors are inversely related.
[b] The student should consult statistics texts (see references for discussions of p values and confidence levels (Ruhlf, 1994; Zav, 1998; Rosner, 1999; Looney, 2002).

[c] Type I errors occur when the null hypothesis is true but rejected.

[d] Type II errors result when the null hypothesis is false but not rejected. Power of a hypothesis test = 1 − P (Type II error) = 1 − B.

[a] Adapted from: appendix A Fey, J.C. (1993). *Biological Data Analysis. A Practical Approach.* Oxford Univeristy Press, Oxford, England and

[b] Wardlaw, A.C. (2000). Provides extensive laboratory examples which use Minitab.

[c] SPSS is high-lighted by Campbell, R.C. (1989).

[d] Young, S.S. (2001). *Computerized Data Acquisition and Analysis for the Life Sciences.* Cambridge Univ. Press, Cambridge, England

[e] Harvard Graphics and Statistica are well-established software packages.

SPREAD SHEET

One of the most common softwares available for data entry and analysis is the spreadsheet. Although the spreadsheet was originally designed for the business world, it is widely used in the scientific community.

A discussion of Microsoft Excel 7 can be found at: *http://blogs.msdn.com/excel/archive/2005/11/08/490502.aspx.* An introduction to excel occurs at: *http://www.webspin.edu/Libr/tut/excel/intor. html.* H.F. Webster (Integrated Laboratory) has an online exercise regarding the use of spreadsheets in physical chemistry.

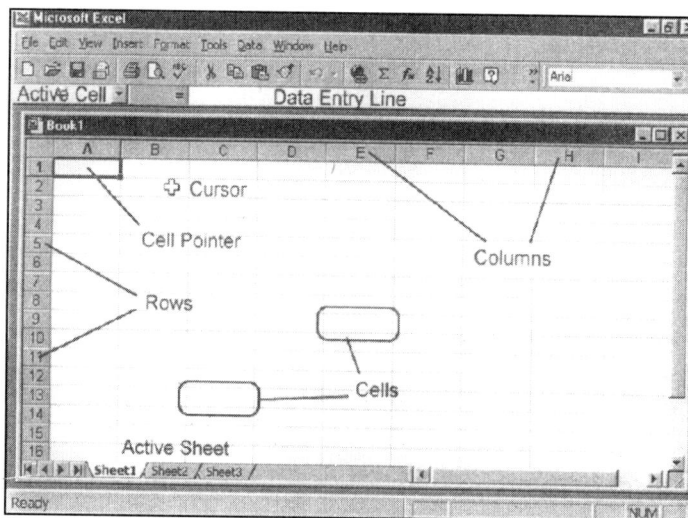

Sample of an excel spreadsheet. From: *http:personal.kent.ed/~rspamdur/instructions.htm.*

STATISTICAL TEST

Summary of Statistical Test Choices [a, b, c]

Variables	Other Parameters	Recommended Choice	Description
One	One group	$\bar{x}$ ±/-SD	= Sum of all the numbers of items in the set, SD = Square root of the variance
	Two groups	T-test	A parametric statistical test
	Three groups	ANOVA	An acronym for a category of tests termed analysis of variance
Two	Both continuous	Correlation$x$	Strength of relationship between variables
	One continuous One discrete	ANOVA	See above
	Both categorical	Chi-Square	Non-parametric alternative top the t-test
Three or More	One group	Multiple regression	Serves to make predictions regarding multiple values.
		Factor analysis	Used to discover simple patterns
		ANOVA repeated measures	Used for repeated measures variables
		Analysis of co-variance	Combines features of simple linear regression with one-way analysis of variance
		Multivariate ANOVA	A set of tests used where there are three or more independent variables with two or more treatments per variable
		Discriminant Function	Employed to determine which variable discriminates between two or more naturally occurring groups

[a] Modified from T. Lee Willoughby – *http://reserchmed.unkc.edu/ttwbiostats/*
[b] Descriptions of these tests can be found in Maxell and Delaney (2004), Myers and Well (2003), Wardlaw (2000), Dytham (1999), Fry (1993 a, b) and Iles (1993 a, b).
[c] see *http://www.cas.lancs.ACUK/glossary*

LABORATORY MANUALS

Arnfiels, E., Bitterman-Wenson, A., Ahuja, A. and Lorber, S. 2003. 'Environmental Science/ A Field and Laboratory Manual'. Kendall Hunt, Dubuque, IW, USA.

Chiras, D.D. 1998. 'Lab Manual to Accompany Environmental Science Action for Sustainable Future'. Benjamin Cummings Publishing, San Francisco, CA, USA.

Colverson, P. and Clemente, R. 1996. 'Environmental Science Lab Manuals'. Kendall Hunt, Dubuque, IW, USA.

Enger, E. and Smith, b. 1997. 'Field and Laboratory Exercises in Environmental Science'. McGraw-Hill, Boston, MA, USA.

Gillipan, M.R. 1991. 'Environmental Science Laboratory. A Manual of Lab and Field Exercises'. Halfmoon Publishing, Halfmoon Bay, CA, USA.

Krumhardt, R.A. and Wirth, D. 1999. 'Environmental Science Lab Manual. Experiences in Environmental Science'. Bellwether Cross, East Dubuque, IL, USA.

Lynn, L.M. 1999. 'Environmental Biology and Ecology Laboratory Manual'. Kendall Hunt, Dubuque, IL, USA.

Rockett, L.C. and VanDellen, J. 1993. 'Laboratory Manual for Miller's Living for The Environment'. Wadsworth, Belmont, CA, USA.

Rostenthal, D.D. 1995. 'Environmental Science Activities'. Wiley and Sons, New York, NY, USA.

Wagner, T.B. and Sanford, R. 2004. 'Laboratory Manual for Environmental Science'. Wiley and Sons, New York, NY, USA.

Wallace, L. 1998. 'Ecology and Environment Quality: LabManual'. Kendall Hunt, Dubuque, IL, USA.

MATHEMATICS REFERENCES

Foster, P.C. 1999. *Easy Mathematics for Biologists*. CRC Press, Boca Raton, FL, USA.

Gauch, H.G. 2002. *Scientific Method in Practice*. Cambridge Univ. Press, Cambridge, England.

Gustavil, B. 2003. *How to Write and Illustrate a Scientific Paper*. Cambridge Univ. Press, Cambridge, England.

Heath, M.T. 2001. *Scientific Computing*. McGraw Hill, New York, NY, USA.

Owen, G.E. 2003. *Fundamentals of Scientific Mathematics*. Dover, Mineola, NY, USA.

Valiela, I. 2000. *Doing Science: Design, Analysis and Communication of Scientific Research*. Oxford Univ. Press, Oxford, UK.

SUGGESTED EXAM QUESTIONS

Chapter 1 – Ecology

1. What is Ecology?
2. Compare biogeochemical cycles. How are they related?
3. Discuss trophic levels.
4. Describe the research methods that ecologists employ?

Chapter 2 – Environmental Science

1. What are the scientific topics that environmental scientists concern themselves with?
2. Compare the statistical methods of environmental scientists with those of the ecologists.

Chapter 3 – Global Warming

1. What are greenhouse gases and what are their sources?
2. What is the evidence that greenhouse gases can affect global temperature?
3. Discuss the factors other than greenhouse gases which can influence global temperature..
4. Describe the effects of elevated greenhouse gases on biological systems.

5. Can the climate of the future be predicted? Include in your answer, global circulation models and feedbacks affecting the greenhouse effects.
6. What is the best estimate of the rise in global temperature that would occur if there is a doubling of atmospheric CO_2 concentration.
7. What is the best estimate of the annual sea-level rise by 2100?
8. What are your ideas to curb global warming?

Chapter 4 – Energy

1. Discuss the future energy requirements of the world? Please include in your answer alternatives to fossil fuels.
2. What are the effects of ionizing radiation on biological systems?

Chapter 5 – Air Pollution

1. Elaborate the types of air pollution and their sources.
2. What atmospheric factors influence the transboundary movement of air pollutants?
3. Compare the chemical reaction resulting in the formation of ozone, nitric acid and acid rain.

4. What is the difference between wet and dry deposition of air pollutants?

5. Compare the effects of air pollutants on plants and animals.

6. Design and experiment to determine the toxicity of mercury to fish.

Chapter 6 – Metals and Volatile Organics

1. What are the mineral elements required for plant and animal growth and development? What are the functions of the mineral elements?

2. Design an experiment to determine at what levels mineral elements become toxic for plant or animal growth and development.

3. What is known about the molecular mechanisms of metal toxicity in plants or animals?

4. Suppose you discover that a company is pumping waste water containing an elevated amount of cadmium into a local river, what legal courses of action are available to curtail the process?

5. What are volatile organics and what are their sources?

6. Prepare an essay regarding the effect of volatile organics on human health.

Chapter 7 – Pesticides

1. Describe the ways that pesticides can be classified?

2. How do natural pesticides differ from conventional pesticides?

3. Compare the modes of actions of fungicides, herbicides, insecticides and rodenticides.

4. What effects do conventional pesticides have on human health?

Chapter 8 – Radioactive Pollution

1. What is radioactivity and what are its sources?

2. Describe the biological effects of radioactivity.

Chapter 9 – Thermal Pollution

1. What is thermal pollution?

2. Discuss the effects of thermal pollution on aquatic organisms.

3. What are the methods for measuring noise?

Chapter 11 – Agricultural Pollution

1. Detail the differences between non-point and point pollution.

2. What are the ways that agricultural runoff can enter the drinking water?

3. Discuss the advantages and disadvantages of applying sludge to farmlands.

4. How can agricultural runoff from swine, poultry and cattle operations be contained?

Chapter 12 – Species

1. What is your understanding of the species concept?

2. Explain the difference between endangered and threatened species?

3. What is meant by the term 'HIPPO'?

4. Prepare an essay describing biodiversity and its value.

Chapter 13 – Population

1. Compare population growth in developed and underdeveloped countries.

2. What is meant by carrying capacity and does it apply to population growth?

Chapter 14 – Environmental Law

1. Describe the approaches available to individuals and groups to protect the environment.

2. Search the Internet to determine International Environmental Laws, Agreements and Regulation. Then, prepare a summary table regarding the results of your search.

Index

DATE DUE